Arithmetic Progression

$$a_n = a_1 + (n - 1)d$$
$$S_n = \frac{n}{2}(a_1 + a_n)$$
$$= \frac{n}{2}[2a_1 + (n - 1)d]$$

Right Lines

$$m = \frac{y_2 - y_1}{x_2 - x_1} \quad \text{(slope formula)}$$
$$y - y_1 = m(x - x_1) \quad \text{(point-slope form)}$$
$$y = mx + b \quad \text{(slope-intercept form)}$$
$$x = \text{constant} \quad \text{(vertical line)}$$
$$y = \text{constant} \quad \text{(horizontal line)}$$

Geometric Progression

$$a_n = a_1 r^{n-1}$$
$$S_n = \frac{a_1(1 - r^n)}{1 - r} \quad (r \neq 1)$$
$$S = \frac{a_1}{1 - r} \quad (|r| < 1)$$

Variation

$$y = kx^n \quad \text{(direct variation)}$$
$$y = \frac{k}{x^n} \quad \text{(inverse variation)}$$

Binomial Theorem

$$(a + b)^n = a^n + \frac{n}{1!}a^{n-1}b + \frac{n(n - 1)}{2!}a^{n-2}b^2$$
$$+ \frac{n(n - 1)(n - 2)}{3!}a^{n-3}b^3 + \cdots + b^n$$

Geometric Formulas

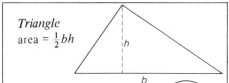

Triangle
area = $\frac{1}{2}bh$

Circle
area = πr^2
circumference = $2\pi r$

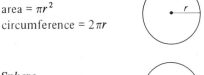

Sphere
volume = $\frac{4}{3}\pi r^3$
surface area = $4\pi r^2$

Right circular cylinder
volume = $\pi r^2 h$
lateral surface area
  $= 2\pi rh$

Right circular cone
volume = $\frac{1}{3}\pi r^2 h$

Greek Alphabet

alpha	A	α		nu	N	ν
beta	B	β		xi	Ξ	ξ
gamma	Γ	γ		omicron	O	o
delta	Δ	δ		pi	Π	π
epsilon	E	ϵ		rho	P	ρ
zeta	Z	ζ		sigma	Σ	σ
eta	H	η		tau	T	τ
theta	Θ	θ		upsilon	Υ	υ
iota	I	ι		phi	Φ	ϕ, φ
kappa	K	κ		chi	X	χ
lambda	Λ	λ		psi	Ψ	ψ
mu	M	μ		omega	Ω	ω

Essentials
of Technical
Mathematics

Essentials of Technical Mathematics

Third Edition

Richard S. Paul

The Department of Mathematics
The Pennsylvania State University

M. Leonard Shaevel

Formerly of the Department of Physics
The Pennsylvania State University

Prentice Hall, Englewood Cliffs, New Jersey 07632

Library of Congress Cataloting-in-Publication Data

Paul, Richard S.
  Essentials of technical mathematics/Richard S. Paul, M. Leonard
Shaevel.--3rd ed.
    p. cm.
  Includes index.
  ISBN 0-13-288812-2
  1. Mathematics--1961- I Shaevel, M. Leonard. II. Title.
QA39.2P39 1989
512'.1--dc19   87-21147
                                                                    CIP

Editorial/production supervision: Pat Walsh
Interior and cover design: Suzanne Behnke
Manufacturing Buyer: Bob Anderson

 © 1989, 1982, 1974 by Prentice-Hall, Inc.
A Division of Simon & Schuster
Englewood Cliffs, New Jersey 07632

Printed in the United States of America

10 9 8 7 6 5 4

ISBN 0-13-288812-2

Prentice-Hall International (UK) Limited, *London*
Prentice-Hall of Australia Pty. Limited, *Sydney*
Prentice-Hall Canada Inc., *Toronto*
Prentice-Hall Hispanoamericana, S. A., *Mexico*
Prentice-Hall of India Private Limited, *New Delhi*
Prentice-Hall of Japan, Inc., *Tokyo*
Simon & Schuster of Southeast Asia Pte. Ltd., *Singapore*
Editora Prentice-Hall do Brasil, Ltda., *Rio de Janeiro*

Contents

Preface   xiii

0
Preliminary Topics   1

0.1 Introduction   **1**
0.2 Operations with Whole Numbers   **1**
0.3 Common Fractions   **3**
0.4 Addition and Subtraction of Fractions   **8**
0.5 Multiplication and Division of Fractions   **13**
0.6 Operations with Decimals   **17**
0.7 Geometrical Concepts and Formulas   **23**

1
Operations with Real Numbers   31

1.1 Real Numbers and Number Lines   **31**
1.2 Inequalities and Absolute Value   **35**
1.3 Basic Laws of Real Numbers   **39**
1.4 The Arithmetic of Real Numbers   **42**

1.5 Fractions   **53**
1.6 Evaluating Formulas   **61**
1.7 Review   **64**

2

Introduction to Exponents and Radicals   69

2.1 Exponents   **69**
2.2 Zero and Negative Exponents   **78**
2.3 Scientific Notation   **83**
2.4 Radicals   **90**
2.5 Rational Exponents   **99**
2.6 Review   **103**

3

Operations with Algebraic Expressions   108

3.1 Algebraic Expressions   **108**
3.2 Addition and Subtraction of Algebraic Expressions   **112**
3.3 Multiplication of Algebraic Expressions   **117**
3.4 Division of Algebraic Expressions   **123**
3.5 SI Units and Conversion of Units   **133**
3.6 Review   **141**

4

Equations   144

4.1 Types of Equations   **144**
4.2 Linear Equations   **148**
4.3 Proportion   **158**
4.4 Translating English to Algebra   **164**
4.5 Word Problems   **168**
4.6 Review   **179**

5

Functions and Graphs   182

5.1 Functions   **182**
5.2 Graphs in Rectangular Coordinates   **194**
5.3 Review   **209**

6

Trigonometric Functions of an Acute Angle   212

6.1 Angles and Angular Measurement   **212**
6.2 Trigonometric Functions of Acute Angles   **224**
6.3 Values of Trigonometric Functions   **233**
6.4 Solution of a Right Triangle   **241**
6.5 Review   **251**

7

Reference Angles and Vectors   256

7.1 Trigonometric Functions of Any Angle   **256**
7.2 Vectors   **269**
7.3 Addition of Vectors—Analytical Method   **279**
7.4 Review   **286**

8

Straight Lines   291

8.1 Slope of a Line   **291**
8.2 Equations of Lines   **297**
8.3 Parallel and Perpendicular Lines   **307**
8.4 Linear Interpolation   **309**
8.5 Review   **313**

9

Systems of Equations   317

9.1 Systems of Linear Equations in Two Variables   **317**
9.2 Methods of Elimination   **321**
9.3 Systems of Linear Equations in Three Variables   **327**
9.4 Determinants of Order 2   **331**
9.5 Determinants of Order 3   **336**
9.6 Properties of Determinants   **342**
9.7 Word Problems   **345**
9.8 Review   **351**

Contents   **vii**

10

Factoring and Fractions   354

10.1 Special Products   **354**
10.2 Factoring   **361**
10.3 Factoring, Continued   **367**
10.4 Reducing Fractions   **370**
10.5 Multiplication and Division of Fractions   **375**
10.6 Addition and Subtraction of Fractions   **380**
10.7 Complex Fractions   **390**
10.8 Fractional Equations   **394**
10.9 Review   **400**

11

Operations with Radicals   407

11.1 Changing the Form of a Radical   **407**
11.2 Addition and Subtraction of Radicals   **416**
11.3 Multiplication of Radicals   **417**
11.4 Division of Radicals   **420**
11.5 Radical Equations   **423**
11.6 Review   **427**

12

Quadratic Equations and Functions   430

12.1 Solution by Factoring and Square Root   **430**
12.2 Completing the Square   **440**
12.3 The Quadratic Formula   **442**
12.4 Quadratic Functions   **448**
12.5 Equations Leading to Quadratic Equations   **457**
12.6 Nonlinear Systems   **465**
12.7 Review   **471**

13

Complex Numbers   475

13.1 The j-Operator and Complex Numbers   **475**
13.2 Operations with Complex Numbers   **481**

13.3 Polar Form of a Complex Number   **490**
13.4 Exponential Form of a Complex Number   **504**
13.5 Operations in Polar Form   **506**
13.6 De Moivre's Theorem   **510**
13.7 Review   **516**

14

Variation   521

14.1 Direct Variation   **521**
14.2 Inverse Variation   **529**
14.3 Review   **534**

15

Exponential and Logarithmic Functions   537

15.1 Exponential Functions   **537**
15.2 Logarithmic Functions   **541**
15.3 Properties of Logarithms   **551**
15.4 Change of Base   **560**
15.5 Exponential and Logarithmic Equations   **561**
15.6 Logarithmic and Semilogarithmic Graph Paper   **565**
15.7 Review   **572**

16

Graphs of the Trigonometric Functions   576

16.1 The Graph of the Sine Function   **576**
16.2 The Graph of the Cosine Function   **587**
16.3 The Sine Curve and Harmonic Motion   **591**
16.4 Graphs of Tangent, Cotangent, Secant, and Cosecant   **593**
16.5 Addition of Ordinates   **596**
16.6 Inverse Trigonometric Functions   **600**
16.7 The Oscilloscope and Lissajous Figures   **604**
16.8 Polar Coordinates   **608**
16.9 Review   **615**

17

Trigonometric Formulas and Equations 619

17.1 Basic Identities **619**
17.2 Addition and Subtraction Formulas **627**
17.3 Double- and Half-Angle Formulas **636**
17.4 Trigonometric Equations **642**
17.5 Review **646**

18

Oblique Triangles, Area, and Angular Speed 650

18.1 The Law of Sines **650**
18.2 The Law of Cosines **659**
18.3 Areas and Angular Speed **665**
18.4 Review **670**

19

Inequalities 673

19.1 Linear Inequalities in One Variable **673**
19.2 Nonlinear Inequalities **679**
19.3 Equations and Inequalities Involving Absolute Value **686**
19.4 Linear Inequalities in Two Variables **690**
19.5 Review **697**

20

Sequences and Series 701

20.1 Sequences and Series **701**
20.2 Arithmetic Progressions **706**
20.3 Geometric Progressions **713**
20.4 Limits of Sequences **718**
20.5 The Infinite Geometric Series **725**
20.6 The Binomial Theorem **730**
20.7 Review **735**

21

Analytic Geometry   739

21.1 The Conic Sections   **739**
21.2 The Distance Formula   **741**
21.3 The Circle   **742**
21.4 The Parabola   **747**
21.5 The Ellipse   **757**
21.6 The Hyperbola   **763**
21.7 Summary of Conic Sections   **770**
21.8 Parametric Equations   **771**
21.9 Review   **777**

A

The Scientific Calculator   A-1

A.1 Introduction   **A-1**
A.2 Data Entry and Scientific Notation   **A-2**
A.3 Arithmetic Operations   **A-4**
A.4 Reciprocal, Square, Square Root, and Factorial   **A-6**
A.5 Powers and Roots   **A-10**
A.6 Trigonometric and Inverse Trigonometric Functions   **A-13**
A.7 Common and Natural Logarithms   **A-16**

B

Answers to Odd-Numbered Problems   B-1

Index   I-1

Preface

The third edition of *Essentials of Technical Mathematics* is designed to meet the mathematical needs of students in the various engineering technologies. Special care has been taken to retain the character and spirit of previous editions. Throughout this edition we continue to include not only the *how*, but also the *why*.

The developments in the text present a good balance between mathematical concepts, manipulative mathematics, and applications. An effort has been made to make the material understandable and meaningful, and to ensure that the language and writing style are appropriate for students having limited mathematical backgrounds.

There are over 1000 examples, which are given in step-by-step detail. An abundance and variety of technical applications are blended throughout the book so that students continually see ways in which the mathematics they are learning can be applied to the engineering technologies. However, the text is virtually self-contained in the sense that it assumes no prior exposure to the concepts on which the applications are based. We point out that these applications simply pose a technical framework for the mathematics problems, and we are not concerned with teaching the nonmathematical aspects of the applications.

The problem sets, which include over 5200 problems, contain a large number of drill problems that the student should be able to do rather easily. This allows the student to gain confidence, develop manipulative skills, and fully understand the basic concepts being studied.

In this edition, uses of calculators are introduced as the need arises. This supplements Appendix A, which covers the most important uses of the scientific calculator in technical mathematics and includes problems (with all answers) for practice.

Besides expanding the material on the calculator, we have made other changes for this edition. A section on translating English to algebra has been added. Much of the exposition has been improved to afford greater clarity and student understanding. Chapter review sections have been expanded. Many problem sets have been revised. The range of applications has been extended both in the illustrative examples and in the exercises.

A general reorganization of material has been made, and the order in which topics appear has been rearranged to allow greater flexibility in planning a course outline that serves the needs of a particular class and curriculum. For example, the chapter on basic trigonometry has been divided into two chapters: one involving right-triangle trigonometry (Chapter 6) and one involving reference angles and vectors (Chapter 7). These occur earlier in the book to benefit those students taking a concurrent physics course. Basic fractions, rational exponents, SI units, and unit conversion appear earlier. Factoring has been divided into two sections. Straight lines appear in a separate chapter. Quadratic equations are discussed before complex numbers are introduced. The chapter on sequences and series precedes analytic geometry. The section on arithmetic and geometric progressions has been divided into two separate sections.

Chapter 0 covers selected topics from arithmetic and geometry. Although this chapter is not necessarily intended to be part of a formal course, it can provide a needed review for many students. With the exception of Chapter 0, each chapter ends with a review section that contains a list of key terms and symbols, a chapter summary of important formulas and rules, a set of programmed-type review questions, and numerous review problems.

Answers to odd-numbered problems appear in the back of the book. Solutions to all problems are available in an Instructor's Manual. Also available is a Student's Solution Manual that contains solutions to every other odd-numbered problem (that is, problems numbered 1, 5, 9, and so on).

I express my appreciation to my colleagues Mylar Giri, who discussed some technical applications with me, and Margaret Stahura, who proofread much of the revised material and offered comments and suggestions for its improvement.

Special thanks go to Rick Williamson, editor, Patrick Walsh, production editor, Linda Thompson, copy editor, and to the entire Prentice-Hall staff for their efficiency, competent assistance, and enthusiastic cooperation.

Finally, special recognition is given to M. Leonard Shaevel. In fond remembrance of a beloved colleague, teacher, and friend.

Richard S. Paul

Preliminary Topics

0.1 INTRODUCTION

This chapter gives a brief review of selected topics in arithmetic and geometry. Even though you may be familiar with much of this material, another exposure to these topics may be beneficial to you. You should devote whatever time is necessary to those topics in which you need review.

0.2 OPERATIONS WITH WHOLE NUMBERS

Whole numbers are numbers used in counting, such as 0, 1, 2, 13, 44, and 610. Whole numbers are either **even** or **odd**, depending, respectively, on whether or not they are divisible by 2. For example, 2, 10, and 64 are even, and 1, 3, and 29 are odd.

Whole numbers are represented by combinations of the ten **digits** 0, 1, 2, 3, 4, 5, 6, 7, 8, and 9. An analysis of the numeral 234 shows that the digit 4 occupies the units place and represents four 1's (4), the digit 3 occupies the tens place and represents three groups of 10 (30), and the digit 2 occupies the hundreds place and represents two groups of 100 (200). Note that the value of each group is ten times that of the group to its immediate right. This system of writing numbers is called the **decimal system**.

We shall assume that you are totally familiar with the basic operations of addition, subtraction, multiplication, and division of whole numbers. It is appropriate, however, to state four important rules that apply to all types of numbers. We give two rules first:

> 1. Numbers can be added in any order.
> 2. Numbers can be multiplied in any order.

For example, by Rule 1 we have

$$2 + 3 + 4 = 4 + 2 + 3 = 4 + 3 + 2 = 9.$$

Rule 2 involves multiplication. We can show a multiplication such as 3 times 2 by 3×2, $3 \cdot 2$, $(3)(2)$, $(3)2$, and $3(2)$. Thus, by Rule 2,

$$(3)(2)(4) = (2)(3)(4) = (4)(3)(2) = 24.$$

Because the numbers 2, 3, and 4 give 24 when multiplied together, 2, 3, and 4 are called **factors** of 24, and 24 is called the **product** of 2, 3, and 4. The next rules are for subtraction and division:

> 3. Subtraction must be performed in the given order.
> 4. Division must be performed in the given order.

For example, $5 - 2$ is *not* the same as $2 - 5$, and $6 \div 3$ is *not* the same as $3 \div 6$. In the statement $6 \div 3 = 2$, 6 is called the **dividend**, 3 is the **divisor**, and 2 is the **quotient**.

In computations involving more than one operation, first perform any multiplications and divisions. Then do any additions and subtractions, as Example 1 shows.

Example 1

a. $6 + (6)(2) = 6 + 12 = 18.$

b. $12 - (2)(2) = 12 - 4 = 8.$

c. $(12)(2) + 6 \div 3 = 24 + 2 = 26.$

When any operations are inside parentheses, do those first, as Example 2 shows.

Example 2

a. $\quad 12 + (2 + 3) - (4)(6 - 3)$

$\quad = 12 + 5 - (4)(3) = 12 + 5 - 12$

$\quad = 17 - 12 = 5.$

b. $16 + (12 \div 3)(4 - 3)$

$= 16 + (4)(1) = 16 + 4$

$= 20.$

c. $(2 - 2)(5) = (0)(5) = 0.$ *Remember*: Zero times any number is zero.

Problem Set 0.2

In Problems **1–32**, *compute the numbers.*

1. $5 + (2)(8).$ **2.** $17 - (3)(4).$ **3.** $(2)(4) - 7.$

4. $(3)(2) - (2)(2).$ **5.** $(8 - 3)(4).$ **6.** $2(5 + 1).$

7. $15 - 3(4).$ **8.** $(6 - 4) + 3.$ **9.** $3(4 + 6).$

10. $(6 - 6)8.$ **11.** $(7 - 2)(3 + 1).$ **12.** $(4 + 5)(2 - 0).$

13. $0(8 - 5).$ **14.** $(3 - 3)(4 - 4).$ **15.** $(6 - 1) - (10 - 7).$

16. $(6 + 7) - (8 - 4).$ **17.** $(8 \div 4) + 3.$ **18.** $(7 - 2)(3) + 2(3 + 1).$

19. $10 + (9 \div 3).$ **20.** $(3 \div 3) + (2)(3 - 1).$

21. $18 \div (6 + 3).$ **22.** $55 \div (17 - 6).$

23. $(6 + 2 - 3) \div (7 - 2).$ **24.** $(8 - 4)(7 - 3) + (16 \div 4).$

25. $13 - 2(5 - 2) + (2 + 3).$ **26.** $8 + (15 \div 3)(3 - 1).$

27. $(36 \div 6)(6 \div 2) - 6.$ **28.** $16 + (12 \div 3)(4 - 3).$

29. $4(4 - 1) - 5(4 \div 2) + 9.$ **30.** $2(3 + 4) + 5 - (2 - 1)(2 + 1).$

31. $3(6 \div 3) + 2 - 4(2 \div 2).$ **32.** $(4 + 3)(3) + 5(5 - 2)(2 - 2).$

33. In $8 \div 2 = 4$, what is the dividend? What is the divisor?

34. In $16 \div 8 = 2$, what is the quotient?

35. In $30 \div (3 + 2) = 6$, what is the divisor?

36. The product of three factors is 30. Two of the factors are 2 and 3. What is the other factor?

37. The product of five factors is 32. Three of the factors are each 2. What is the *product* of the other factors?

38. The numbers 2 and 3 are factors of 6. Find two other whole numbers that are also factors of 6.

39. The tens and units digits of the number 862 are reversed and the new number is subtracted from 862. What is the result?

40. What digit must be changed in 6257 to make it an even number?

41. In the number 62,417, what place does the 6 occupy?

0.3   COMMON FRACTIONS

The division $6 \div 3$ can be written as the fraction $\frac{6}{3}$, where 6 is the **numerator** and 3 is the **denominator**. The fraction is also called the **quotient** of 6 divided by 3, or simply the quotient $\frac{6}{3}$. You may recall that $\frac{6}{3}$ means the number which when multiplied by 3 gives 6. Because $(3)(2) = 6$, then $\frac{6}{3} = 2$.

Let us quickly do away with one point of confusion—*division by zero is not defined*. For example, the symbol $\frac{1}{0}$ means a number which when multiplied by 0 gives 1. But there is no such number, since any number times 0 is 0. Thus $\frac{1}{0}$ has no meaning. Sometimes students write $\frac{1}{0} = 0$, which is *false*. However, it *is* true that $\frac{0}{1} = 0$, because $\frac{0}{1}$ is the number which when multiplied by 1 gives 0: That number is 0. Similarly, $\frac{0}{5} = 0$ because $(5)(0) = 0$.

Fractions are of two types, depending on the relationship between the numerator and denominator. If the numerator is *less* than the denominator, as in $\frac{3}{8}$, the fraction is called a **proper fraction** and its value is less than 1. If the numerator is *equal to or greater than* the denominator, as in $\frac{3}{3}$ and $\frac{3}{2}$, the fraction has a value of at least 1 and is called an **improper fraction**.

Any improper fraction can be written either as a whole number or as the sum of a whole number and a proper fraction. This last form is called a **mixed number**. For example, because $23 \div 6$ gives 3 with a remainder of 5, then

$$\frac{23}{6} = 3 + \frac{5}{6} = 3\frac{5}{6}.$$

The improper fraction $\frac{23}{6}$ has the same value as the mixed number $3\frac{5}{6}$, which means 3 *plus* $\frac{5}{6}$.

Example 1

---

a. $\dfrac{16}{13} = 1 + \dfrac{3}{13} = 1\dfrac{3}{13}.$

b. $\dfrac{14}{3} = 4 + \dfrac{2}{3} = 4\dfrac{2}{3}.$

c. $\dfrac{27}{4} = 6 + \dfrac{3}{4} = 6\dfrac{3}{4}.$

Fractions that have the same value (or represent the same number) are called **equivalent fractions**. The basic rule used in obtaining equivalent fractions is called the **fundamental principle of fractions**:

Multiplying or dividing the numerator and denominator of a fraction by the same number, except zero, results in a fraction that is equivalent to the original fraction.

For example,

$$\frac{2}{5} = \frac{(2)(4)}{(5)(4)} = \frac{8}{20} \quad \text{and} \quad \frac{14}{21} = \frac{14 \div 7}{21 \div 7} = \frac{2}{3}.$$

By using the fundamental principle, we can write a fraction as an equivalent fraction with a specified numerator or denominator. For example, suppose we write $\frac{5}{8}$ as an equivalent fraction with denominator 24. Because the original denominator 8 must be multiplied by 3, so must the numerator:

$$\frac{5}{8} = \frac{(5)(3)}{(8)(3)} = \frac{15}{24}.$$

Example 2

a. Write $\frac{3}{8}$ as an equivalent fraction with denominator 32.

Solution

$$\frac{3}{8} = \frac{(3)(4)}{(8)(4)} = \frac{12}{32}.$$

b. Write 12 as an equivalent fraction with denominator 2.

Solution

$$12 = \frac{12}{1} = \frac{(12)(2)}{(1)(2)} = \frac{24}{2}.$$

c. Write $\frac{12}{20}$ as an equivalent fraction with numerator 3.

Solution

$$\frac{12}{20} = \frac{12 \div 4}{20 \div 4} = \frac{3}{5}.$$

A fraction is in **lowest terms**, or is (completely) **reduced**, when its numerator and denominator have no whole-number factors in common except 1. For example, $\frac{18}{12}$ is not in lowest terms because 18 and 12 have a common factor of 6:

$$18 = (6)(3) \quad \text{and} \quad 12 = (6)(2).$$

To reduce $\frac{18}{12}$, we use the fundamental principle by dividing the numerator and denominator by the common factor 6.

$$\frac{18}{12} = \frac{\frac{18}{6}}{\frac{12}{6}} = \frac{3}{2}.$$

Because 3 and 2 have no factors in common except 1, then $\frac{3}{2}$ is the reduced form of $\frac{18}{12}$. More simply, to reduce $\frac{18}{12}$ we usually write

$$\frac{\overset{3}{\cancel{18}}}{\underset{2}{\cancel{12}}} = \frac{3}{2},$$

where the slashes indicate the division by a common factor. Sometimes this division is called **cancellation**. You can think of this problem in another way by showing the common factor 6:

$$\frac{18}{12} = \frac{\overset{1}{(\cancel{6})}(3)}{\underset{1}{(\cancel{6})}(2)} = \frac{3}{2}.$$

The fraction $\frac{18}{12}$ can also be reduced by repeated cancellation. Watch how we cancel a factor of 2 and then 3:

$$\frac{\overset{9}{\cancel{18}}}{\underset{6}{\cancel{12}}} = \frac{\overset{3}{\cancel{9}}}{\underset{2}{\cancel{6}}} = \frac{3}{2}.$$

In the next example we reduce a more complicated fraction in a similar manner.

Example 3

$$\frac{(12)(6)(\overset{1}{\cancel{3}})(14)}{(22)(14)(\underset{7}{\cancel{21}})(18)} = \frac{(12)(6)(\overset{2}{\cancel{14}})}{(22)(14)(\underset{1}{\cancel{7}})(18)} = \frac{(12)(6)(\overset{1}{\cancel{2}})}{(\underset{11}{\cancel{22}})(14)(18)}$$

$$= \frac{(12)(\overset{1}{\cancel{6}})}{(11)(14)(\underset{3}{\cancel{18}})} = \frac{\overset{4}{\cancel{12}}}{(11)(14)(\underset{1}{\cancel{3}})}$$

$$= \frac{\overset{2}{\cancel{4}}}{(11)(\underset{7}{\cancel{14}})} = \frac{2}{77}.$$

This can be written more compactly as

$$\frac{(\overset{\overset{2}{\cancel{4}}}{\cancel{12}})(6)(\overset{1}{\cancel{3}})(\overset{\overset{1}{\cancel{2}}}{\cancel{14}})}{(\underset{11}{\cancel{22}})(\underset{7}{\cancel{14}})(\underset{\underset{1}{\cancel{7}}}{\cancel{21}})(\underset{\underset{1}{\cancel{8}}}{\cancel{18}})} = \frac{2}{77}.$$

In Example 3 we used canceling of common factors to reduce a fraction in which the numerator and denominator were written as a *product* of numbers. In a fraction where the numerator or the denominator is a *sum* or *difference* of numbers, this type of cancellation *cannot* be done. For example, $\frac{6}{2}$, which is 3, can be written $(4 + 2)/2$. But

$$\frac{4 + \overset{1}{\cancel{2}}}{\underset{1}{\cancel{2}}} = \frac{5}{1} \qquad \text{is } false.$$

Problem Set 0.3

1. Fill in the blank: $\frac{24}{8}$ is the number which when multiplied by _____ gives 24.
2. In the fraction $\frac{6}{11}$, what is the numerator?
3. In the fraction $\frac{5}{6}$, what is the denominator?
4. Which of the following are proper fractions?

$$\frac{3}{5}, \quad \frac{8}{7}, \quad \frac{2}{9}, \quad \frac{7}{11}, \quad \frac{8}{8}.$$

5. Which of the following are improper fractions?

$$\frac{8}{7}, \quad \frac{12}{13}, \quad \frac{7}{7}, \quad \frac{8}{1}.$$

6. Which of the following fractions are in lowest terms?

$$\frac{9}{27}, \quad \frac{4}{7}, \quad \frac{13}{3}, \quad \frac{14}{21}.$$

*In Problems **7–14**, write the fractions as mixed numbers.*

7. $\dfrac{23}{3}$.

8. $\dfrac{37}{6}$.

9. $\dfrac{18}{5}$.

10. $\dfrac{74}{9}$.

11. $\dfrac{3}{2}$.

12. $\dfrac{251}{24}$.

13. $\dfrac{13}{4}$.

14. $\dfrac{5003}{1000}$.

15. Write $\frac{2}{3}$ as an equivalent fraction with denominator 18.
16. Write $\frac{3}{7}$ as an equivalent fraction with numerator 21.

*In Problems **17–28**, fill in the missing numbers.*

17. $\dfrac{4}{5} = \dfrac{?}{25}$.

18. $\dfrac{15}{24} = \dfrac{5}{?}$.

19. $\dfrac{2}{3} = \dfrac{22}{?}$.

20. $\dfrac{12}{32} = \dfrac{?}{8}$.

21. $\dfrac{8}{24} = \dfrac{1}{?}$.

22. $\dfrac{100}{5} = \dfrac{?}{1}$.

23. $\dfrac{18}{12} = \dfrac{?}{2}$.

24. $7 = \dfrac{?}{3}$.

25. $5 = \dfrac{?}{3}$.

26. $3 = \dfrac{?}{12}$.

27. $\dfrac{1}{7} = \dfrac{7}{?}$.

28. $\dfrac{6}{6} = \dfrac{?}{12}$.

*In Problems **29–50**, completely reduce the fractions.*

29. $\dfrac{28}{12}$.

30. $\dfrac{210}{26}$.

31. $\dfrac{60}{45}$.

32. $\dfrac{18}{99}$.

33. $\dfrac{24}{30}$.

34. $\dfrac{275}{1000}$.

35. $\dfrac{66}{144}$.

36. $\dfrac{7+9}{2}$.

37. $\dfrac{4}{8+8}$.

38. $\dfrac{(10)(3)}{5}$.

39. $\dfrac{6}{(12)(5)}$.

40. $\dfrac{(9)(10)}{(6)(8)}$.

41. $\dfrac{(3)(16)}{(4)(9)}$.

42. $\dfrac{(4)(14)}{(21)(2)}$.

43. $\dfrac{(21)(15)}{(5)(14)}$.

44. $\dfrac{(4)(9)(6)}{(6)(18)}$.

45. $\dfrac{(8)(12)(15)}{(3)(4)(6)}$.

46. $\dfrac{(100)(12)(5)}{(6)(15)(25)}$.

47. $\dfrac{(12)(6)(3)(14)}{(22)(14)(21)(18)}$.

48. $\dfrac{(4)(12)(10)(7)}{(14)(8)(5)(6)}$.

49. $\dfrac{(81)(5)(3)(7)}{(27)(18)(4)}$.

50. $\dfrac{(24)(50)(3)}{(75)(6)(16)(5)}$.

0.4 ADDITION AND SUBTRACTION OF FRACTIONS

The sum (or difference) of two fractions with the *same* denominator is a fraction which has the same (common) denominator and a numerator which is the sum (or difference) of the numerators of the original fractions.

Example 1

a. $\dfrac{2}{3} + \dfrac{5}{3} = \dfrac{2+5}{3} = \dfrac{7}{3}$.

b. $\dfrac{8}{9} - \dfrac{6}{9} = \dfrac{8-6}{9} = \dfrac{2}{9}$.

c. $\dfrac{7}{18} + \dfrac{3}{18} - \dfrac{5}{18} = \dfrac{7+3-5}{18} = \dfrac{5}{18}$.

d. $\dfrac{12}{27} - \dfrac{6}{27} + \dfrac{4}{27} - \dfrac{1}{27} = \dfrac{12-6+4-1}{27} = \dfrac{9}{27} = \dfrac{1}{3}$.   *Always reduce your answer, if possible.*

When adding fractions, we *do not* add the denominators. For example,

$$\frac{1}{2} + \frac{1}{2} = \frac{1+1}{2+2} \quad \text{is } \textit{false:}$$

$$\frac{1}{2} + \frac{1}{2} = \frac{1+1}{2} = \frac{2}{2} = 1, \quad \text{and} \quad \frac{1+1}{2+2} = \frac{2}{4} = \frac{1}{2}.$$

To add or subtract fractions with denominators that are not the same, the fractions must first be rewritten as equivalent fractions that *do* have a common denominator. The simplest common denominator to use is the least whole number such that each of the given denominators is a factor of it. This is called the **least common denominator**, or L.C.D., of the fractions. For example, the denominators of $\frac{1}{3}$, $\frac{5}{8}$, and $\frac{7}{12}$ are 3, 8, and 12. These denominators divide 24, and 24 is the *least* such number that has 3, 8, and 12 as factors. Thus 24 is the L.C.D.

Basically, finding an L.C.D. involves using *prime numbers*. A **prime number** is a whole number greater than 1 that has only two whole-number factors: itself and 1. Thus 2, 3, 5, 7, 11, 13, and 17 are prime numbers. However, 15 is *not* a prime, because 15 has 3 and 5 as well as 15 and 1 as factors. In Example 2 we illustrate the following fact:

> Every whole number greater than 1 can be written as a product of primes.

Example 2

Write each number as a product of primes.

a. $20 = (2)(10) = (2)(2)(5)$.   [Because $10 = (2)(5)$.]

b. $35 = (5)(7)$.

c. $42 = (2)(21) = (2)(3)(7)$.

d. $24 = (2)(12) = (2)(2)(6) = (2)(2)(2)(3)$.

How to Find an L.C.D.

In general, to find the L.C.D. of a group of fractions, first write each denominator as a product of prime factors. Next, from the different prime factors that occur, form a product in which the number of times each different prime appears is the greatest number of times that it appears in any *single* denominator. This product is the L.C.D. For example, to determine the L.C.D. of $\frac{1}{3}$, $\frac{5}{12}$, and $\frac{7}{10}$, we first write each denominator as a product of primes:

$$3 = 3,$$

$$12 = (2)(6) = (2)(2)(3),$$

$$10 = (5)(2).$$

There are three different prime factors involved: 3, 2, and 5. The greatest number of times that 3 appears in any *single* denominator is once. For 2, it is twice. For 5, it is once. Thus

$$\text{L.C.D.} = (3)(2)(2)(5) = 60.$$

Similarly, the L.C.D. of $\frac{1}{6}$, $\frac{2}{9}$, and $\frac{5}{12}$ is determined as follows:

$$6 = (3)(2),$$
$$9 = (3)(3),$$
$$12 = (2)(2)(3).$$

There are two different prime factors involved: 3 and 2. In any single denominator, the most often that 3 occurs is twice. The most often that 2 occurs is also twice. Thus

$$L.C.D. = (3)(3)(2)(2) = 36.$$

To show how to add (or subtract) fractions with different denominators, we shall find

$$\frac{1}{6} + \frac{2}{9} + \frac{5}{12}.$$

First we find the L.C.D. As shown before, it is 36. Next we write each fraction as an equivalent fraction with a denominator of 36.

$$\frac{1}{6} = \frac{(1)(6)}{(6)(6)} = \frac{6}{36}, \qquad \frac{2}{9} = \frac{(2)(4)}{(9)(4)} = \frac{8}{36}, \qquad \frac{5}{12} = \frac{(5)(3)}{(12)(3)} = \frac{15}{36}.$$

Now that all denominators are the same, we add the fractions directly as in Example 1.

$$\frac{1}{6} + \frac{2}{9} + \frac{5}{12} = \frac{6}{36} + \frac{8}{36} + \frac{15}{36} = \frac{6+8+15}{36} = \frac{29}{36}.$$

Example 3

Find $\dfrac{2}{3} + \dfrac{7}{8} + \dfrac{3}{16}$.

Solution Since $3 = 3$, $8 = (2)(2)(2)$, and $16 = (2)(2)(2)(2)$, then

$$L.C.D. = (3)(2)(2)(2)(2) = 48.$$

Thus

$$\frac{2}{3} + \frac{7}{8} + \frac{3}{16} = \frac{(2)(16)}{(3)(16)} + \frac{(7)(6)}{(8)(6)} + \frac{(3)(3)}{(16)(3)}$$

$$= \frac{32}{48} + \frac{42}{48} + \frac{9}{48} = \frac{32+42+9}{48} = \frac{83}{48}.$$

Example 4

Perform the operations and simplify.

a. $\dfrac{3}{5} - \dfrac{1}{2} + \dfrac{1}{4} = \dfrac{12}{20} - \dfrac{10}{20} + \dfrac{5}{20} = \dfrac{12 - 10 + 5}{20} = \dfrac{7}{20}.$

b. $2 + \dfrac{1}{6} + \dfrac{1}{3} = \dfrac{2}{1} + \dfrac{1}{6} + \dfrac{1}{3} = \dfrac{12}{6} + \dfrac{1}{6} + \dfrac{2}{6} = \dfrac{15}{6} = \dfrac{5}{2}.$

Just as we can write an improper fraction as a mixed number, we can also write a mixed number as an improper fraction. To do this, we write the whole-number part as a fraction and add it to the fractional part.

Example 5

a. $3\dfrac{3}{4} = 3 + \dfrac{3}{4} = \dfrac{3}{1} + \dfrac{3}{4} = \dfrac{12}{4} + \dfrac{3}{4} = \dfrac{15}{4}.$

b. $5\dfrac{2}{7} = \dfrac{5}{1} + \dfrac{2}{7} = \dfrac{35}{7} + \dfrac{2}{7} = \dfrac{37}{7}.$

To add mixed numbers, we can add the whole-number parts first and treat the fractional parts as before. For example,

$$1\dfrac{3}{4} + 2\dfrac{2}{3} = (1 + 2) + \left(\dfrac{3}{4} + \dfrac{2}{3}\right)$$

$$= 3 + \left(\dfrac{9}{12} + \dfrac{8}{12}\right)$$

$$= 3 + \dfrac{17}{12} = 3 + 1\dfrac{5}{12}$$

$$= 4\dfrac{5}{12}.$$

To subtract one mixed number from another, a slight modification in the above procedure may be necessary, as shown in the following example:

$$6\dfrac{1}{3} - 2\dfrac{2}{3}.$$

We cannot subtract $\frac{2}{3}$ from $\frac{1}{3}$ (without introducing negative numbers). However, we can express $6\frac{1}{3}$ as $5 + 1 + \frac{1}{3} = 5 + \frac{3}{3} + \frac{1}{3} = 5\frac{4}{3}$. Thus we can rewrite the problem as

$$6\dfrac{1}{3} - 2\dfrac{2}{3} = 5\dfrac{4}{3} - 2\dfrac{2}{3} = 5 - 2 + \dfrac{4}{3} - \dfrac{2}{3} = 3\dfrac{2}{3}.$$

Alternatively, we can express the given fractions as improper fractions and subtract. (This method also applies to addition.)

$$6\frac{1}{3} - 2\frac{2}{3} = \frac{19}{3} - \frac{8}{3} = \frac{11}{3} = 3\frac{2}{3}.$$

Example 6

a. $3\frac{3}{4} + 4\frac{2}{3} = 3\frac{9}{12} + 4\frac{8}{12} = 7\frac{17}{12} = 8\frac{5}{12}.$

b. $4\frac{3}{4} - 2\frac{7}{8} = 4\frac{6}{8} - 2\frac{7}{8} = 3\frac{14}{8} - 2\frac{7}{8} = 1\frac{7}{8}.$

Problem Set 0.4

In Problems 1–8, perform the operations and simplify.

1. $\frac{7}{8} + \frac{17}{8}.$

2. $\frac{8}{9} - \frac{6}{9}.$

3. $\frac{7}{12} - \frac{5}{12}.$

4. $\frac{7}{18} + \frac{3}{18} - \frac{5}{18}.$

5. $\frac{5}{13} + \frac{11}{13} - \frac{4}{13}.$

6. $\frac{12}{27} - \frac{6}{27} + \frac{4}{27} - \frac{1}{27}.$

7. $\frac{2}{8 + 8}.$

8. $\frac{6 + 6}{3 + 3}.$

In Problems 9–16, find the L.C.D. of the fractions.

9. $\frac{5}{6}, \frac{2}{9}.$

10. $\frac{4}{21}, \frac{3}{14}.$

11. $\frac{7}{4}, \frac{3}{2}, \frac{1}{5}.$

12. $\frac{1}{15}, \frac{3}{8}, \frac{7}{10}.$

13. $\frac{1}{3}, \frac{7}{18}, \frac{5}{12}.$

14. $\frac{5}{6}, \frac{9}{20}, \frac{2}{15}.$

15. $\frac{7}{30}, \frac{5}{12}, \frac{11}{20}.$

16. $\frac{3}{20}, \frac{14}{25}, \frac{27}{50}.$

In Problems 17–20, write the mixed numbers as fractions.

17. $4\frac{3}{5}.$

18. $3\frac{2}{3}.$

19. $7\frac{2}{7}.$

20. $10\frac{1}{10}.$

In Problems 21–38, perform the operations and simplify.

21. $\frac{3}{4} + \frac{5}{6}.$

22. $\frac{3}{5} + \frac{2}{3}.$

23. $\frac{3}{8} - \frac{5}{14}.$

24. $\frac{5}{6} - \frac{2}{21}.$

25. $\frac{1}{3} + \frac{8}{9} + \frac{5}{12}.$

26. $\frac{2}{3} + \frac{3}{4} + \frac{5}{8}.$

27. $\dfrac{7}{4} - \dfrac{3}{2} + \dfrac{1}{5}.$ 28. $\dfrac{1}{5} + \dfrac{1}{8} + \dfrac{1}{12}.$ 29. $\dfrac{5}{6} + \dfrac{2}{15} - \dfrac{1}{3}.$

30. $\dfrac{7}{12} - \dfrac{5}{18} + \dfrac{3}{8}.$ 31. $7\dfrac{2}{3} - 4\dfrac{3}{4}.$ 32. $2\dfrac{1}{12} - 1\dfrac{7}{8}.$

33. $3\dfrac{2}{3} + 4\dfrac{1}{8} - 6\dfrac{1}{4}.$ 34. $4\dfrac{2}{7} - 2\dfrac{2}{5} + 1\dfrac{2}{35}.$ 35. $3 - \dfrac{7}{15} + \dfrac{3}{10}.$

36. $2 + \dfrac{3}{8} - \dfrac{1}{6}.$ 37. $\dfrac{2}{5} + 2 - \dfrac{3}{8}.$ 38. $\dfrac{5}{12} + \dfrac{11}{18} - 1.$

0.5 MULTIPLICATION AND DIVISION OF FRACTIONS

The product of two or more fractions is a fraction whose numerator and denominator are obtained by multiplying the numerators and denominators of the original fractions, respectively. For example,

$$\frac{2}{3} \cdot \frac{5}{7} = \frac{(2)(5)}{(3)(7)} = \frac{10}{21}.$$

Example 1

a. $\dfrac{3}{5} \cdot \dfrac{8}{4} \cdot \dfrac{2}{9} = \dfrac{(3)(8)(2)}{(5)(4)(9)} = \dfrac{\overset{1}{(\cancel{3})}\overset{2}{(\cancel{8})}(2)}{(5)(\cancel{4})(\cancel{9})} = \dfrac{4}{15}.$

We could have canceled immediately, as the following shows:

$$\frac{\overset{1}{\cancel{3}}}{5} \cdot \frac{\overset{2}{\cancel{8}}}{\underset{1}{\cancel{4}}} \cdot \frac{2}{\underset{3}{\cancel{9}}} = \frac{4}{15}.$$

b. $\dfrac{17}{3} \cdot \dfrac{11}{4} \cdot \dfrac{2}{17} = \dfrac{\overset{1}{\cancel{17}}}{3} \cdot \dfrac{11}{\underset{2}{\cancel{4}}} \cdot \dfrac{\overset{1}{\cancel{2}}}{\underset{1}{\cancel{17}}} = \dfrac{11}{6}.$

c. $3\left(\dfrac{1}{2}\right) = \left(\dfrac{3}{1}\right)\left(\dfrac{1}{2}\right) = \dfrac{3}{2}.$

d. $5 \cdot \dfrac{2}{5} = \dfrac{\overset{1}{\cancel{5}}}{1} \cdot \dfrac{2}{\underset{1}{\cancel{5}}} = \dfrac{2}{1} = 2.$

Usually we simply write

$$\cancel{5} \cdot \frac{2}{\cancel{5}} = 2.$$

e. $\dfrac{4}{9} \cdot 18 = \dfrac{4}{\underset{1}{9}} \cdot \overset{2}{18} = 8.$

f. $\left(2\dfrac{1}{2}\right)\left(\dfrac{4}{5}\right) = \dfrac{\overset{1}{\cancel{5}}}{\underset{1}{2}} \cdot \dfrac{\overset{2}{\cancel{4}}}{\underset{1}{\cancel{5}}} = \dfrac{2}{1} = 2.$

Note that we first wrote the mixed number $2\frac{1}{2}$ as the improper fraction $\frac{5}{2}$.

Turning to division of fractions, we consider $\frac{2}{3} \div \frac{3}{4}$ or

$$\dfrac{\dfrac{2}{3}}{\dfrac{3}{4}}.$$

If we use the fundamental principle and multiply both $\frac{2}{3}$ and $\frac{3}{4}$ by $\frac{4}{3}$, we obtain

$$\dfrac{\dfrac{2}{3}}{\dfrac{3}{4}} = \dfrac{\dfrac{2}{3} \cdot \dfrac{4}{3}}{\dfrac{3}{\cancel{4}} \cdot \dfrac{\cancel{4}}{\cancel{3}}} = \dfrac{\dfrac{2}{3} \cdot \dfrac{4}{3}}{1} = \dfrac{2}{3} \cdot \dfrac{4}{3} = \dfrac{8}{9}.$$

Notice that to *divide* $\frac{2}{3}$ by $\frac{3}{4}$, we can simply interchange the numerator and denominator of the divisor, $\frac{3}{4}$, and *multiply* $\frac{2}{3}$ by the result (look above at the next-to-last step). That is,

$$\dfrac{2}{3} \div \dfrac{3}{4} = \dfrac{2}{3} \cdot \dfrac{4}{3} = \dfrac{8}{9}.$$

The interchanging of numerator and denominator is called **inverting**. Thus to divide one fraction by another, we *invert the divisor and multiply.*

Division of fractions can be described in terms of the *reciprocal* of a number. The **reciprocal** of a number, *other than zero*, is 1 divided by that number. For example, the reciprocal of 2 is $\frac{1}{2}$ and the reciprocal of $\frac{3}{4}$ is

$$\dfrac{1}{\dfrac{3}{4}} = 1 \div \dfrac{3}{4} = 1 \cdot \dfrac{4}{3} = \dfrac{4}{3}.$$

From this last example it is clear that *the reciprocal of a fraction can be found by interchanging its numerator with its denominator.* Thus we can say that $\frac{2}{3}$ divided by $\frac{3}{4}$ is equal to $\frac{2}{3}$ multiplied by the reciprocal of $\frac{3}{4}$, or $\frac{4}{3}$.

Example 2

a. $\dfrac{7}{8} \div \dfrac{5}{4} = \dfrac{7}{8} \cdot \dfrac{\overset{1}{\cancel{4}}}{5} = \dfrac{7}{10}.$

b. $\dfrac{\frac{3}{5}}{\frac{9}{10}} = \dfrac{3}{5} \div \dfrac{9}{10} = \dfrac{\overset{1}{\cancel{3}}}{\cancel{5}} \cdot \dfrac{\overset{2}{\cancel{10}}}{\cancel{9}} = \dfrac{2}{3}.$

c. $\dfrac{4}{\frac{3}{2}} = 4 \div \dfrac{3}{2} = 4 \cdot \dfrac{2}{3} = \dfrac{4}{1} \cdot \dfrac{2}{3} = \dfrac{8}{3}.$

d. $\dfrac{\frac{3}{8}}{4} = \dfrac{3}{8} \div 4 = \dfrac{3}{8} \div \dfrac{4}{1} = \dfrac{3}{8} \cdot \dfrac{1}{4} = \dfrac{3}{32}.$

e. $4\dfrac{1}{3} \div \dfrac{5}{3} = \dfrac{13}{\cancel{3}} \cdot \dfrac{\overset{1}{\cancel{3}}}{5} = \dfrac{13}{5}.$

Example 3

Simplify.

a. $\dfrac{2 - \frac{3}{4}}{3 + \frac{1}{8}}.$

We first separately simplify the numerator $2 - \frac{3}{4}$ and the denominator $3 + \frac{1}{8}$.

$$\dfrac{2 - \dfrac{3}{4}}{3 + \dfrac{1}{8}} = \dfrac{\dfrac{8}{4} - \dfrac{3}{4}}{\dfrac{24}{8} + \dfrac{1}{8}} = \dfrac{\dfrac{5}{4}}{\dfrac{25}{8}} = \dfrac{\overset{1}{\cancel{5}}}{\cancel{4}} \cdot \dfrac{\overset{2}{\cancel{8}}}{\cancel{25}} = \dfrac{2}{5}.$$

b. $\dfrac{\frac{1}{5} + \frac{3}{4}}{\frac{1}{8}} = \dfrac{\frac{4}{20} + \frac{15}{20}}{\frac{1}{8}} = \dfrac{\frac{19}{20}}{\frac{1}{8}} = \dfrac{19}{\cancel{20}} \cdot \dfrac{\overset{2}{\cancel{8}}}{1} = \dfrac{38}{5}.$

c. $\dfrac{\left(\frac{3}{5}\right)\left(\frac{4}{7}\right)}{\frac{6}{11}} = \dfrac{\overset{1}{\cancel{3}}}{5} \cdot \dfrac{\overset{2}{\cancel{4}}}{7} \cdot \dfrac{11}{\underset{1}{\underset{\cancel{2}}{\cancel{6}}}} = \dfrac{22}{35}.$

Problem Set 0.5

*In Problems **1–42**, compute and give answers in reduced form.*

1. $\left(\dfrac{3}{5}\right)\left(\dfrac{25}{9}\right).$

2. $\left(\dfrac{8}{3}\right)\left(\dfrac{15}{4}\right).$

3. $\dfrac{14}{15}\cdot\dfrac{25}{24}.$

4. $\dfrac{7}{12}\cdot 9.$

5. $\dfrac{6}{7}\cdot\dfrac{0}{3}.$

6. $0\cdot\dfrac{2}{15}.$

7. $(7)\left(\dfrac{6}{21}\right).$

8. $\left(\dfrac{2}{3}\right)(5).$

9. $\dfrac{1}{9}\cdot\dfrac{10}{3}.$

10. $\dfrac{1}{4}\cdot\dfrac{3}{1}.$

11. $\dfrac{3}{4}\cdot\dfrac{8}{5}\cdot\dfrac{4}{9}.$

12. $\dfrac{2}{5}\cdot\dfrac{3}{6}\cdot\dfrac{7}{5}.$

13. $\dfrac{3}{5}\cdot\dfrac{4}{11}\cdot\dfrac{7}{3}\cdot\dfrac{25}{4}.$

14. $7\cdot\dfrac{8}{5}\cdot\dfrac{6}{12}\cdot\dfrac{10}{49}.$

15. $\left(2\dfrac{2}{5}\right)\left(1\dfrac{1}{8}\right).$

16. $(8)\left(3\dfrac{1}{8}\right).$

17. $\left(5\dfrac{2}{3}\right)\left(2\dfrac{3}{4}\right)\left(\dfrac{2}{17}\right).$

18. $\left(2\dfrac{2}{3}\right)\left(3\dfrac{1}{2}\right)(4).$

19. $\dfrac{8}{3}\div\dfrac{5}{4}.$

20. $\dfrac{7}{5}\div\dfrac{3}{10}.$

21. $16\div\dfrac{12}{5}.$

22. $0\div\dfrac{8}{4}.$

23. $\dfrac{\frac{7}{10}}{\frac{21}{5}}.$

24. $\dfrac{\frac{14}{3}}{\frac{6}{15}}.$

25. $\dfrac{\frac{18}{11}}{\frac{8}{33}}.$

26. $\dfrac{\frac{2}{5}}{\frac{2}{5}}.$

27. $\dfrac{\frac{3}{5}}{\frac{2}{2}}.$

28. $\dfrac{\frac{7}{1}}{\frac{1}{4}}.$

29. $\dfrac{\frac{4}{1}}{\frac{1}{5}}.$

30. $\dfrac{\frac{3}{5}}{\frac{6}{6}}.$

31. $\dfrac{\frac{4}{8}}{\frac{8}{9}}.$

32. $\dfrac{\frac{1}{2}}{\frac{2}{3}}.$

33. $\dfrac{\frac{12}{25}\cdot\frac{15}{7}}{20}.$

34. $\dfrac{\frac{4}{9}}{\frac{2}{3}\cdot 8}.$

35. $\dfrac{6+\frac{1}{3}}{7}.$

36. $\dfrac{\frac{3}{4}-\frac{3}{16}}{\frac{1}{3}}.$

37. $\dfrac{7-\frac{2}{3}}{15-\frac{1}{3}}.$

38. $\dfrac{1+\frac{1}{2}}{1-\frac{1}{2}}.$

39. $\dfrac{\frac{8}{5}+\frac{2}{3}}{2+\frac{4}{7}}.$

40. $\dfrac{\dfrac{6}{7} - \dfrac{6}{7}}{\dfrac{4}{3} + \dfrac{5}{3}}.$

41. $\dfrac{\dfrac{1}{2} - \dfrac{1}{3}}{\dfrac{1}{4} + \dfrac{1}{5}}.$

42. $\dfrac{\left(\dfrac{2}{3}\right)\left(\dfrac{4}{5}\right) + 1}{2 + \dfrac{1}{15}}.$

In Problems **43–48**, *find the reciprocals.*

43. 6.

44. $\dfrac{3}{2}.$

45. $\dfrac{3}{5}.$

46. 1.

47. $\dfrac{1}{9}.$

48. $\dfrac{1}{2}.$

0.6 OPERATIONS WITH DECIMALS

A **decimal fraction** is a fraction whose denominator is a power of 10, that is, 10, 100, 1000, and so on. A decimal fraction, such as $\frac{6}{10}$, $\frac{15}{100}$, $\frac{121}{1000}$, or $\frac{62}{1000}$, can be written as an equivalent *decimal* number if we remove the denominator and place a *decimal point* in the numerator. We position the decimal point so that the number of digits to its right is equal to the number of zeros in the denominator of the original fraction. This means that to the right of the decimal point are units of tenths, hundredths, thousandths, and so on. For example, we shall convert the following fractions to decimal form:

$$\frac{6}{10} = 0.6 \qquad \text{one digit to the right,}$$

$$\frac{15}{100} = 0.15 \qquad \text{two digits to the right,}$$

$$\frac{121}{1000} = 0.121 \qquad \text{three digits to the right,}$$

$$\frac{1624}{100} = 16.24 \qquad \text{two digits to the right,}$$

$$\frac{62}{1000} = 0.062 \qquad \text{three digits to the right.}$$

Because $\frac{6}{10} = 0.6$ and $\frac{6}{10} = \frac{60}{100} = 0.60$, you can see that inserting zeros to the right of the decimal point after the last digit of a decimal does not change its *value*.

To add or subtract decimals, we place each number so that the decimal points are aligned, one under another, and then add or subtract the numbers as if there were no decimal points. In the result, we insert a decimal point that is aligned with the others.

Example 1

a. Find $0.123 + 1.624 + 0.0621 + 0.1$.

Solution

$$
\begin{array}{r}
0.1230 \\
1.6240 \\
0.0621 \\
0.1000 \\
\hline
1.9091
\end{array}
$$

b. Find $0.726 - 0.0246$.

Solution

$$
\begin{array}{r}
0.7260 \\
-0.0246 \\
\hline
0.7014
\end{array}
$$

In multiplying two decimals, the decimal point can, as usual, be disregarded as far as the calculation is concerned. After obtaining the product, however, we insert a decimal point in it so that the number of digits to the right of the decimal point is the sum of the number of digits to the right of the decimal point in each of the numbers to be multiplied. If the number of digits in the product is not sufficient to do this, additional zeros can be inserted to the left of the leftmost digit. For example:

$$
\begin{array}{r}
6\,2.4 \\
\times\ 1.2\,3 \\
\hline
1\,8\,7\,2 \\
1\,2\,4\,8\ \ \\
6\,2\,4\ \ \ \ \\
\hline
7\,6.7\,5\,2
\end{array}
$$

The factor 62.4 has one digit to the right of the decimal point and the factor 1.23 has two digits to the right. Their sum, 3, is the number of digits to the right of the decimal point in the product. As an additional example, we have:

$$
\begin{array}{r}
0.1\,2\,3 \\
\times\ 0.0\,1\,2 \\
\hline
2\,4\,6 \\
1\,2\,3\ \ \\
\hline
0.0\,0\,1\,4\,7\,6
\end{array}
$$

Here, two zeros were properly inserted in the answer to agree with the requirement that there be six digits to the right of the decimal point. The additional zeros are necessary to determine the correct value of the decimal. We say that 0.001476 has six *decimal places* (to the right of the decimal point).

It should not be difficult for you to realize that to multiply any decimal by 10, we need only to move the decimal point one position to the right. Similarly, we move the decimal point two, three, and four places to the right to multiply by 100, 1000, and 10,000, respectively. For example:

$$(12.12)(10) = 121.2,$$

$$(12.12)(100) = 1212. = 1212,$$

$$(12.12)(10,000) = 121,200.$$

To divide a decimal by a power of 10, we move the decimal point to the left the same number of places as there are zeros in the power of ten. For example:

$$\frac{26.24}{10} = 2.624,$$

$$\frac{0.08624}{100} = 0.0008624,$$

$$\frac{1183.421}{1000} = 1.183421.$$

We can write a decimal as a common fraction by replacing the decimal point by the appropriate power of 10 in the denominator. For example,

$$62.5 = \frac{625}{10} = \frac{125}{2},$$

$$2.432 = \frac{2432}{1000} = \frac{304}{125}.$$

When using long division to divide a decimal by a decimal, we can think of the problem as a fraction and multiply both the numerator and denominator by a power of 10 that will make the denominator a whole number. That is, the decimal points in the numerator and denominator are shifted. For example, in the division 62.314 ÷ 72.62,

$$\frac{62.314}{72.62} \quad \text{becomes} \quad \frac{6231.4}{7262.}$$

or

$$72.62\overline{)62.314} \quad \text{becomes} \quad 7262\overline{)6231.4}.$$

The division is then performed without regard to the decimal point. However, a decimal point must be inserted in the quotient in such a manner that it is aligned with the decimal point in the dividend. For example:

```
                0.8 5 8 0 8
      7 2 6 2)6 2 3 1.4 0 0 0 0
             5 8 0 9 6
               4 2 1 8 0
               3 6 3 1 0
                 5 8 7 0 0
                 5 8 0 9 6
                   6 0 4 0 0
                   5 8 0 9 6
                     2 3 0 4
```

In this case the answer is not an exact quotient; to approximate it, we round the quotient by disregarding all digits beyond a desired decimal place. If the first digit to the right of the last retained digit is less than five, the last retained digit remains unchanged. If the first digit to the right of the last retained digit is five or more, the last retained digit is increased by one. For example, rounding 0.85808 to four, three, two, and one decimal places gives 0.8581, 0.858, 0.86, and 0.9, respectively. Similarly, to one decimal place:

$$48.329 \quad becomes \quad 48.3;$$

$$48.392 \quad becomes \quad 48.4;$$

$$48.076 \quad becomes \quad 48.1;$$

$$48.95 \quad becomes \quad 49.0.$$

The digits used to represent the accuracy of an approximate number, such as a number obtained in a physical measurement, are said to be **significant figures**, or **significant digits**. If an approximate number is not a whole number, the digits beginning with the leftmost nonzero digit and ending with the rightmost digit are significant.

Number	Significant figures
0.023	2
0.00062141	5
0.00010	2
2.43	3
2.430	4
2.04300	6
89.620	5
18.0	3

Unless stated otherwise, if an approximate number is a whole number, the digits beginning with the leftmost nonzero digit and ending with the rightmost nonzero digit are significant.

Number	Significant figures
35,000	2
8,020	3
7,890	3
284,000,000	3
284,001,000	6

In a problem involving multiplication or division of approximate numbers, the result usually should not have more significant figures than the number of significant figures in the least accurate number in the problem. To see why this is reasonable, suppose that the length and width of a rectangle are estimated to be 8.32 and 5.27 centimeters, respectively. Here the 8.32 actually signifies a number between 8.315 and 8.325. The 5.27 signifies a number between 5.265 and 5.275. You might be tempted to say that the area of the rectangle is $8.32 \times 5.27 = 43.8464$ square centimeters. However, since the length and width are measured only to an accuracy of three significant figures, we cannot expect to compute the area to an accuracy of six significant figures. Since both numbers have three-figure accuracy, so should our answer, which we give as 43.8 square centimeters. If the given numbers were exact, then we could use the value 43.8464.

Example 2

Assume all numbers are approximate and compute.

a. $22.3 \times 13.65 = 304$, not 304.4, because the least accurate number has three significant figures.

b. $\dfrac{85.32}{37.624} = 2.268$, not 2.2677.

c. $\dfrac{0.00349}{0.051} = 0.068$, not 0.0684, because the least accurate number has two significant figures.

In addition or subtraction of approximate numbers, the result should be no more accurate than the least precise of the numbers involved. For example, in the problem $21.31 + 100.6 = 121.91$, we round 121.91 to 121.9 because 100.6 is measured to the nearest tenth and thus is less precise than 21.31, which is measured to the nearest hundredth.

Sometimes we can express a common fraction as an equivalent decimal by placing a decimal point in the numerator and then dividing. For example:

$$\frac{5}{8} = \frac{5.}{8} = 0.625,$$

$$\frac{43}{64} = 0.671875.$$

We conclude this section with a brief discussion about *percentage*. The term **percentage** means *per one hundred*. For example, 5 percent means $5(\frac{1}{100})$, or $\frac{5}{100}$. Using the percent symbol (%), we have

$$5\% = \frac{5}{100}.$$

To write a percentage as a decimal, we move the decimal point two places to the left and remove the percent symbol. For example,

$$5\% = 0.05$$

$$12.2\% = 0.122.$$

$$9.6\% = 0.096.$$

To write a decimal as a percentage, we move the decimal point two places to the right and attach a percent symbol. For example,

$$0.08 = 8\%,$$

$$0.004 = 0.4\%,$$

$$0.23 = 23\%.$$

To find a percentage of a number, we write the percentage as a decimal and multiply the number by the decimal. For example,

$$25\% \text{ of } 80 \quad \text{is} \quad (0.25)(80) = 20.$$

Thus 25% of 80 is 20. Similarly,

$$4\% \text{ of } 250 \quad \text{is} \quad (0.04)(250) = 10.$$

Problem Set 0.6

In Problems **1–8**, *compute the given sums and differences. Assume all numbers are exact.*

1. $0.021 + 0.1206 + 12.6 + 123.$

2. $1.006 + 1.0 + 0.629 + 0.4.$

3. $71.62 + 0.01 + 0.0006 + 4.1.$

4. $84.0264 + 0.621 + 0.006 + 71.$

5. $0.6241 - 0.5968.$

6. $0.0026 - 0.00094.$

7. $1.006 - 0.99.$

8. $71.6241 - 12.345.$

*In Problems **9–14**, state the number of significant digits in each number.*

9. 0.0023.

10. 8.620.

11. 92,500.

12. 23.00.

13. 14.620.

14. 5000.

*In Problems **15–20**, round each number to three significant figures.*

15. 0.3256.

16. 84162.

17. 0.0026514.

18. 72.6001.

19. 983.7.

20. 7.698.

*In Problems **21–28**, compute the given approximate numbers.*

21. (72.64)(0.023).

22. (861.4)(0.6241).

23. (0.00601)(1.005)(0.04).

24. (0.621)(0.101)(0.8)(1000).

25. 24.530688 ÷ 0.310.

26. 246.21 ÷ 8.60.

27. 0.00241 ÷ 1.6.

28. 0.624 ÷ 1.002.

*In Problems **29–34**, change the given decimals to fractions.*

29. 0.6241.

30. 1.0241.

31. 0.006.

32. 22.01.

33. 0.62415.

34. 11.6.

*In Problems **35–38**, change the given fractions to decimals.*

35. $\dfrac{1}{8}$.

36. $\dfrac{63}{64}$.

37. $\dfrac{3}{32}$.

38. $\dfrac{160.7}{2500}$.

39. Find 4% of 67.

40. Find 1.1% of 550.

41. Find 15% of 4.

42. Find 125% of 50.

0.7 GEOMETRICAL CONCEPTS AND FORMULAS

At times we shall refer to basic results of plane geometry, some of which are now listed for you:

1. **A right angle** is an angle of 90°. The angle in Fig. 0.1 is a right angle, as indicated by the right-angle symbol ⌐.

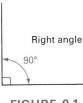

Right angle

90°

FIGURE 0.1

2. Two lines are **perpendicular** to each other if they intersect at right angles. The lines in Fig. 0.2 are (mutually) perpendicular, as indicated by the right-angle symbol.

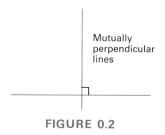

Mutually perpendicular lines

FIGURE 0.2

3. Two angles are **complementary** if their sum is 90°. Two angles are **supplementary** if their sum is 180°. In Fig. 0.3(a), angles A and B are complementary angles; in (b), angles A and B are supplementary.

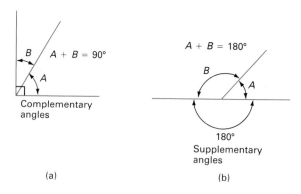

(a)   (b)

FIGURE 0.3

4. The **bisector** of an angle is the line that divides the angle into two angles of equal measure. In Fig. 0.4, line L bisects angle A.

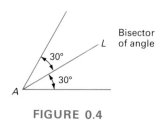

FIGURE 0.4

5. The **perpendicular bisector** of a line segment is the line that is perpendicular to the segment at its midpoint. In Fig. 0.5, line L is the perpendicular bisector of line segment AB.

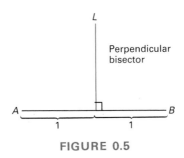

FIGURE 0.5

6. If two parallel lines are cut by another line, called a **transversal**, any pair of **alternate interior angles** formed are equal. In Fig. 0.6, the interior angles are A, B, C, and D. Thus angle A equals angle D and angle B equals angle C. Moreover, angles A and C are supplementary angles: $A + C = 180°$. Also, $B + D = 180°$. Angles E and B are **vertical angles**, which are equal. Thus $E = B$ and, similarly, $C = F$.

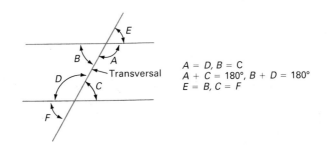

FIGURE 0.6

7. The sum of the angles of a triangle is $180°$ (see Fig. 0.7). The sum of the lengths of any two sides of a triangle is greater than the length of the third side. For example, $a + c$ is greater than b.

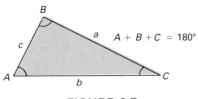

FIGURE 0.7

8. In an **isosceles** triangle (one in which at least two sides are equal), the angles opposite the equal sides have equal measure. In Fig. 0.8, $a = b$, so $A = B$.

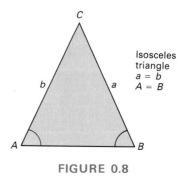

FIGURE 0.8

9. An **equilateral** triangle is one with three equal sides and, therefore, with three equal angles of 60°. The triangle in Fig. 0.9 is equilateral.

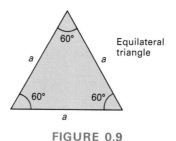

FIGURE 0.9

10. Two triangles are **similar** if their corresponding angles are equal. The triangles in Fig. 0.10 are similar: $A = A'$, $B = B'$, and $C = C'$. If two triangles are similar,

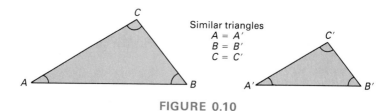

FIGURE 0.10

then the lengths of their corresponding sides are proportional. That is, for example,

$$\frac{\overline{AB}}{\overline{A'B'}} = \frac{\overline{CB}}{\overline{C'B'}},$$

where the *length* of the line segment from A to B is denoted by $\overline{AB}$ and similarly for the other sides.

11. Two triangles are **congruent** if they can be made to coincide, that is, if they have equal corresponding angles and sides. Two triangles are congruent if any of the following conditions holds.

    a. Two sides and the included angle of one triangle are equal, respectively, to two sides and the included angle of the other.

    b. Two angles and the included side of one triangle are equal, respectively, to two angles and the included side of the other.

    c. Three sides of one triangle are equal, respectively, to three sides of the other.

12. A **right triangle** has one angle of 90° (see Fig. 0.11). The side opposite the right angle is called the **hypotenuse**, and the other sides are called the **legs**. The

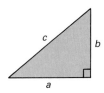

FIGURE 0.11

Pythagorean theorem states that the square of the length of the hypotenuse of a right triangle is equal to the sum of the squares of the lengths of the legs of the triangle. Symbolically,

$$c^2 = a^2 + b^2.$$

It follows that

$$c = \sqrt{a^2 + b^2}.$$

13. The **area of a triangle** is equal to one-half the product of the base and height. In Fig. 0.12, the area is $\frac{1}{2}bh$.

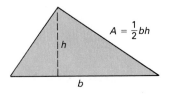

FIGURE 0.12

14. A **central angle** of a circle is an angle formed by two radii of the circle. **In Fig.** 0.13, angle A is a central angle.

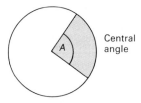

Central angle

FIGURE 0.13

15. In the following useful formulas, r denotes a radius and h denotes an altitude (or height).
    a. Circle:

$$\text{Area} = \pi r^2;$$
$$\text{Circumference} = 2\pi r.$$

    b. Right circular cylinder:

$$\text{Volume} = \pi r^2 h;$$
$$\text{Lateral surface area} = 2\pi rh;$$
$$\text{Total surface area} = 2\pi rh + 2\pi r^2.$$

    c. Right circular cone:

$$\text{Volume} = \frac{1}{3}\pi r^2 h.$$

    d. Sphere:

$$\text{Volume} = \frac{4}{3}\pi r^3;$$
$$\text{Surface area} = 4\pi r^2.$$

e. Trapezoid (a four-sided figure with a pair of parallel sides):

$$\text{Area} = \frac{1}{2}h(a + b).$$

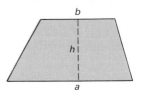

16. The **perimeter** of a geometric figure is the measure of the boundary of the figure. In Fig. 0.11, the perimeter is $a + b + c$.

Problem Set 0.7

1. Find the area of the shaded region in Fig. 0.14 if the radius of the inner circle is 8.72 centimeters and the radius of the outer circle is 9.88 centimeters. Use $\pi = 3.1416$ and give your answer to two decimal places.

FIGURE 0.14

2. Find the area of the region in Fig. 0.15. Assume that all units are centimeters.

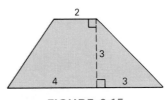

FIGURE 0.15

3. Find the hypotenuse of a right triangle if its other sides have lengths 6 and 8 units.

4. In Fig. 0.16, find angle E if $A = 20°$, $C = 90°$, and segments AB and CD are parallel.

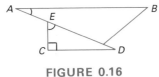

FIGURE 0.16

5. Find the length of the belt around the pulleys in Fig. 0.17 if the centers of the pulleys are 28 centimeters apart and the radius of each pulley is 5 centimeters. Use $\pi = 3.14$.

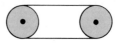

FIGURE 0.17

6. Find the area and perimeter of the metal plate in Fig. 0.18. Assume that all units are centimeters.

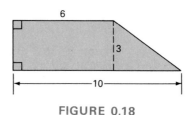

FIGURE 0.18

7. A crate is at rest on the inclined plane shown in Fig. 0.19. Find the angle A between the weight **W** and the perpendicular to the incline.

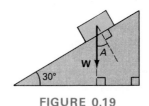

FIGURE 0.19

8. A cylindrical concrete pipe, which is 100 centimeters long, has an inner radius of 4 centimeters and an outer radius of 5 centimeters. Find the volume of concrete that was used to make the pipe. Use $\pi = 3.1416$.

9. Given a circle of radius 3 units, show that if the radius is doubled, then the area is four times as great.

10. Given a sphere of radius 3 units, show that if the radius is doubled, then the volume is eight times as great.

Operations with Real Numbers

1.1 REAL NUMBERS AND NUMBER LINES

At the foundation of any study of technical mathematics is a good working knowledge of basic algebra. Thus our immediate goal is to help you develop that background. In some ways, beginning the study of algebra is like beginning the study of a language. There are words, phrases, concepts, and symbols with meanings to be learned. As a first step, let us look at the types of numbers with which we deal.

Probably the numbers most familiar to you are the counting numbers 1, 2, 3, ... (the three dots mean *and so on*). These are called the **positive integers** or *natural numbers*. The numbers $-1, -2, -3, \ldots$ are the **negative integers**. The positive and negative integers together with the number zero make up the **integers**. Thus 3, -5, and 0 are integers.

A **rational number** is a number that can be written as an integer divided by a nonzero integer, that is, as the ratio of two integers. (Think of the word *rational* as *ratio*-nal.) Some rational numbers are

$$\frac{1}{2}, \quad \frac{5}{3}, \quad \frac{0}{6} \ (= 0), \quad \text{and} \quad \frac{-3}{7}.$$

Every integer is also a rational number. For example, 3 can be written as $\frac{3}{1}$, so 3 is rational. We emphasize that expressions like $\frac{1}{0}$ and $\frac{0}{0}$ do not satisfy the definition of a rational number. In fact, they actually have no meaning because *division by zero is not defined* (see Sec. 1.4 for the reason). So $\frac{1}{0}$ and $\frac{0}{0}$ are not numbers at all.

31

Any rational number can be written as a decimal that is either *terminating*, such as $\frac{3}{4} = 0.75$, or *repeating* (from some point on, a block of one or more digits repeats without end), such as

$$\tfrac{1}{3} = 0.333\ldots \quad \text{and} \quad \tfrac{15}{11} = 1.3636\ldots$$

(in the second case, the block of digits 36 repeats endlessly). It is also a fact that every terminating or repeating decimal represents a rational number. For example, 2.7 is $\frac{27}{10}$, which is rational.

A decimal that is nonterminating and nonrepeating is not a rational number and is called an **irrational number**. A well-known example of an irrational number is π (a Greek letter read *pi*), which is the value of the ratio of the circumference of a circle to the diameter of the circle:

$$\pi = 3.14159\ldots,$$

In a calculation, π is sometimes replaced by $\frac{22}{7}$ or 3.14, which are rational. However, these numbers are just approximations of π. We indicate this by writing $\pi \approx \frac{22}{7}$ or $\pi \approx 3.14$, where the symbol $\approx$ is read *is approximately equal to*. Another irrational number is $\sqrt{2}$ (read *the square root of* 2), which represents a number whose square is 2:

$$\sqrt{2} \approx 1.41421.$$

Together, the rational and irrational numbers make up the **real numbers**. In other words:

> The real numbers consist of all decimal numbers.

Thus all of the following are real numbers:

$$4, \qquad -2.35, \qquad \tfrac{1}{3}, \qquad \pi, \quad \text{and} \quad \sqrt{2}.$$

For the time being, whenever we use the word *number* we shall mean *real number*. (Numbers that are not real numbers are discussed in Chapter 13.)

Real numbers can be thought of as points on a line, like the markings on a thermometer. First we consider one direction along the line to be *positive* and the other direction to be *negative*. Usually the line is horizontal with the positive direction to the *right* and the negative direction to the *left*. The positive direction is indicated by an arrow (see Fig. 1.1). Next we select a point O, called the **origin**, to represent the number 0. Then we choose some unit of length, called a *unit distance*,

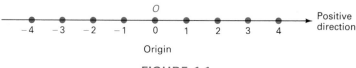

FIGURE 1.1

and use it to mark off successive points to the right and to the left of the origin. These points are identified with the positive and negative integers, as shown in Fig. 1.1.

The other points on the line are handled as follows. With *any* point on this line we associate a real number that depends on the position of the point with respect to the origin. For example, with the point $\frac{1}{2}$ unit from the origin in the *positive* direction, we associate the real number $\frac{1}{2}$ (see Fig. 1.2). The number $\frac{1}{2}$ is called the **coordinate** of

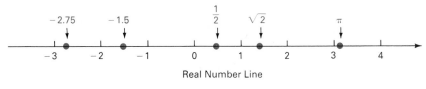

Real Number Line

FIGURE 1.2

that point, and the point is called the **graph** of the number. Similarly, the point 1.5 units to the *left* of the origin has coordinate -1.5. Figure 1.2 shows other points along with their coordinates. (Of course, the locations of some points are approximate, like that of the point with coordinate π.)

Each point on the line corresponds to exactly one real number, and each real number corresponds to exactly one point on the line. Because of these correspondences, we may think of real numbers as points, and vice versa. For example, we shall feel free to speak interchangeably of the *point* 2 and the *number* 2. For obvious reasons we call this line a **(real) number line** or a **coordinate line**.

Numbers to the right of the origin are called **positive numbers**, and numbers to the left of the origin are **negative numbers**. Thus real numbers (except zero) are *signed numbers*: that is, either positive (or plus) or negative (or minus). In fact, we could, for example, write 3 as $+3$ when we want to emphasize its sign.

In algebra it is typical to use letters like a, b, x, and y to stand for unspecified numbers (or quantities). Letters used in this way are called **literal numbers**. **Algebraic expressions** are numbers, including literal numbers, or numbers combined by operations. Examples are 6, $x + 3$, and x/y.

One powerful advantage of using letters together with operation symbols is that such a practice allows us to work with expressions that involve arbitrary numbers instead of specific ones. For example, the arithmetic sum $5 + 9$ is just a special case of the more general algebraic sum $a + b$, where a and b are *any* pair of numbers. Here a and b are called **terms** of the sum. The result of multiplying a and b is called the **product** of a and b, written $a \times b$ or $a \cdot b$, and a and b are called **factors** of the product. Other notations for the product of a and b are ab, $a(b)$, $(a)b$, and $(a)(b)$. For example, the product of 2 and a may be written as $2a$.

Another advantage of using algebraic symbols is that it makes it easy to express general rules. For example, just as $0 + 6 = 6$ and $1 \cdot 6 = 6$, we can state the more general rules that for any number a,

$$0 + a = a \qquad \text{and} \qquad 1 \cdot a = a.$$

In technical work, by using symbols we are able to express statements in a very simple and concise way. For example, an important relationship in physics is:

Force is equal to mass times acceleration.

Using mathematical shorthand, we can let the letter F denote force (this seems natural), m denote mass, and a denote acceleration and simply write

$$F = ma.$$

This symbolic form is compact and is certainly a convenient way to express the relationship.

A letter that represents a fixed number throughout a discussion is called a **constant**. Specific numbers, such as 2 and π, are also constants. On the other hand, a letter that can take on different values is called a **variable**, as illustrated in the next example.

Example 1

When a weight is attached to the lower end of a spring that is suspended from a support, there is a resulting elongation (or stretch) of the spring. The formula

$$F = ks \qquad (F \text{ equals } k \text{ times } s),$$

called *Hooke's law*, gives the relationship between the weight (or applied force) F and the resulting elongation s of the spring. Here k is a fixed number that depends on the nature of the particular spring, so k is a constant. However, hanging different weights results in different elongations of the spring. Thus F and s can each take on different values, so they are variables.

From time to time we shall refer to the notion of a *set*. A **set** is a collection of objects, called **members** or **elements** of that set. For example, the set of even numbers between 5 and 11 has as its elements the numbers 6, 8, and 10.

One way of specifying a set is to list its elements between braces. For example, the set of numbers 6, 8, 10 may be written

$$\{6, 8, 10\}.$$

The order in which the elements are listed is not important. We could have written the preceding set as $\{10, 6, 8\}$. There are four other ways to write it—can you find them?

Problem Set 1.1

*In Problems **1–6**, complete the statements.*

1. The numbers 1, 2, 3, ... are called _____ integers.
2. The integers consist of the positive integers, the negative integers, and _____ .

3. Real numbers that are not rational must be _____ numbers.

4. Every real number is a signed number except for the number _____ .

5. A collection of objects is called a _____ .

6. The number of elements in the set $\{1, 3, 0, -6\}$ is _____ .

In Problems 7–16, classify the statement as either true or false.

7. 15 is a positive number.

8. $\frac{2}{3}$ is an integer.

9. 0 is not rational.

10. π is a real number.

11. $\frac{5}{2}$ is a rational number.

12. -32 is an integer.

13. $\frac{4}{2}$ is not an integer.

14. An integer is positive or negative.

15. The sets $\{2, 4, 6\}$ and $\{4, 6, 2\}$ are not the same.

16. On the real number line, -6 is to the left of -5.

In Problems 17 and 18, draw a number line and plot (that is, locate) the points having the given coordinates.

17. (a) 0.   (b) -2.   (c) 1.5.   (d) $-3\frac{1}{2}$.

18. (a) 4.   (b) -2.5.   (c) $\frac{3}{2}$.   (d) $-\sqrt{2}$.

In Problems 19–22, determine which of the given two numbers is farther from zero on a number line.

19. 3, 4.   20. 7, -8.   21. -3, 1.   22. -5, -8.

In Problems 23–26, determine which of the given two numbers is to the right of the other number on a number line.

23. 6, -5.   24. -3, -4.   25. -13, -12.   26. 0, -2.

27. The formula $C = 2\pi r$ gives the circumference C of a circle with radius r. The formula applies to all possible circles. (a) Is π a constant or a variable? (b) Is r a constant or a variable?

1.2   INEQUALITIES AND ABSOLUTE VALUE

There is a geometrical way to compare two numbers in the sense that one number is "smaller" than the other. Let a and b be two real numbers. If a is to the *left* of b on a number line, we say that a is **less than** b, written $a < b$ (see Fig. 1.3). For example, 1 is

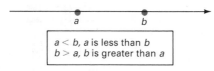

FIGURE 1.3

to the left of 3, so $1 < 3$. Another way to express that a is less than b is to say that b is **greater than** a, written $b > a$. This means that b is to the *right* of a on a number line (see Fig. 1.3). Thus $3 > 1$.

Statements involving inequality symbols like $<$ and $>$ are called **inequalities**. If you think of an inequality symbol as a pointer, it should always point to the "smaller" number.

Example 1

From Fig. 1.4 it is clear that

$$3 < 7, \qquad -2 < 1, \qquad 0 < 3, \qquad -5 < -2,$$
$$7 > 0, \qquad 0 > -2, \qquad 3 > -5, \qquad -2 > -5.$$

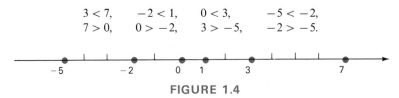

FIGURE 1.4

Because all positive numbers lie to the right of the origin, the statements *a is positive* and $a > 0$ mean the same. Similarly, *a is negative* means $a < 0$.

Two more inequality symbols are $\leq$ and $\geq$. Each plays a double role. The notation

$$a \leq b \quad \text{is read} \quad a \text{ is } \textbf{\textit{less than or equal to}} \text{ } b$$

and means that either $a < b$ or $a = b$. Similarly,

$$a \geq b \quad \text{is read} \quad a \text{ is } \textbf{\textit{greater than or equal to}} \text{ } b$$

and means that either $a > b$ or $a = b$.

Example 2

a. $3.4 \leq 3.5$ because $3.4 < 3.5$. Similarly, $-3 \leq -2$.

b. $7 \geq 5$ because $7 > 5$. Similarly, $-4 \geq -6$.

c. $3 \leq 3$ because $3 = 3$. Similarly, $\frac{1}{2} \leq 0.5$.

d. If $a \geq 0$, then $a > 0$ or $a = 0$, and a is said to be *nonnegative*.

If $a < b$ and c lies between a and b, then not only is $a < c$, but $c < b$ also. We indicate this by writing $a < c < b$. For example, $0 < 7 < 9$ and $-2 < -1 < 2$. Similarly, $-4 \leq x \leq 1$ means that $-4 \leq x$ and $x \leq 1$ *simultaneously*.

The notion of *absolute value* involves *distance* on a number line. The distance of a number a from the origin is called the **absolute value** (or *magnitude*) of a and is denoted by $|a|$.

$$|a| \text{ denotes the absolute value of } a.$$

For example, the numbers 3 and −3 are each three units from 0 (see Fig. 1.5), so

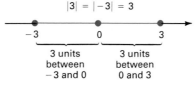

FIGURE 1.5

$|3| = 3$ and $|-3| = 3$. This illustrates that

$$|a| = |-a|.$$

Numbers that have the same absolute values, like 3 and −3, are said to be **numerically equal** or **equal in magnitude**. Other examples of absolute value are

$$|6| = 6, \qquad |-8| = 8, \qquad |0| = 0, \qquad |\tfrac{1}{2}| = \tfrac{1}{2}, \qquad |-\pi| = \pi.$$

Note that $|x|$ is *never* a negative number; it is always positive or 0:

$$|a| \geq 0.$$

Example 3

a. $|5 - 3| = |2| = 2$.

b. $|-2| > -4$ because $|-2| = 2$ and $2 > -4$.

c. $-5 < 3$, but $|-5| > |3|$ because $|-5| = 5$, $|3| = 3$ and $5 > 3$.

Problem Set 1.2

In Problems **1–10**, *write each statement by using inequality symbols.*

1. 4 is less than 5.
2. y is greater than 0.
3. −3 is greater than −5.
4. −2 is negative.
5. x is between 0 and 1.
6. x is less than −3.
7. x is positive.
8. x is greater than y.
9. y is greater than or equal to 4.
10. x is less than or equal to 1.

In Problems **11–14**, *complete the statements.*

11. Using the symbol $>$, we can write $3 < 4$ as _____.
12. Because 4 and −4 are both four units from 0, they have the same _____ _____.

13. The absolute value of a number must be either positive or _____ . It is never _____ .

14. If $a \geq 0$, then a is a _____ number.

In Problems 15–20, put $<$ or $>$ in the blank so that the inequality is true.

15. 6 ____ 12.

16. 5 ____ 0.

17. -4 ____ -6.

18. -2 ____ 1.

19. 0 ____ -3.

20. -5 ____ -2.

In Problems 21–26, find the value.

21. $|-4|$.

22. $|319|$.

23. $|-\frac{2}{3}|$.

24. $|0|$.

25. $|-0.14|$.

26. $|15 - 13|$.

In Problems 27–29, determine which of the two numbers has the greater absolute value.

27. 5, 3.

28. -5, -3.

29. -4, 2.

30. Arrange the numbers -6, -3, 0, 4, 5 in a list so that each number has a smaller absolute value than the numbers that follow it in the list.

In Problems 31–34, state whether the given inequality is true or false.

31. $|-2| > |-1|$.

32. $|-1| \leq 0$.

33. $|-9| < -8$.

34. $-4 < -1 < 2$.

35. A *resistor* is a device that provides opposition to the current in an electrical circuit. The unit of resistance is the *ohm*, which is denoted by Ω (the Greek letter omega). Suppose that a 50-Ω resistor is actually as much as 20% lower or as much as 20% higher than the stated value. (We say that the *tolerance* is 20%.)
   (a) Set up an inequality that indicates the values between which the actual resistance R lies.
   (b) Suppose that the current I (in units of amperes) through a resistor R is given by the formula

$$I = \frac{V}{R} \quad \text{[Ohm's law]},$$

where V is potential difference (loosely called voltage), in volts. If $V = 20$ volts in this formula, use the range of values of R from Part a to set up an inequality that indicates the values between which the actual current I lies.

1.3 BASIC LAWS OF REAL NUMBERS

The real numbers have three special properties that involve the operations of addition and multiplication. They are the *commutative*, *associative*, and *distributive laws*. These are powerful laws because many procedures in algebra are based on them. Therefore, it is important that you understand their meaning and make effective use of them.

The commutative laws state that we can add or multiply two numbers in any *order.* (In stating these and other laws, all letters represent arbitrary real numbers and can be replaced by any expression representing a real number.)

$$\boxed{\begin{array}{c} \text{COMMUTATIVE LAWS} \\ a + b = b + a, \qquad ab = ba. \end{array}}$$

For example,

$$2 + 3 = 3 + 2 \quad \text{because} \quad 2 + 3 = 5 \quad \text{and} \quad 3 + 2 = 5.$$

Similarly,

$$2 \cdot 3 = 6 = 3 \cdot 2.$$

Example 1

Using the commutative laws.

a. $3 + x = x + 3$.

b. $xy = yx$.

c. $x \cdot \frac{2}{3} = \frac{2}{3}x$.

The associative laws state that when finding the sum or product of three numbers, we can *group* the numbers in any way.

$$\boxed{\begin{array}{c} \text{ASSOCIATIVE LAWS} \\ (a + b) + c = a + (b + c), \\ a(bc) = (ab)c. \end{array}}$$

For example,

$$(3 + 4) + 5 = 3 + (4 + 5)$$

because

$$(3 + 4) + 5 = 7 + 5 = 12$$
$$\text{and} \quad 3 + (4 + 5) = 3 + 9 = 12.$$

Thus when we find the sum $3 + 4 + 5$, it does not matter whether we begin by adding the first two numbers or the last two numbers. Either way gives the same result.
Similarly,

$$(3 \cdot 4) \cdot 5 = 3 \cdot (4 \cdot 5)$$

because

$$(3 \cdot 4) \cdot 5 = 12 \cdot 5 = 60 \quad \text{and} \quad 3 \cdot (4 \cdot 5) = 3 \cdot 20 = 60.$$

By writing $3 \cdot 4 \cdot 5$ we are actually denoting either $(3 \cdot 4) \cdot 5$ or $3 \cdot (4 \cdot 5)$. Similarly, when we write the number $2xy$, we mean either $(2x)y$ or $2(xy)$, which are equal.

Example 2

Simplify by using the associative laws.

a. $(x + 5) + 9 = x + (5 + 9) = x + 14$.

b. $2(4y) = (2 \cdot 4)y = 8y$.

Example 3

Show that $x(2y) = 2xy$.

Solution

$$\begin{aligned} x(2y) &= (x \cdot 2)y && \text{[associative law]} \\ &= (2x)y && \text{[commutative law]} \\ &= 2xy. \end{aligned}$$

Example 3 illustrated an important general fact that is based on the commutative and associative laws. *We can rearrange or group the numbers in a sum or product in any way we want.*

Example 4

a. $2 + y + x + 4 = (2 + 4) + x + y = 6 + x + y$.

b. $2 \cdot \frac{1}{7} \cdot 50 \cdot 7 = (2 \cdot 50) \cdot (\frac{1}{7} \cdot 7) = 100 \cdot 1 = 100$.

c. $7 \cdot p \cdot 2 = (7 \cdot 2)p = 14p$.

d. $abcde = abedc = (be)(adc)$, and so on.

The distributive laws involve *both* addition and multiplication. They deal with a product in which one factor is a sum.

> **DISTRIBUTIVE LAWS**
>
> $a(b + c) = ab + ac,$
>
> $(a + b)c = ac + bc.$

We say that multiplication distributes over addition. For example, to find $7(2 + 3)$ we can multiply 7 by 5, or we can multiply 7 by *each* term in the second factor, $2 + 3$, and then add the results:

$$7(2 + 3) = 7 \cdot 2 + 7 \cdot 3.$$

To verify this we have

$$7(2 + 3) = 7 \cdot 5 = 35$$
$$\text{and} \quad 7 \cdot 2 + 7 \cdot 3 = 14 + 21 = 35.$$

Similarly, by the distributive law $(a + b)c = ac + bc$, we have

$$(2 + 3) \cdot 5 = 2 \cdot 5 + 3 \cdot 5$$

because

$$(2 + 3) \cdot 5 = 5 \cdot 5 = 25$$
$$\text{and} \quad 2 \cdot 5 + 3 \cdot 5 = 10 + 15 = 25.$$

Example 5

Perform the indicated operations by using the distributive laws.

a. $2(x + 5) = 2 \cdot x + 2 \cdot 5 = 2x + 10$. Note that

$$2(x + 5) \neq 2x + 5$$

because you must also multiply the 5 by 2. (The symbol $\neq$ is read *is not equal to*.)

b. $15(12) = 15(10 + 2) = 15 \cdot 10 + 15 \cdot 2 = 150 + 30 = 180$. You can do this problem mentally!

c. $(7 + x)y = 7 \cdot y + x \cdot y = 7y + xy$.

The distributive laws can be made more general to take care of cases where one factor has more than two terms. For example,

$$a(b + c + d) = ab + ac + ad.$$

Again we multiply the first factor, a, by each term of the second factor, $b + c + d$. Thus

$$3(4 + x + y) = 3 \cdot 4 + 3 \cdot x + 3 \cdot y = 12 + 3x + 3y.$$

Here are two other useful properties that are also called distributive laws.

$$a(b - c) = ab - ac,$$
$$(a - b)c = ac - bc.$$

By combining the distributive laws we can write such things as

$$x(y + z - w) = xy + xz - xw.$$

Example 6

Perform the indicated operations.

a. $x(y - 3) = xy - x(3)$ [distributive law]

   $\quad\quad\quad = xy - 3x$ [commutative law].

b. $(2 - x)y = 2y - xy$ [distributive law].

Problem Set 1.3

In Problems **1–10**, *name the law that is used in the given statement.*

1. $x \cdot 5 = 5 \cdot x$.
2. $x(zy) = x(yz)$.
3. $3(x + y) = 3x + 3y$.
4. $(x + y) + 1 = 1 + (x + y)$.
5. $3(2x) = (3 \cdot 2)x$.
6. $x + (x + y) = (x + x) + y$.
7. $4(x - y) = (x - y)4$.
8. $5(3 + 9) = 5(9 + 3)$.
9. $(s + 5)t = st + 5t$.
10. $(-1)[-3 + 4] = (-1)(-3) + (-1)(4)$.

In Problems **11–18**, *determine whether the given statement is true or false.*

11. $(x + y) + z = (y + x) + z$.
12. $7 + (4 + y) = (7 + 4) + y$.
13. $4(x + 4) = 4x + 4$.
14. $(x - 2)y = y(x - 2)$.
15. $3(x - 4) = 3x - 4$.
16. $x + 3y = 3y + x$.
17. $2(3 \cdot 6) = 2 \cdot 3 + 2 \cdot 6$.
18. $(3 + x) + 2 = 3 + (x + 2)$.

In Problems **19–36**, *perform the indicated operations.*

19. $(x + 3) + 5$.
20. $7 + (4 + y)$.
21. $4(3x)$.
22. $(2x)6$.
23. $x(5y)$.
24. $4 + x + 8$.
25. $4 + w + 2 + t$.
26. $3 \cdot x \cdot 5$.
27. $4(2 + x)$.
28. $(x + 2)y$.
29. $(5 - y)z$.
30. $6(7 + t)$.
31. $x(7 - y)$.
32. $(a - 3)2$.
33. $x(y + z - w)$.
34. $4(x - z + y)$.
35. $2(2 \cdot y \cdot 4)$.
36. $x(3 \cdot y \cdot 2)$.

1.4   THE ARITHMETIC OF REAL NUMBERS

We now look at the rules that deal with addition, subtraction, multiplication, and division of real numbers. Recall that real numbers (except 0) are signed numbers; each has a sign of either + or − . There are two rules of signs for addition. The first is:

To add two numbers having the same sign, add their absolute values and prefix the result with their common sign.

Thus the sum of two positive numbers is positive, and the sum of two negative numbers is negative.

Let us apply the above rule to find $(-2) + (-7)$. The sum of the absolute values is $2 + 7$ (or 9). We prefix this with a minus sign because the given numbers are negative.

$$(-2) + (-7) = -(2 + 7) = -9.$$

Example 1

a. $(+2) + (+7) = +(2 + 7) = 9$ (usually the $+$ sign is omitted).

b. $(-7) + (-\pi) = -(7 + \pi)$.

c. $\underbrace{(-2) + (-9)}_{-11} + (-3) = (-11) + (-3) = -14.$

The second rule for addition is:

ADDITION WITH UNLIKE SIGNS

To add a positive number and a negative number, subtract the smaller absolute value from the larger absolute value and prefix the result with the sign of the number having the larger absolute value. When two numbers with unlike signs have the same absolute value, their sum is zero.

Example 2

Perform the additions.

a. $3 + (-11)$.

Solution $|3| = 3, |-11| = 11$, and $|-11| > |3|$. Thus we subtract 3 from 11 and prefix the result with the sign of -11:

$$3 + (-11) = -(11 - 3) = -8.$$

b. $-4 + 9 = +(9 - 4) = 5$ (again, the $+$ sign is usually not written).

c. $-4 + 2 = -(4 - 2) = -2.$

d. $3 + (-3) = 0$, because 3 and -3 have unlike signs but have the same absolute value. In general,

$$a + (-a) = 0.$$

For each real number a, there is an *opposite*. The **opposite** (or **negative**) of a, written $-a$, is a number which when added to a gives 0. That is, $a + (-a) = 0$. For example,

The opposite of 4 is -4

because $4 + (-4) = 0$. Similarly,

The opposite of -4 is 4

because $-4 + 4 = 0$. The opposite of -4 is represented symbolically by $-(-4)$, so we have just shown that $-(-4) = 4$. In general, we have the rule

$$-(-a) = a.$$

That is, the opposite of the opposite of a number is the number itself. Moreover, the statement $-(-4) = 4$ shows that the negative of a number does not have to be a negative number. That is, the number $-a$ may be a positive number (as it is if $a = -4$). Finally, the opposite of 0 is 0 because $0 + 0 = 0$.

Example 3

a. $-(-5) = 5$.

b. To simplify $-(-xy)$, we let xy play the role of a in the rule $-(-a) = a$. Thus

$$-(-xy) = xy.$$

(We remind you that letters appearing in the rules can be replaced by any expression representing a real number.)

Using the idea of the opposite of a number, we can algebraically define absolute value without referring to the geometric notion of a number line:

ABSOLUTE VALUE

$$|a| = \begin{cases} a, & \text{if } a \geq 0, \\ -a, & \text{if } a < 0. \end{cases}$$

(We remind you that $-a$ is a positive number when $a < 0$. The minus sign indicates that we have changed the sign of a.) For example, $|4| = 4$ because $4 > 0$, and $|-4| = -(-4) = 4$ because $-4 < 0$. That is, the absolute value of -4 is the negative of -4.

We define subtraction in terms of addition and an opposite:

SUBTRACTION

The **difference** of a and b, denoted by $a - b$, is defined by

$$a - b = a + (-b).$$

That is, *to subtract b from a, **add** the opposite of b to a.*

Note: In the statement $a - b = a + (-b)$, the first minus sign refers to subtraction, but the second refers to an opposite. For example,

$$5 - (7) = 5 + (-7) = -2.$$

Subtraction is *not* commutative: $a - b \neq b - a$. For example,

$$5 - 7 = -2, \quad \text{but} \quad 7 - 5 = 2.$$

Example 4

Perform the subtractions.

a. $2 - (-7)$.

    Solution   The opposite of -7 is 7, so

$$2 - (-7) = 2 + (7) = 9.$$

    This illustrates that

$$a - (-b) = a + b.$$

b. $-12 - 10 = -12 + (-10) = -22.$

c. $-6 - (-10) = -6 + 10 = 4.$

d. $0 - 3 = 0 + (-3) = -3.$

Some problems involve a "string" of additions and subtractions, such as

$$11 - 5 + 4 - 9.$$

This is actually the *sum*

$$11 + (-5) + 4 + (-9).$$

The commutative and associative laws allow us to rearrange and group the terms so that positive numbers are together and negative numbers are together. Then the positive and negative numbers can be added separately:

$$\underbrace{11 + 4}_{} + \underbrace{(-5) + (-9)}_{} = 15 + (-14) = 1.$$

More simply, we write

$$11 - 5 + 4 - 9 = 11 + 4 - 5 - 9 = 15 - 14 = 1.$$

Similarly,

$$-12 + 4 - 6 + 3 = 7 - 18 = -11.$$

We now turn to multiplication. Whenever a number a is multiplied by 0, the result is 0:

$$\boxed{a \cdot 0 = 0.}$$

The rules of signs for multiplication are:

MULTIPLICATION

To multiply two positive numbers or two negative numbers, multiply their absolute values. To multiply a positive number and a negative number, multiply their absolute values and prefix the result with a minus sign.

These rules imply that the product of two positive numbers or two negative numbers is positive. The product of a positive number and a negative number is negative.

Example 5

a. $(+2)(+6) = 12.$

b. $(-3)(-6) = 3 \cdot 6 = 18.$

c. $(-2)(4) = -(2 \cdot 4) = -8.$

d. $(5)(-10) = -(5 \cdot 10) = -50.$

e. $(-1)(3) = -(1 \cdot 3) = -3.$

The result of Example 5(e) can be made more general:

$$(-1)a = -a.$$

Thus $(-1)x = -x$ and $-4 = (-1)(4)$.

Example 6

a. $\underbrace{(-6)(-2)}_{12}(-1) = 12(-1) = -12$. Here there is an odd number (3) of negative factors and

the product is negative. In general, *the product of an **odd** number of negative numbers is negative.*

b. $\underbrace{(-2)(-1)}_{2}\underbrace{(-3)(-4)}_{12} = (2)(12) = 24$. Here there is an even number (4) of negative factors.

In general, *the product of an **even** number of negative numbers is positive.*

The operation of division is linked to that of multiplication. If $b \neq 0$, then the **quotient** of a by b is denoted by $a \div b$ and is that number c by which b must be multiplied to give a. We call a the **dividend** and b the **divisor**. For example, $6 \div 3$ is 2 because $3 \cdot 2 = 6$. The dividend is 6, the divisor is 3, and the quotient is 2. The quotient $a \div b$ can also be denoted by the fraction a/b, where we call a the **numerator** and b the **denominator**. Thus we have:

DIVISION

$$\frac{a}{b} = c \quad \text{means} \quad a = bc.$$

---

*Division of a number by 0 is **not** defined.* For example, assuming that $\frac{5}{0}$ were some number c would mean that $5 = 0 \cdot c$. But 0 times any number is 0, so $0 \cdot c$ cannot be 5. Thus division by 0 is not possible.

The rules of signs for division are similar to those for multiplication:

DIVISION

To divide two positive numbers or two negative numbers, divide their absolute values. To divide a positive number and a negative number, divide their absolute values and prefix the result with a minus sign.

Thus the quotient of two positive numbers or two negative numbers is positive, and the quotient of a positive number and a negative number is negative.

Example 7

a. $\dfrac{-6}{-3} = \dfrac{6}{3} = 2.$

b. $\dfrac{-4}{5} = -\dfrac{4}{5}.$

c. $\dfrac{18}{-6} = -\dfrac{18}{6} = -3.$

d. $-\dfrac{-24}{3} = -\left(-\dfrac{24}{3}\right) = -(-8) = 8.$ The last step follows from the rule $-(-a) = a.$

Example 8

a. $\dfrac{7(-4)}{3} = \dfrac{-28}{3} = -\dfrac{28}{3}.$

b. $\dfrac{6(-2)}{-3} = \dfrac{-12}{-3} = \dfrac{12}{3} = 4.$

c. $\dfrac{2}{3(-5)} = \dfrac{2}{-15} = -\dfrac{2}{15}.$

d. $\dfrac{-6}{(-5)(-7)} = \dfrac{-6}{35} = -\dfrac{6}{35}.$

In symbols we can state the rules of signs for multiplication and division as follows:

MULTIPLICATION	DIVISION
$(-a)(-b) = ab.$	$\dfrac{-a}{-b} = \dfrac{a}{b}.$
$a(-b) = (-a)(b) = -(ab) = -ab.$	$\dfrac{-a}{b} = -\dfrac{a}{b} = \dfrac{a}{-b}.$

For example, $(-2)(-3) = 2 \cdot 3 = 6$ illustrates $(-a)(-b) = ab.$ Similarly,

$$3(-4) = (-3)(4) = -(3 \cdot 4) = -12.$$

Example 9

a. $(-2)(-x) = 2x$ because $(-a)(-b) = ab.$

b. $(-3)x = -3x$ because $(-a)(b) = -ab.$

c. $\dfrac{(-x)y}{z} = \dfrac{-(xy)}{z} = -\dfrac{xy}{z}$.

d. $-\dfrac{x}{y(-z)} = -\dfrac{x}{-(yz)} = -\left(-\dfrac{x}{yz}\right) = \dfrac{x}{yz}$.

Example 10

a. $2(-4x) = -(2 \cdot 4x)$ [since $a(-b) = -(ab)$]

       $= -8x$.

b. $-4(3x) = -(4 \cdot 3x)$ [since $-ab = -(ab)$]

       $= -12x$.

c. $-3(x - 5) = (-3)(x - 5)$ [since $-ab = (-a)b$]

       $= (-3)(x) - (-3)(5)$ [distributive law]

       $= -3x - (-15)$

       $= -3x + 15$.

The following table gives the rules for operations with the numbers 0 and 1.

OPERATIONS WITH 0 AND 1	
Rule	*Example*
1. $0 + a = a$.	1. $0 + (-3) = -3$.
2. $0 - a = -a$.	2. $0 - 4 = -4$.
3. $a - 0 = a$.	3. $xy - 0 = xy$.
4. $a \cdot 0 = 0$.	4. $(-4) \cdot 0 = 0$.
5. $\dfrac{0}{a} = 0$ if $a \neq 0$.	5. $\dfrac{0}{3} = 0$.
6. $\dfrac{a}{0}$ is not defined.	6. $\dfrac{3}{0}$ is not defined.
7. $1 \cdot a = a$.	7. $x = 1 \cdot x$.
8. $\dfrac{a}{1} = a$.	8. $xy = \dfrac{xy}{1}$.

With some expressions involving two or more operations, parentheses (or other grouping symbols) may not be present to indicate the order in which the operations are to be performed. For example, to find

$$2 + 3 \times 4,$$

you might be confused as to whether you should first compute $2 + 3$ or first compute 3×4. In cases such as this, common practice is to use the following **order of operations**.

ORDER OF OPERATIONS

1. Any operations indicated inside parentheses should be done before performing any operations immediately outside the parentheses.
2. First do any multiplications and divisions. Work from left to right and do them in the order in which they appear.
3. Then do any additions and subtractions in the order in which they appear from left to right.

Thus to compute

$$2 + 3 \times 4,$$

we *do not* add 2 to 3 and multiply by 4:

$$2 + 3 \times 4 \neq 5 \cdot 4.$$

Based on the preceding rules, the correct procedure is to first multiply 3 by 4 and then add the result to 2:

$$2 + 3 \times 4 = 2 + (3 \times 4) = 2 + 12 = 14.$$

Similarly, to compute

$$-1 + 10 \div 2 \times (3 - 9)$$

we first take care of the subtraction inside the parentheses. Then we do the division because, from left to right, it precedes the multiplication. Addition is performed last.

$$-1 + 10 \div 2 \times \underbrace{(3 - 9)}_{\text{Divide next}} = -1 + \underbrace{10 \div 2}_{} \times (-6)$$

$$= -1 + \underbrace{5 \times (-6)}_{\text{Multiply next}}$$

$$= -1 + (-30)$$

$$= -31.$$

In a fraction, the fraction bar acts as a grouping symbol, so we simplify the numerator and denominator separately. Thus

$$\frac{-2 \times 7 + 6}{14 \div 7} = \frac{-14 + 6}{2} = \frac{-8}{2} = -4.$$

Example 11

a. $-3 - 4(-2) = -3 - (-8) = -3 + 8 = 5.$

b. $(-4 + 3)(7 - 5) = (-1)(2) = -2.$

c. $\dfrac{1 - (2 - 8)}{-3} = \dfrac{1 - (-6)}{-3} = \dfrac{7}{-3} = -\dfrac{7}{3}.$

Appendix A covers the use of the scientific calculator and discusses (in Sec. A.1) the two common types of logic systems used in the design of such calculators. They are *algebraic logic* (ALG) and *reverse Polish notation* (RPN). Section A.3 covers arithmetic operations with these systems; it is suggested that you read it at this time.

Basically, algebraic logic follows the hierarchy of operations given by the preceding rules. That is, multiplications and divisions occur before additions and subtractions. For example, with algebraic logic, the keystroke sequence

$$\boxed{2}\ \boxed{+}\ \boxed{3}\ \boxed{\times}\ \boxed{4}\ \boxed{=}$$

correctly results in $2 + (3 \times 4) = 14$ because the multiplication has priority over the addition. With RPN an operation is executed as soon as that operation key is pressed. Thus entering the preceding sequence of numbers and operations gives $(2 + 3) \times 4 = 20$. To get the correct answer, the multiplication should be done first (use $3 \times 4 + 2$), or the parentheses keys should be used if available so that the expression inside the parentheses is treated as a single number. In fact, with either system you should use parentheses if there is any doubt as to how the calculator will treat a given expression.

Throughout the rest of the book, our comments regarding the calculator will focus on ALG logic, and we shall give rounded displays for our answers. (RPN logic is discussed more fully in the appendix.) For example, to calculate

$$35.6 \times (-6.13) \div 7.412$$

we may use the following keystroke sequence:

$\boxed{35.6}\ \boxed{\times}\ \boxed{6.13}\ \boxed{+/-}\ \boxed{\div}\ \boxed{7.412}\ \boxed{=}$ Display: -29.4425

↑
Change-sign
key

Note that the change-sign key is used to enter a negative number. It is pressed *after* entering 6.13. If the numbers in this problem were approximate, we would round our answer to three significant digits and give it as -29.4. (See Sec. 0.6 for a discussion of approximate numbers and significant digits.)

Problem Set 1.4

In Problems 1-58, perform the indicated operations.

1. $(+4) + (+8)$.
2. $(-2) + (-3)$.
3. $(-7) + (-2)$.
4. $-5 + 0$.
5. $6 + (-4)$.
6. $-2 + 4$.
7. $-5 + 2$.
8. $8 + (-9)$.
9. $8 - 10$.
10. $6 - 9$.
11. $-2 - 8$.
12. $3 - (-4)$.
13. $-6 - (-5)$.
14. $(-8) - 7$.
15. $-14 - (-14)$.
16. $5 - 6 + 1$.
17. $-2 - 5 + 9 - 1$.
18. $3 - 5 + 6 - 4$.
19. $(-3)(-4)$.
20. $(+3)(+4)$.
21. $7(-3)$.
22. $(5)(-2)$.
23. $-(-4)$.
24. $(-0.34) \cdot 0$.
25. $(-2)(2)(-1)(-2)(-1)$.
26. $(-2)(-1)(-2)(-1)(-1)$.

27. $\dfrac{-6}{-3}$.
28. $\dfrac{+4}{+2}$.
29. $\dfrac{-10}{2}$.

30. $\dfrac{25}{-5}$.
31. $\dfrac{0}{-3}$.
32. $-\dfrac{7}{-14}$.

33. $-\dfrac{-6}{7}$.
34. $0 - 0.12$.
35. $(2 - 3)4 + 5$.

36. $(-6 - 7) + 8$.
37. $(-6)(-7) + 8$.
38. $3(5 + 1)$.

39. $4 - 8 + 6(-2)$.
40. $(8 + 1)(-7)$.
41. $\dfrac{3 - 5}{4 - 6}$.

42. $\dfrac{-6}{3(-2)}$.
43. $\dfrac{12}{-6 + 6}$.
44. $\dfrac{-7 - 5}{4}$.

45. $(-2)(4)(-3) + 6(-2)$.
46. $-7 - (-3 - 2)$.

47. $(-2)(-3) - (-4)(2)$.
48. $\dfrac{-4 - (2 - 2)}{(-3)(2) - (-2)(4)}$.

49. $\dfrac{12}{2(-2 + 1)(3 - 4)}$.
50. $\dfrac{(-9)(3 - 5)}{(2 - 5) - 6}$.

51. $16 - \dfrac{4(-2)}{2} - \dfrac{32(-4)}{-(-1)}$.
52. $\dfrac{(-3)(2) - (-2)(4)}{-4 - (-4)}$.

53. $|-2| - |2 - 3(-1)|$.
54. $-|-2| - |2(-4)|$.
55. $4 \times 3 + 6 \div 2 - 1$.
56. $7 + 3 \times 4 - 9 \div 3$.
57. $7 + 9 \div 3 + 4 \times 6$.
58. $8 \times 4 + 2 \times 6 \div 3$.

In Problems 59-72, perform the indicated operations.

59. $-(-5xy)$.
60. $(-1)(-bc)$.
61. $(-1)(-x)$.
62. $(-7)(x)$.
63. $(-6)(-x)$.
64. $(-8x)(-2)$.
65. $6(-4x)$.
66. $-5(4x)$.
67. $-4(x - 2)$.

68. $0 - ab$.
69. $\dfrac{x}{(-y)(z)}$.
70. $-\dfrac{x}{(-y)z}$.

71. $-\dfrac{-x}{(-y)(-z)}$.
72. $\dfrac{(-x)(-y)}{(-w)(-z)}$.

73. At 3 P.M. a television weatherperson gave the current temperature as 15°C and predicted that the low temperature that evening would be −3°C. Assuming that the prediction comes true, set up an expression involving subtraction of signed numbers that gives the temperature drop, and simplify your result.

1.5 FRACTIONS

In this section we shall discuss some basic rules for working with fractions. (Throughout we assume that all denominators are nonzero.) The first two rules are:

Rule	Example
1. $\dfrac{a}{a} = 1.$	$\dfrac{7}{7} = 1.$
2. $\dfrac{a}{b} \cdot \dfrac{c}{d} = \dfrac{ac}{bd}.$	$\dfrac{2}{3} \cdot \dfrac{4}{5} = \dfrac{2 \cdot 4}{3 \cdot 5} = \dfrac{8}{15}.$

Rule 1 simply states that a (nonzero) number divided by itself is 1. Rule 2 gives the procedure for multiplying two fractions: Multiply their numerators and multiply their denominators.

Example 1

a. $\dfrac{-5x}{-5x} = 1$ [by Rule 1 with $-5x$ replacing a].

b. $\dfrac{3}{y} \cdot \dfrac{x}{2} = \dfrac{3 \cdot x}{y \cdot 2} = \dfrac{3x}{2y}$ [by Rule 2].

c. $\left(-\dfrac{3}{5}\right)\left(-\dfrac{x}{7}\right) = \dfrac{3}{5} \cdot \dfrac{x}{7}$ [since $(-a)(-b) = ab$]

         $= \dfrac{3 \cdot x}{5 \cdot 7}$ [by Rule 2]

         $= \dfrac{3x}{35}.$

d. $\dfrac{-2}{3} \cdot \dfrac{-x}{-y} = \dfrac{(-2)(-x)}{3(-y)} = \dfrac{2x}{-3y}$

         $= -\dfrac{2x}{3y}$ $\left[\text{since } \dfrac{a}{-b} = -\dfrac{a}{b}\right].$

e. $2 \cdot \dfrac{y}{3} = \dfrac{2}{1} \cdot \dfrac{y}{3}$ $\left[\text{since } a = \dfrac{a}{1} \right]$

$\qquad = \dfrac{2y}{3}.$

The result of Example 1(e) can be obtained directly from the following rule:

Rule	Example
3. $a \cdot \dfrac{b}{c} = \dfrac{ab}{c} = \dfrac{a}{c} \cdot b.$	$2 \cdot \dfrac{7}{3} = \dfrac{2 \cdot 7}{3} = \dfrac{2}{3} \cdot 7.$

Example 2

a. $\dfrac{4x}{y} = 4\left(\dfrac{x}{y}\right)$, by Rule 3.

b. $\dfrac{9}{5}(-4) = -\left(\dfrac{9}{5} \cdot 4\right) = -\dfrac{9 \cdot 4}{5} = -\dfrac{36}{5}.$

c. $\dfrac{\frac{2}{3}b}{c} = \dfrac{2}{3}\left(\dfrac{b}{c}\right) = \dfrac{2b}{3c}$, by Rules 3 and 2.

To obtain our next rule, called the *cancellation property*, we shall simplify the fraction $\dfrac{ac}{bc}$. Notice that the numerator and denominator have a factor of c in common.

$$\dfrac{ac}{bc} = \dfrac{a}{b} \cdot \dfrac{c}{c} \qquad \text{[Rule 2]}$$

$$= \dfrac{a}{b} \cdot 1 \qquad \text{[Rule 1]}$$

$$= \dfrac{a}{b}.$$

CANCELLATION PROPERTY

4. $\dfrac{ac}{bc} = \dfrac{a}{b}.$

This means that removing (or *canceling*) a common factor of *both* the numerator and denominator of a fraction results in an *equivalent fraction* (that is, a fraction that has the same value as the original fraction). For example,

$$\frac{2 \cdot 3}{5 \cdot 3} = \frac{2}{5} \quad \text{and} \quad \frac{5 \cdot x}{5 \cdot y} = \frac{x}{y}.$$

We can indicate cancellation by using slashes:

$$\frac{a \cdot \cancel{c}}{b \cdot \cancel{c}} = \frac{a}{b}.$$

Example 3

Simplify.

a. $\dfrac{3xy}{3yz} = \dfrac{\cancel{3}\cancel{x}\cancel{y}}{\cancel{3}\cancel{y}z} = \dfrac{x}{z}.$

b. $\dfrac{-2x}{-2} = \dfrac{(\cancel{-2}) \cdot x}{(\cancel{-2}) \cdot 1} = \dfrac{x}{1} = x.$

c. $\dfrac{-12x}{8y} = -\dfrac{12x}{8y} = -\dfrac{4 \cdot 3x}{4 \cdot 2y} = -\dfrac{3x}{2y}.$

Actually, we could have mentally canceled the common factor 4 and written

$$\frac{-12x}{8y} = -\frac{\overset{3}{\cancel{12}}x}{\underset{2}{\cancel{8}}y} = -\frac{3x}{2y}.$$

d. $49 \cdot \dfrac{x}{7} = \dfrac{\overset{7}{\cancel{49}}x}{\cancel{7}} = 7x.$

e. $\dfrac{3}{y} \cdot \dfrac{y}{6} = \dfrac{\cancel{3}\cancel{y}}{\underset{2}{\cancel{6}}\cancel{y}} = \dfrac{1}{2}.$

To simplify the work we can cancel common factors *before* multiplying the fractions:

$$\frac{\cancel{3}}{\cancel{y}} \cdot \frac{\cancel{y}}{\underset{2}{\cancel{6}}} = \frac{1}{2}.$$

Similarly, part (d) can be done as follows.

$$\overset{7}{\cancel{49}} \cdot \frac{x}{\cancel{7}} = 7x.$$

By writing Rule 4 as $\dfrac{a}{b} = \dfrac{ac}{bc}$, you can see that multiplying both the numerator and denominator of a fraction by a nonzero number c results in an equivalent fraction. For example,

$$\frac{2}{3} = \frac{2 \cdot 4}{3 \cdot 4} = \frac{8}{12}.$$

Dividing both the numerator and denominator of a fraction by a nonzero number also results in an equivalent fraction. For example,

$$\frac{4}{10} = \frac{4/2}{10/2} = \frac{2}{5}.$$

These facts are together called the *fundamental principle of fractions*.

FUNDAMENTAL PRINCIPLE OF FRACTIONS

5. $\dfrac{a}{b} = \dfrac{ac}{bc};$ $\dfrac{a}{b} = \dfrac{a \div c}{b \div c}.$

Example 4

a. Write $\dfrac{5x}{8}$ as an equivalent fraction with denominator 16.

Solution   To get a denominator of 16, we multiply the original denominator (and the numerator) by 2.

$$\frac{5x}{8} = \frac{(5x) \cdot 2}{8 \cdot 2} = \frac{10x}{16}.$$

b. Write $\dfrac{4}{10y}$ as an equivalent fraction with denominator $5y$.

Solution   To change the $10y$ to $5y$, we must divide it by 2. Thus

$$\frac{4}{10y} = \frac{4/2}{10y/2} = \frac{2}{5y}.$$

Observe that dividing the numerator and denominator by 2 has the same effect as canceling the common factor 2 in the numerator and denominator.

To divide the fraction a/b by the fraction c/d, we can apply the fundamental principle:

$$\frac{\dfrac{a}{b}}{\dfrac{c}{d}} = \frac{\dfrac{a}{b}\cdot\dfrac{d}{c}}{\dfrac{c}{d}\cdot\dfrac{d}{c}} = \frac{\dfrac{a}{b}\cdot\dfrac{d}{c}}{cd} = \frac{\dfrac{a}{b}\cdot\dfrac{d}{c}}{1} = \frac{a}{b}\cdot\frac{d}{c}.$$

> **DIVISION OF FRACTIONS**
>
> 6. $\dfrac{a}{b} \div \dfrac{c}{d} = \dfrac{a}{b}\cdot\dfrac{d}{c} = \dfrac{ad}{bc}.$

Observe that to divide a/b by c/d, we multiply a/b by the fraction obtained by interchanging the numerator and denominator of the divisor c/d. This is commonly referred to as *inverting* the divisor and multiplying. For example,

$$\frac{\dfrac{4}{7}}{\dfrac{3}{5}} = \frac{4}{7}\cdot\frac{5}{3} = \frac{20}{21}.$$

Example 5

Perform the operations and simplify.

a. $\dfrac{x}{3} \div \dfrac{4}{5} = \dfrac{x}{3}\cdot\dfrac{5}{4}$ [Rule 6]

$= \dfrac{5x}{12}.$

b. $\dfrac{\dfrac{5x}{2}}{\dfrac{x}{1}} = \dfrac{\dfrac{5x}{2}}{x} = \dfrac{5x}{2}\cdot\dfrac{1}{x} = \dfrac{5x}{2x} = \dfrac{5}{2}$, by cancellation.

c. $\dfrac{-4}{\dfrac{y}{3}} = (-4)\left(\dfrac{3}{y}\right) = -\left(4\cdot\dfrac{3}{y}\right) = -\dfrac{12}{y}.$

If a is a nonzero real number, the number $1/a$ is called the *reciprocal* of a:

> The **reciprocal** of a is $\dfrac{1}{a}$.

For example, the reciprocal of 2 is $\frac{1}{2}$. There is no reciprocal of 0 because division by 0 is not defined. The reciprocal of $\frac{2}{3}$ is

$$\frac{1}{\frac{2}{3}} = 1 \cdot \frac{3}{2} = \frac{3}{2}.$$

Our result illustrates that the reciprocal of a fraction can be found by simply interchanging its numerator and denominator. That is, the reciprocal of c/d is d/c. With this fact we can restate Rule 6 as follows: To divide a/b by c/d, multiply a/b by the reciprocal of c/d.

The product of a nonzero number a and its reciprocal is

$$a \cdot \frac{1}{a} = \frac{a}{a} = 1.$$

$$7. \quad a \cdot \frac{1}{a} = 1.$$

For example, $\frac{2}{3} \cdot \frac{3}{2} = 1$. Moreover, because

$$a \cdot \frac{1}{b} = \frac{a \cdot 1}{b} = \frac{a}{b},$$

division can be expressed in terms of *multiplication*:

$$8. \quad \frac{a}{b} = a \cdot \frac{1}{b}.$$

That is, *dividing a by b is the same as multiplying a by the reciprocal of b.*

Example 6

a. $\dfrac{3}{4} = 3 \cdot \dfrac{1}{4}.$

b. $\dfrac{x}{3} = x \cdot \dfrac{1}{3} = \dfrac{1}{3}x.$

c. $\dfrac{-\frac{1}{2}}{8} = \left(-\dfrac{1}{2}\right)\dfrac{1}{8} = -\dfrac{1}{16}.$

d. $7 \cdot \dfrac{1}{8y} = \dfrac{7}{8y}.$

Reciprocals can be found with a calculator by using the reciprocal key (see Sec. A.4). For example, to compute $\frac{1}{8}$ we have

$$\boxed{8}\ \boxed{1/x} \qquad \text{Display:}\quad 0.125$$

$\uparrow$
Reciprocal
key

Problem Set 1.5

*In Problems **1–6**, find the reciprocals of the given numbers.*

1. 8.

2. 0.5.

3. $\dfrac{4}{5}$.

4. $\dfrac{1}{y}$.

5. $\dfrac{x}{x+1}$.

6. $-\dfrac{x}{4}$.

*In Problems **7–56**, perform the indicated operations.*

7. $\dfrac{2x}{2x}$.

8. $-\dfrac{6y}{6y}$.

9. $\dfrac{8}{3}\cdot\dfrac{5}{7}$.

10. $\left(-\dfrac{4}{3}\right)\left(-\dfrac{11}{5}\right)$.

11. $\left(-\dfrac{3}{4}\right)\left(\dfrac{7}{2}\right)$.

12. $\dfrac{4}{3}\cdot\dfrac{2}{-5}$.

13. $\dfrac{2}{5}\cdot\dfrac{x}{y}$.

14. $\dfrac{2x}{3}\cdot\dfrac{4}{9y}$.

15. $\dfrac{-2}{x}\cdot\dfrac{3y}{7}$.

16. $\dfrac{-6x}{5s}\cdot\dfrac{-3}{7t}$.

17. $\dfrac{s}{8}\cdot\dfrac{2}{t}$.

18. $\dfrac{14}{x}\cdot\dfrac{y}{21}$.

19. $x\cdot\dfrac{3}{4}$.

20. $6\cdot\dfrac{x}{5}$.

21. $(-3x)\left(\dfrac{y}{2}\right)$.

22. $\dfrac{x}{3}\cdot 5$.

23. $\dfrac{3}{4}(-6)$.

24. $(-x)\left(\dfrac{-y}{-z}\right)$.

25. $\dfrac{2yz}{2xz}$.

26. $\dfrac{3x}{12y}$.

27. $\dfrac{18q}{12pq}$.

28. $-\dfrac{-27y}{-6}$.

29. $\dfrac{15xy}{25z}$.

30. $\dfrac{7x}{14}$.

31. $\dfrac{24z}{-18z}$.

32. $\dfrac{-4x}{2}$.

33. $\dfrac{25}{x}\cdot\dfrac{x}{15}$.

34. $\left(-\dfrac{4}{9}\right)(-18)$.

35. $5\left(-\dfrac{3}{5}\right)$.

36. $\left(-\dfrac{4}{5}\right)\left(\dfrac{-25}{8}\right)$.

37. $\dfrac{-2x}{21}\cdot\dfrac{3y}{-4x}$.

38. $\dfrac{1}{5}(5a)$.

39. $\dfrac{2}{-3}(9x)$.

40. $(-6)\left(-\dfrac{3}{8}x\right)$.

41. $6a\cdot\dfrac{1}{6a}$.

42. $-\dfrac{1}{a+b}\cdot(a+b)$.

43. $\dfrac{3}{2} \div \dfrac{5}{7}$.

44. $\dfrac{-4}{9} \div \dfrac{3}{8}$.

45. $\dfrac{5}{6} \div \dfrac{10}{-9}$.

46. $\left(-\dfrac{1}{16}\right) \div \dfrac{6}{-20}$.

47. $\dfrac{2}{x} \div \dfrac{3}{y}$.

48. $\dfrac{x}{y} \div \dfrac{-2}{3}$.

49. $\dfrac{-4x}{3} \div \dfrac{x}{-24}$.

50. $\dfrac{-14y}{-6z} \div \dfrac{-21}{12z}$.

51. $\dfrac{\frac{8}{3}}{-2}$.

52. $\dfrac{\frac{3}{5}xy}{z}$.

53. $\dfrac{12}{\frac{x}{6}}$.

54. $\dfrac{-3y}{\frac{y}{6}}$.

55. $-\dfrac{\frac{4x}{-y}}{10}$.

56. $\dfrac{-\frac{xz}{8}}{2xy}$.

57. Write $\dfrac{2}{x}$ as an equivalent fraction with denominator xy.

58. Write $\dfrac{x}{y}$ as an equivalent fraction with denominator $4y$.

59. Write $\dfrac{3x}{4y}$ as an equivalent fraction with denominator $24yz$.

60. Write $\dfrac{14zt}{36xzw}$ as an equivalent fraction with denominator $18xw$.

*In Problems **61–64**, find the reciprocal of the given approximate number. Round each answer to the number of significant digits contained in the given number.*

61. 548.1.   62. 6.008.   63. 0.00316.   64. 776,000.

*In Problems **65–72**, perform the indicated operations. Round your answers to three significant digits.*

65. $\dfrac{1}{73.4} + \dfrac{1}{82.67}$.

66. $73.7\left(\dfrac{84.3}{96.7}\right)$.

67. $\dfrac{\frac{83.45}{6.87}}{4.62}$.

68. $\dfrac{\frac{1}{1.21}}{0.0367}$.

69. $\dfrac{3.21}{5.02}\left(\dfrac{3.62 - 4.73}{8.31}\right)$.

70. $\dfrac{94.3}{45.7} \div \dfrac{35.6 - 46.9}{2.956}$.

71. $\dfrac{-\frac{6.42}{-1.12}}{8.73 - 4.14}$.

72. $\dfrac{8.62}{\frac{-4.10(7.21 + 2.91)}{6.84}}$.

73. The power of a lens (in units of diopters) is the reciprocal of the focal length of the lens (in meters). If a lens has a focal length of 0.2 meter, find its power.

1.6 EVALUATING FORMULAS

A knowledge of the arithmetic of real numbers is essential when you are evaluating a formula. This involves replacing each letter in the formula by the number it represents and then performing the indicated operations.

Example 1

In the study of straight lines, the following formula (called the *slope formula*) occurs:

$$m = \frac{y_2 - y_1}{x_2 - x_1}.$$

The numbers to the right and just below the letters are called **subscripts**. Subscripts are commonly used in science and technical work to create variables in a convenient way. We read x_1 as *x-sub-one*, x_2 as *x-sub-two*, and so on. Find the value of m when $x_1 = 5, x_2 = 3, y_1 = -5$, and $y_2 = 7$.

Solution In the formula we substitute 5 for x_1, 3 for x_2, and so on. This gives

$$m = \frac{y_2 - y_1}{x_2 - x_1} = \frac{7 - (-5)}{3 - 5} = \frac{12}{-2} = -6.$$

Example 2

To describe statistical data, the terms **average value**, or **(arithmetic) mean**, are used. Given the set of n values $x_1, x_2, x_3, \ldots, x_n$, the mean $\bar{x}$ (read *x-bar*) is their sum divided by n, the number of values. That is,

$$\bar{x} = \frac{x_1 + x_2 + x_3 + \cdots + x_n}{n}. \tag{1}$$

Suppose that the Celsius temperature of methyl alcohol used in a manufacturing process was measured at different times. The readings are given in Table 1.1.

Table 1.1

Reading	Temperature (°C)
1	4.2
2	−3.0
3	−2.6
4	1.7
5	2.2

Find the mean temperature, denoted by $\bar{T}$ (read *T-bar*).

Solution   Because there are five readings here, $n = 5$. Replacing the x's in Eq. 1 by T's, we have

$$\bar{T} = \frac{T_1 + T_2 + T_3 + T_4 + T_5}{5}.$$

When $T_1 = 4.2$, $T_2 = -3.0$, and so on, we have

$$\bar{T} = \frac{4.2 + (-3.0) + (-2.6) + 1.7 + 2.2}{5} = \frac{2.5}{5} = 0.5°C.$$

Example 3

The relationships between the Celsius, Fahrenheit, and Kelvin (or absolute) temperature scales are familiar to scientists and engineers:

$$T_C = \frac{5}{9}(T_F - 32);$$

$$T_F = \frac{9}{5}T_C + 32;$$

$$T_K = T_C + 273.$$

Here T_C, T_F, and T_K represent the respective temperatures on the three scales. As before, the small letters to the right of and just below the T's are subscripts. For example, T_C is read *T-sub-C* and represents a temperature on the Celsius scale. On the Celsius temperature scale, the normal melting point of mercury is $-39°C$. What is the normal melting point of mercury when expressed on the Kelvin and Fahrenheit temperature scales?

Solution   Here $T_C = -39°C$. The melting point of mercury on the Kelvin* scale is

$$T_K = T_C + 273 = -39 + 273 = 234 \text{ K}.$$

On the Fahrenheit scale the corresponding temperature is

$$T_F = \frac{9}{5}T_C + 32 = \frac{9}{5}(-39) + 32 = -70.2 + 32 = -38.2°F.$$

Example 4

The **percentage error** is an indication of the accuracy of a measurement and is given by

$$\% \text{ERR} = \frac{|\text{standard value} - \text{measured value}|}{\text{standard value}} \times 100.$$

To determine the accuracy of a frequency meter, a frequency *standard* of 60 hertz is measured. If the measured value is 58.5 hertz, find the percentage error.

---

* In modern notation, an absolute (or Kelvin) temperature is indicated by using the symbol K without any degree symbol.

Solution   In this case,

$$\% \,\mathrm{ERR} = \frac{|60 - 58.5|}{60} \times 100 = \frac{1.5}{60} \times 100 = 2.5\%.$$

Problem Set 1.6

In Problems 1 and 2, use the slope formula in Example 1 to find m from the given information.

1. $x_1 = 5$, $x_2 = -8$, $y_1 = -1$, $y_2 = -2$.
2. $x_1 = 3$, $x_2 = 5$, $y_1 = 4$, $y_2 = 1$.
3. A math student taking calculus must compute y' (read y *prime*), where

$$y' = \frac{(x)(1) - (x - 5)(1)}{(x)(x)}$$

and $x = -2$. What is this value?

4. Repeat Problem 3 if $x = 3$.

The formula

$$d = |a - b|$$

gives the distance d between the points a and b on a number line. In Problems 5 and 6, find d for the points having the given coordinates.

5. $a = -3$, $b = 5$.   6. $a = -4$, $b = -9$.

7. During an experiment with mercury, both its volume, in cubic centimeters (cm^3), and Celsius temperature were measured. The data are given in Table 1.2.

Table 1.2

Reading	Temperature (°C)	Volume (cm³)
1	3.6	460.0
2	2.8	450.0
3	1.3	448.0
4	−0.8	435.0
5	−1.4	419.0
6	−1.6	415.0

(a) Find the mean temperature $\bar{T}$. (Refer to Example 2.)
(b) Find the mean volume $\bar{V}$.

8. At room temperature the horizontal velocities of five oxygen molecules are −475, 478, −472, 482, and 483 meters per second. Find the mean velocity, $\bar{V}$. (Refer to Example 2.)

9. In technical situations, room temperature is often taken to be 20°C. Express this temperature on the Fahrenheit and Kelvin temperature scales. (Refer to Example 3.)

10. The normal melting point of nitrogen is about −210°C. Express this temperature on the Fahrenheit and Kelvin temperature scales. (Refer to Example 3.)

For an arrangement of three particles on a number line, the location $\bar{x}$ of the center of mass is given by the formula

$$\bar{x} = \frac{m_1 x_1 + m_2 x_2 + m_3 x_3}{m_1 + m_2 + m_3}.$$

Here m_1, m_2, and m_3 are the masses of the particles, and x_1, x_2, and x_3 are their locations, respectively. In Problems 11 and 12, find $\bar{x}$ from the given information.

11. $m_1 = 2$, $m_2 = 3$, $m_3 = 4$, $x_1 = -2$, $x_2 = -3$, $x_3 = 8$.

12. $x_1 = -4$, $x_2 = 4$, $x_3 = -4$, $m_1 = 2$, $m_2 = 2$, $m_3 = 2$.

13. If **speed** is defined to be the absolute value of velocity, find the speeds of the molecules in Problem 8 and use them to find the mean speed.

14. A 2000-ohm resistor is found to have a measured value of 2050 ohms. Find the percentage error. (Refer to Example 4.)

15. A voltage standard of 50 volts is measured with two different instruments. One gives a reading of 49 volts and the other, a reading of 51 volts. Find the percentage error in each case. (Refer to Example 4.)

16. The ends of a certain copper rod are kept at different temperatures. The rate H, in calories per second, at which heat is conducted through the rod is given by

$$H = 0.14(T_h - T_l),$$

where T_h is the higher temperature, T_l is the lower temperature, and both T_h and T_l are in degrees Celsius. Find H for each of the following situations.

(a) $T_h = 820°C$, $T_l = 200°C$.   (b) $T_h = 50°C$, $T_l = -50°C$.

(c) $T_h = 0°C$, $T_l = -10°C$.   (d) $T_h = -10°C$, $T_l = -20°C$.

17. The formula

$$R = \frac{1}{\dfrac{1}{R_1} + \dfrac{1}{R_2}}$$

gives the effective (or equivalent) resistance R of two resistors R_1 and R_2 connected in parallel. Evaluate R if $R_1 = 2.5$ and $R_2 = 3.8$. (All units are ohms.)

1.7   REVIEW

Important Terms and Symbols

Section 1.1   Positive integer, natural number, negative integer, integer, rational number, terminating decimal, repeating decimal, irrational number, $\approx$, real number, origin, coordinate, real number line, positive number, negative number, literal number, algebraic expression, term, product, factor, constant, variable, set, element, member.

Section 1.2 Less than, greater than, inequality symbols, $<$, $>$, $\leq$, $\geq$, $a < c < b$, inequality, nonnegative, absolute value (geometrically), $|a|$, magnitude, numerically equal, equal in magnitude.

Section 1.3 Commutative laws, associative laws, distributive laws, $\neq$.

Section 1.4 Rules of signs, opposite (or negative), absolute value (algebraically), difference, quotient, dividend, divisor, numerator, denominator, order of operations.

Section 1.5 Cancellation property, equivalent fractions, fundamental principle of fractions, inverting, reciprocal.

Section 1.6 Subscript.

Formula Summary

Commutative Laws

$a + b = b + a.$

$ab = ba.$

Associative Laws

$(a + b) + c = a + (b + c).$

$(ab)c = a(bc).$

Distributive Laws

$a(b + c) = ab + ac.$

$(a + b)c = ac + bc.$

Subtraction

$a - b = a + (-b).$

Division

$\dfrac{a}{b} = a \cdot \dfrac{1}{b};$ $\dfrac{a}{b} = c$ means $a = bc.$

Negatives

$a + (-a) = 0.$ $-(-a) = a.$ $a - (-b) = a + b.$

$-a = (-1)a.$ $(-a)(-b) = ab.$ $(-a)(b) = -(ab) = a(-b).$

Zero and One

$0 + a = a.$ $0 - a = -a.$ $a - 0 = a.$ $a \cdot 0 = 0.$

$1 \cdot a = a.$ $\dfrac{a}{1} = a.$ $\dfrac{0}{a} = 0$ if $a \neq 0.$ $\dfrac{a}{0}$ is undefined.

Fractions

$\dfrac{-a}{-b} = \dfrac{a}{b}.$ $\dfrac{-a}{b} = -\dfrac{a}{b} = \dfrac{a}{-b}.$ $\dfrac{a}{b} = \dfrac{ac}{bc}.$ $a \cdot \dfrac{b}{c} = \dfrac{ab}{c} = \dfrac{a}{c} \cdot b.$

$\dfrac{a}{b} \cdot \dfrac{c}{d} = \dfrac{ac}{bd}.$ $\dfrac{a}{b} \div \dfrac{c}{d} = \dfrac{a}{b} \cdot \dfrac{d}{c} = \dfrac{ad}{bc}.$

Absolute Value

$|a| = \begin{cases} a, & \text{if } a \geq 0, \\ -a, & \text{if } a < 0. \end{cases}$ $|a| = |-a|.$ $|a| \geq 0.$

Review Questions

1. A real number that cannot be expressed as a ratio of two integers is called a(n) _____ number.

2. The number 7 is $\underline{\text{(rational)(irrational)}}$.

3. True or false: If $x = -4$, then $|x| = -x$. _____

4. The statement $a < b$ means that b lies to the _____ of a on the real number line.

5. The product of an odd number of negative numbers is $\underset{\text{(a)}}{\underline{\text{(positive)(negative)}}}$, and the product of an even number of negative numbers is $\underset{\text{(b)}}{\underline{\text{(positive)(negative)}}}$.

6. By the definition of division, $a/b = c$ if $a = \underline{\text{(a)}}$, or a/b means a times $\underline{\text{(b)}}$.

7. The _____ law of addition states that $a + b = b + a$.

8. The result of subtracting -7 from 0 is _____ .

9. The distributive law states that $a(b + c) = $ _____ .

10. $|-6| = \underline{\text{(a)}}$, $|0| = \underline{\text{(b)}}$, and $|6| = \underline{\text{(c)}}$.

11. The associative law states that $a(bc) = $ _____ .

12. The sets $A = \{1, 2, 3, 4\}$ and $B = \{1, 3, 2, 4\}$ $\underline{\text{(are)(are not)}}$ equal.

13. True or false: $|x| > 0$ for any real number x. _____

14. The sum of four negative numbers is a _____ number.

15. $\dfrac{8}{5}$ means _____ times $\dfrac{1}{5}$.

16. If -6 is subtracted from 6, the result is _____ .

17. $\dfrac{-7}{1}$ in a simpler form is just _____ .

18. The opposite of 5 is $\underline{\text{(a)}}$, and the reciprocal of 5 is $\underline{\text{(b)}}$.

19. The sum of a number and its opposite is _____ .

20. The product of a nonzero number and its reciprocal is _____ .

21. The reciprocal of $\dfrac{3}{5}$ is _____ .

22. The negative of -4 is _____ .

23. All numbers have a reciprocal except _____ .

24. The product of $\dfrac{x}{2}$ and _____ equals 1.

25. $\dfrac{7}{0}$ has no meaning, but $\dfrac{0}{7} = $ _____ .

26. $8 - 5$ means $8 + $ _____ .

27. $5 + $ _____ $= 0$.

28. What rule allows you to multiply the numerator and denominator of a fraction by the same nonzero number? _____

Answers to Review Questions

1. Irrational. 2. Rational. 3. True. 4. Right. 5. (a) Negative, (b) Positive.
6. (a) bc, (b) $1/b$. 7. Commutative. 8. 7. 9. $ab + ac$. 10. (a) 6, (b) 0, (c) 6.
11. $(ab)c$. 12. Are. 13. False. 14. Negative. 15. 8. 16. 12. 17. -7.
18. (a) -5, (b) $\frac{1}{5}$. 19. 0. 20. 1. 21. $\frac{5}{3}$. 22. 4. 23. 0. 24. $2/x$. 25. 0.
26. -5. 27. -5. 28. Fundamental principle of fractions.

Review Problems

In Problems **1-8**, *name the law used in the given statement.*

1. $8 + y = y + 8$.
2. $2(x + 3y) = 2x + 6y$.
3. $2x + (x + y) = (2x + x) + y$.
4. $2(4x) = (2 \cdot 4)x$.
5. $5(x + 4) = (x + 4)5$.
6. $5(x + 4) = 5(4 + x)$.
7. $(a - 3)b = ab - 3b$.
8. $(3x)(y) = 3(xy)$.

In Problems **9-16**, *determine whether each statement is true or false.*

9. $-3 < -2$.
10. $3 > 0$.
11. $|-6| = 6$.
12. $-5 > 1$.
13. $|-3| < |-2|$.
14. $-|-4| = 4$.
15. $2(3x) = (2 \cdot 3)(2 \cdot x)$.
16. $2(3 - x) = 6 - x$.

In Problems **17-61**, *perform the indicated operations.*

17. $\dfrac{(-2)(-4)}{-16}$.
18. $(-3)(-4 + 7)$.
19. $x(y - 5)$.

20. $2(x - y)$.
21. $2(-5y)$.
22. $\dfrac{7}{2 - 2}$.

23. $-6 - (-5)$.
24. $\dfrac{-2}{6}$.
25. $3(-8 + 4)$.

26. $(-2)(-4)(-1)$.
27. $\dfrac{14}{(-2)(-7)}$.
28. $\dfrac{6(-3)}{-2}$.

29. $-4(5x - 3)$.
30. $8\left(\dfrac{-5}{12}x\right)$.
31. $(7 + x) - 8$.

32. $\dfrac{6 - 8}{-4}$.
33. $2(-6) + (-4)(-1)$.
34. $\dfrac{8 - 9}{7 - 6}$.

35. $-(-3xz)$.
36. $-6 + 4 - 9(2)$.
37. $\left(-\dfrac{3}{4}\right)(-20)$.

38. $-9(-12 - 8)$.
39. $(-8)\left(\dfrac{-5}{8}\right)$.
40. $(a + b)c$.

41. $(4x)7$.
42. $-\dfrac{-8}{-64}$.
43. $\dfrac{7 - 9}{(7)(-9)}$.

44. $(8 - 8)(8x)$.
45. $\dfrac{6(-4)}{-2}$.
46. $\dfrac{(-2)(4)(-6)}{0 - 3}$.

47. $\dfrac{(-2) - (-1)(0)}{-2}$.
48. $\dfrac{8(-2) - (-2)(-8)}{3(-2) - 2}$.
49. $-4(3x)$.

50. $x(1 - y + z)$.
51. $(8 - x)y$.
52. $(x + 7) - (7 - 4)$.

53. $\dfrac{-\frac{2}{3}}{6}$.
54. $-\dfrac{-12xy}{8xz}$.
55. $\dfrac{2x}{-y} \cdot \dfrac{3y}{14}$.

56. $\dfrac{6}{x} \cdot \dfrac{-3}{14}$.
57. $15x\left(-\dfrac{y}{5}\right)$.
58. $\left(-\dfrac{18w}{15}\right) \div \dfrac{2w}{x}$.

59. $\dfrac{25}{-9x} \div \dfrac{15}{-3}.$

60. $\dfrac{5xy}{\dfrac{10x}{3}}.$

61. $\dfrac{\dfrac{2x}{3}}{-18x}.$

62. Find the reciprocals of the following numbers. Round your results to four significant figures.

    (a) 0.006932.   (b) 297.5.   (c) 6.094π.

63. Compute $\dfrac{a - 2b + c}{ad}$ in each case.

    (a) $a = 2, b = -3, c = 4, d = -1.$ (b) $a = -3, b = 4, c = -2, d = -2.$

*In Problems **64** and **65**, for the given values of x and y, find (a) $|x + y|$, (b) $|x - y|$, (c) $|x| + |y|$, and (d) $|x| - |y|$.*

64. $x = 2, y = 5.$ **65.** $x = -3, y = -8.$

66. The formula $C = \frac{5}{9}(F - 32)$ is used to convert a Fahrenheit temperature F to a Celsius temperature C. According to the *Guinness Book of World Records*, the highest recorded temperature in the United States is $134°F$ (in the shade) at Death Valley, California on July 10, 1913. Express this temperature in degrees Celsius.

67. Convert a temperature of $-40°F$ to degrees Celsius. (Refer to Problem 66.)

Introduction to Exponents and Radicals

2.1   EXPONENTS

We can use mathematical shorthand to indicate the product of four 2's:

$$2 \cdot 2 \cdot 2 \cdot 2 \quad \text{is abbreviated} \quad 2^4$$

and is called the *fourth power of* 2. The symbol 2^4 is read 2 (*raised*) *to the fourth power*. We call 2 the *base* and 4 the *exponent*. In general, the abbreviation a^n indicates that the number a is used as a factor n times, where n is a positive integer. The expression a^n is called the **nth power of a**, where n is the **exponent** and a is the **base**:

> The nth power of a:
>
> $$a^n = \underbrace{a \cdot a \cdots a.}_{n \text{ factors of } a}$$

If a number does not have an exponent indicated, the exponent is understood to be 1. Thus $5 = 5^1$.

Example 1

a. $6^2 = 6 \cdot 6 = 36$ (the *second power* of 6 or 6 *squared*). The base is 6 and the exponent is 2. Do *not* confuse 6^2 with $6 + 6$:

$$6^2 \neq 6 + 6 \quad \text{because} \quad 36 \neq 12.$$

b. $x^3 = x \cdot x \cdot x$ (x to the third, or x cubed).

c. $4^2 x^4 = 4 \cdot 4 \cdot x \cdot x \cdot x \cdot x$.

d. $(x - 2)^1 = x - 2$. The base is $x - 2$.

Example 2

a. $(-1)^4 = (-1)(-1)(-1)(-1) = 1$. In general, *an even power of a negative number is positive.*

b. $(-6)^3 = (-6)(-6)(-6) = -216$. *An odd power of a negative number is negative.*

We emphasize that *an exponent applies only to the quantity immediately to the left and below it.* For example,

Expression	Base	Value
$(-3)^2$	-3	$(-3)^2 = (-3)(-3) = 9$.
-3^2	3	$-3^2 = -(3 \cdot 3) = -9$.
$(3x)^2$	$3x$	$(3x)^2 = (3x)(3x)$.
$3x^2$	x	$3x^2 = 3 \cdot x \cdot x$.

We turn now to the basic rules of exponents. The product $2^3 \cdot 2^2$ is $8 \cdot 4 = 32$. However, we can also write the product as

$$2^3 \cdot 2^2 = (2 \cdot 2 \cdot 2)(2 \cdot 2) = 2 \cdot 2 \cdot 2 \cdot 2 \cdot 2 = 2^5.$$

In the result, notice that the exponent 5 is the sum of the exponents 3 and 2 in the original expression. That is,

$$2^3 2^2 = 2^5 = 2^{3+2}.$$

This illustrates the general rule that to **multiply** numbers with the same base, we may **add** the exponents and keep the base the same:

$$1. \quad a^m a^n = a^{m+n}.$$

Example 3

a. $x^5 \cdot x^4 = x^{5+4} = x^9$.

b. $y^8 y^6 = y^{8+6} = y^{14}$.

c. $(x - 2)^3 (x - 2)^5 = (x - 2)^8$.

d. $x(x^n) = x^1 x^n = x^{1+n} = x^{n+1}$.

e. $t^2 t^5 t^4 = t^{2+5+4} = t^{11}$. (The pattern of Rule 1 extends to a product of more than two powers having the same base.)

Example 4

a. $\dfrac{x^3 x^5}{y^2 y^4} = \dfrac{x^{3+5}}{y^{2+4}} = \dfrac{x^8}{y^6}$.

b. $(3x^4)(9x^2) = 3 \cdot x^4 \cdot 9 \cdot x^2 = 27x^{4+2} = 27x^6$.

Keep in mind that Rule 1 requires that both powers have the same base before exponents are added. For example,

$$(-2)^2(-2^4) \neq (-2)^6 \quad \text{but} \quad (-2)^2(-2^4) = 4(-16) = -64.$$

↑	↑
Base	Base
-2	2

Furthermore, in Rule 1 the base remains the same after exponents are added. Do not multiply the bases! Thus

$$3^4 \cdot 3^2 \neq 9^6 \quad \text{but} \quad 3^4 \cdot 3^2 = 3^6 = 729.$$

Another rule of exponents deals with taking a power of an expression that is itself a power of a number. For example, consider taking the second power of the third power of 2:

$$(2^3)^2 = (8)^2 = 64.$$

On the other hand, we can look at $(2^3)^2$ another way. Because 2^3 is the base for the exponent 2, we have

$$(2^3)^2 = 2^3 \cdot 2^3 = 2^{3+3} = 2^6.$$

Observe that the exponent 6 in the result is the product of the exponents 3 and 2 in the original expression. That is,

$$(2^3)^2 = 2^6 = 2^{3 \cdot 2}.$$

In general, *to find a power of a power, we may **multiply** the exponents and keep the base the same*:

$$\boxed{2. \quad (a^m)^n = a^{mn}.}$$

Example 5

a. $(10^2)^6 = 10^{2 \cdot 6} = 10^{12}$.

b. $(x^4)^3 = x^{4 \cdot 3} = x^{12}$.

c. $(y^8)^5 = y^{8 \cdot 5} = y^{40}$.

d. $[(x + 7)^3]^5 = (x + 7)^{15}$.

We caution you not to confuse $(x^4)^3$ with $x^4 x^3$. The first expression is a power of a power, so we *multiply* the exponents. The second expression is a product of powers with the same base, so we *add* the exponents.

$$(x^4)^3 = x^{12}, \quad \text{but} \quad x^4 x^3 = x^7.$$

Now let us look at division (here, as elsewhere in this book, we assume that all denominators are nonzero). In the fraction $4^5/4^3$, the numerator has more factors of 4 than does the denominator because $5 > 3$. By cancellation we have

$$\frac{4^5}{4^3} = \frac{\cancel{4} \cdot \cancel{4} \cdot \cancel{4} \cdot 4 \cdot 4}{\cancel{4} \cdot \cancel{4} \cdot \cancel{4}} = 4^2.$$

Notice that the exponent in the result is the *difference* of the exponents in the original expression: $2 = 5 - 3$. Similarly, in the fraction $4^3/4^5$, the exponent in the denominator is greater than that in the numerator. Cancellation gives two factors of a in the denominator.

$$\frac{4^3}{4^5} = \frac{\cancel{4} \cdot \cancel{4} \cdot \cancel{4}}{\cancel{4} \cdot \cancel{4} \cdot \cancel{4} \cdot 4 \cdot 4} = \frac{1}{4^2} = \frac{1}{4^{5-3}}.$$

More generally we have the following rules.

3.	$\dfrac{a^m}{a^n} = a^{m-n}$	for $m > n$,
4.	$\dfrac{a^m}{a^n} = \dfrac{1}{a^{n-m}}$	for $n > m$,
5.	$\dfrac{a^n}{a^n} = 1.$	

Example 6

a. $\dfrac{x^{11}}{x^7} = x^{11-7} = x^4.$

b. $\dfrac{t^6}{t^9} = \dfrac{1}{t^{9-6}} = \dfrac{1}{t^3}.$

c. $\dfrac{y^{10}}{y^{10}} = 1.$

d. $\dfrac{-x^8}{x} = -\dfrac{x^8}{x} = -x^{8-1} = -x^7.$

e. $\dfrac{(x^2+1)^4}{(x^2+1)^{12}} = \dfrac{1}{(x^2+1)^{12-4}} = \dfrac{1}{(x^2+1)^8}.$

f. $\dfrac{(-2)^3}{(-2)^4} = \dfrac{1}{(-2)^{4-3}} = \dfrac{1}{(-2)^1} = \dfrac{1}{-2} = -\dfrac{1}{2}.$

We now turn to a power of a product, such as raising the product $2 \cdot 3$ to the second power:

$$(2 \cdot 3)^2 = 6^2 = 36.$$

However, we can also write

$$(2 \cdot 3)^2 = (2 \cdot 3)(2 \cdot 3) = (2 \cdot 2)(3 \cdot 3) = 2^2 \cdot 3^2.$$

Notice that the result is the product of the second powers of the original factors 2 and 3. That is,

$$(2 \cdot 3)^2 = 2^2 \cdot 3^2 = 4 \cdot 9 = 36.$$

There is a similar pattern for a power of a quotient:

$$\left(\frac{2}{3}\right)^2 = \frac{2}{3} \cdot \frac{2}{3} = \frac{2 \cdot 2}{3 \cdot 3} = \frac{2^2}{3^2}.$$

More generally, *to raise a product to a power, we may raise each factor of the product to that power. To raise a quotient to a power, we may raise both the numerator and denominator to that power.*

6. $(ab)^n = a^n b^n.$

7. $\left(\dfrac{a}{b}\right)^n = \dfrac{a^n}{b^n}.$

Example 7

a. $(xy)^4 = x^4 y^4.$

b. $(abc)^6 = a^6 b^6 c^6.$

c. $\left(\dfrac{x}{y}\right)^{12} = \dfrac{x^{12}}{y^{12}}.$

d. $\left(\dfrac{2}{z}\right)^4 = \dfrac{2^4}{z^4} = \dfrac{16}{z^4}.$

e. $(3x)^3 = 3^3 x^3 = 27x^3.$ Note that $(3x)^3 \neq 3x^3.$

f. $\dfrac{20^3}{5^3} = \left(\dfrac{20}{5}\right)^3 = 4^3 = 64,$ by Rule 7.

Rule 6 applies to a power of a *product*, not the power of a *sum*. For example,

$$(3 + 4)^2 \neq 3^2 + 4^2 \quad \text{because} \quad 49 \neq 25.$$

The next examples show various ways in which Rules 1–7 can be used. These rules are now summarized.

1. $a^m a^n = a^{m+n}$.
2. $(a^m)^n = a^{mn}$.
3. $\dfrac{a^m}{a^n} = a^{m-n}$ for $m > n$.
4. $\dfrac{a^m}{a^n} = \dfrac{1}{a^{n-m}}$ for $n > m$.
5. $\dfrac{a^n}{a^n} = 1$.
6. $(ab)^n = a^n b^n$.
7. $\left(\dfrac{a}{b}\right)^n = \dfrac{a^n}{b^n}$.

Example 8

Find $(x^4 y^2)^6$.

Solution   The expression is a power of a product, so we first use the rule $(ab)^n = a^n b^n$ with x^4 replacing a and y^2 replacing b.

$$(x^4 y^2)^6 = (x^4)^6 (y^2)^6$$
$$= x^{24} y^{12} \quad \text{[Rule 2]}.$$

Example 9

In each of the following, the numbers to the right refer to the rules used.

a. $\dfrac{(x^6)^3}{(x^3)^4} = \dfrac{x^{18}}{x^{12}} = x^6$ [2, 3].

b. $\dfrac{y^2(y^6)^3}{y^{24}} = \dfrac{y^2(y^{18})}{y^{24}} = \dfrac{y^{20}}{y^{24}} = \dfrac{1}{y^4}$ [2, 1, 4].

c. $(2s^2 t^3)^3 = 2^3(s^2)^3(t^3)^3 = 8s^6 t^9$ [6, 2].

d. $\left(\dfrac{x}{y^4}\right)^2 = \dfrac{x^2}{(y^4)^2} = \dfrac{x^2}{y^8}$ [7, 2].

e. $\left(\dfrac{2b^2}{c^3d^4}\right)^6 = \dfrac{(2b^2)^6}{(c^3d^4)^6} = \dfrac{2^6(b^2)^6}{(c^3)^6(d^4)^6} = \dfrac{64b^{12}}{c^{18}d^{24}}$ [7, 6, 2].

The next example shows you how to deal with minus signs involved with powers.

Example 10

a. $(-x^2y)^9 = [(-1)x^2y]^9 = (-1)^9(x^2)^9y^9 = (-1)x^{18}y^9 = -x^{18}y^9.$
(Note the use of brackets for grouping.)

b. $(-x)^{10} = [(-1)x]^{10} = (-1)^{10}x^{10} = 1 \cdot x^{10} = x^{10}.$

Example 11

The energy $\mathcal{E}$ stored in the electric field of a charged capacitor is given by the formula

$$\mathcal{E} = \tfrac{1}{2}CV^2,$$

where C is the capacitance and V is the voltage across the terminals of the capacitor. What happens to the energy if the voltage is doubled?

Solution   The formula states that when the voltage is V, the energy $\mathcal{E}$ is $\tfrac{1}{2}CV^2$. Let $\mathcal{E}_1$ be the energy when the voltage is doubled, that is, after the voltage is changed from V to $2V$. Then, replacing $\mathcal{E}$ by $\mathcal{E}_1$ and V by $2V$ in the preceding formula gives

$$\mathcal{E}_1 = \tfrac{1}{2}C(2V)^2 = \tfrac{1}{2}C(4V^2) = 4(\tfrac{1}{2}CV^2).$$

But the factor $\tfrac{1}{2}CV^2$ is $\mathcal{E}$, the original energy. Thus

$$\mathcal{E}_1 = 4\mathcal{E},$$

so the energy increases to four times its original value. That is, doubling the voltage increases the energy by a factor of 4.

In the hierarchy of operations, raising to powers has priority over multiplication and division. Thus

$$4 \times 2^3 = 4 \times (2^3) = 4 \times 8 = 32.$$

To calculate powers with a calculator, two keys are available. The square key $\boxed{x^2}$ is used to compute the square of a number x (see Sec. A.4).

Example	Keystroke Sequence	Display
$(6.5)^2$	$\boxed{6.5}$ $\boxed{x^2}$	42.25

The power key $\boxed{y^x}$ is used to compute the xth power of y (see Sec. A.5). First the base y is keyed in, followed by $\boxed{y^x}$, the exponent x, and $\boxed{=}$.

Example	Keystroke Sequence	Display
$(4.56)^2$	$\boxed{4.56}\ \boxed{y^x}\ \boxed{2}\ \boxed{=}$	20.7936
$(2.6)^4$	$\boxed{2.6}\ \boxed{y^x}\ \boxed{4}\ \boxed{=}$	45.6976

Problem Set 2.1

In Problems 1–12, evaluate the numbers without using a calculator.

1. 2^3.

2. $(-3)^2$.

3. $2^4 - 2^5$.

4. $(-7)^1$.

5. $-(-2)^4$.

6. $\dfrac{-2^2}{(-2)^3}$.

7. $(-2^3)(-3)^2$.

8. $(3-5)^2$.

9. $\dfrac{-(-3)^2}{(-3)^3}$.

10. $2^3 \cdot 3^2 - 6^2$.

11. $(10^3)^2$.

12. $(-2+3^2)^2$.

In Problems 13–82, simplify.

13. $x^3 x^8$.

14. $x^4 x^4$.

15. $y^5 y^4$.

16. $t^2 t$.

17. $x^2 x^4 x$.

18. $x^a x^b$.

19. $(x-2)^5(x-2)^3$.

20. $y^9 y^{91} y^2$.

21. $\dfrac{x^5 x^2}{y^2 y^3 y^4}$.

22. $\dfrac{x(x^2)}{y^2 y^3}$.

23. $(2x^2)(7x^6)$.

24. $x^4(3x^2)$.

25. $(-3x)(4x^3)$.

26. $(-2x^5)(-3x^4)$.

27. $(x^8)^2$.

28. $(x^4)^3$.

29. $(x^3)^3$.

30. $(x^5)^7$.

31. $(t^2)^n$.

32. $(x^b)^c$.

33. $(x^4)^2(x^3)^7$.

34. $x^6(x^4)^2$.

35. $\dfrac{x^7}{x^3}$.

36. $\dfrac{x^8}{x^{12}}$.

37. $\dfrac{x^{21}}{x^{22}}$.

38. $\dfrac{(a+b)^{16}}{(a+b)^{12}}$.

39. $\dfrac{y^{14}}{-y^8}$.

40. $\dfrac{-y^2}{-y^5}$.

41. $\dfrac{x^2 x^8}{x^{16}}$.

42. $\dfrac{x^{18}}{x^{10} x^{10}}$.

43. $\dfrac{(x^5)^3}{x^2}$.

44. $\dfrac{(x^4)^2}{(x^5)^3}$.

45. $\dfrac{(x^4)^2}{x(x^6)}$.

46. $\dfrac{t^{12}(t^6)}{(w^5)^3}$.

47. $\dfrac{(x^2)^4(x^4)^2}{(x^3)^7}$.

48. $\dfrac{1}{(x^4)^5}(x^4)^5$.

49. $(ab)^6$.

50. $(xy)^4$.

51. $(2x)^4$.

52. $(3x)^3$.

53. $(2x^4 y^2)^4$.

54. $(x^2 yz^3)^2$.

55. $(3y)^2(5y^4)$.

56. $(3y^2)(2y^2)^4$.

57. $\left(\dfrac{a}{b}\right)^3$.

58. $\left(\dfrac{x}{2}\right)^4$.

59. $\left(\dfrac{3}{x}\right)^4$.

60. $\left(\dfrac{x}{y}\right)^2$.

61. $(xy^2)^4$.

62. $(3x^2)^3$.

63. $\left(\dfrac{2y}{z}\right)^3$.

64. $\left(\dfrac{1}{x^2y^3}\right)^5$.

65. $\left(\dfrac{2}{3}a^2b^3c^6\right)^2$.

66. $(-4)(2x^2)^2$.

67. $\left(\dfrac{x^2}{y^5}\right)^3$.

68. $\left(\dfrac{2x^2}{y^2}\right)^4$.

69. $\left(\dfrac{x^2y^3}{2z^4}\right)^4$.

70. $\left(\dfrac{2x^4}{5y^2}\right)^3$.

71. $(-x)^{13}$.

72. $(-3x)^4$.

73. $(-2x^2y)^4$.

74. $(-3)^2(-x)^3$.

75. $\dfrac{(-xy)^5}{(-t)^4}$.

76. $\dfrac{-y^3}{(-z)^2}$.

77. $(-3x)^3(-x)^7$.

78. $\dfrac{5^{100}}{5^{99}}$.

79. $\dfrac{2^6 2^{11}}{(2^5)^3}$.

80. $\dfrac{(80)^6}{(40)^6}$.

81. $(x^ay^b)^c$.

82. $\left(\dfrac{x^a}{y^b}\right)^a$.

*In Problems **83-92**, perform the indicated operations. Round your results to three significant digits.*

83. $(5.64)^2$.

84. $(0.00456)^2$.

85. $(17.6)^3$.

86. $(1.05)^4$.

87. $\dfrac{1}{(1.06)^5}$.

88. $\left(\dfrac{5.89}{2.57}\right)^4$.

89. $(0.0287)^2(46.3)^3$.

90. $(2.86^3)^2$.

91. $\dfrac{(34.57)^3}{(23.9)^4}$.

92. $\dfrac{(1.27)^{12}(310.6)}{(1.895)^8(228)}$.

93. The power P (in watts) dissipated in a resistance R (in ohms) is given by $P = I^2R$, where I is the current (in amperes). Find the power in each of the following cases.
    (a) $I = 3$ amperes and $R = 2$ ohms.
    (b) $I = -2$ amperes and $R = 2$ ohms.
    (c) $I = 0.1$ ampere and $R = 5$ ohms.

94. Under certain conditions, the pressure p of a vacuum system is given by the formula

$$p = \frac{ah^2}{V_0}.$$

Find p if (a) $h = 3d$, and (b) $h = \frac{2}{3}d$.

95. In addition to the formula in Problem 93, another formula for the power P dissipated in a resistance R is $P = V^2/R$, where V is the voltage across the resistor terminals. What happens to the power if the voltage is (a) doubled, (b) tripled, and (c) halved?

96. Compute the value of

$$\frac{(x_1 - \bar{x})^2 + (x_2 - \bar{x})^2 + (x_3 - \bar{x})^2 + (x_4 - \bar{x})^2 + (x_5 - \bar{x})^2}{4}$$

if $x_1 = 6$, $x_2 = 8$, $x_3 = 5$, $x_4 = 3$, $x_5 = 3$, and $\bar{x} = 5$.

97. Repeat Problem 96 if $x_1 = 3.45$, $x_2 = 4.81$, $x_3 = 9.84$, $x_4 = 7.11$, $x_5 = 3.84$, and $\bar{x} = 5.81$. Give your answer to two decimal places.

98. The volume V of a sphere of radius r is given by $V = \frac{4}{3}\pi r^3$. What is the effect of doubling the radius?

99. The volume V of a cylinder with radius r and height h is given by $V = \pi r^2 h$.
    (a) Find the volume (in cubic meters) of a cylinder with $r = 0.850$ meters and $h = 1.50$ meters. (Use the $\boxed{\pi}$ key on your calculator and round your result to three significant digits.)
    (b) If both the radius and height of a cylinder are doubled, by what factor does the volume increase?

2.2 ZERO AND NEGATIVE EXPONENTS

Up to now we have worked only with exponents that are positive integers. Now we shall attach a meaning to a^0, where $a \neq 0$ (the symbol 0^0 is not defined). We want to define a^0 so that the rules for exponents hold. For instance, by the rule $a^m a^n = a^{m+n}$, we must have

$$a^m \cdot a^0 = a^{m+0} = a^m.$$

The result is identical to the first factor in the original expression, so multiplying by a^0 should have the same effect in multiplication as does the number 1. Thus it is reasonable to define a^0 to be 1.

$$\boxed{8. \quad a^0 = 1 \quad \text{if} \quad a \neq 0.}$$

With this definition, the other rules of exponents also hold if an exponent is 0. To illustrate, for the rule $(ab)^n = a^n b^n$, we have

$$(ab)^0 = 1 = 1 \cdot 1 = a^0 b^0.$$

Example 1

a. $2^0 = 1$.

b. $(-3)^0 = 1$.

c. $(x^2 + 3)^0 = 1$.

d. $5(\frac{3}{2})^0 = 5(1) = 5$.

We now give meaning to a negative exponent, as in a^{-n}, where n is positive. Again we want the rules of exponents to hold. From the rule $a^m a^n = a^{m+n}$ we must have

$$a^n \cdot a^{-n} = a^{n+(-n)} = a^0 = 1.$$

Thus the effect of multiplying a^n by a^{-n} is the same as multiplying a^n by its reciprocal, because the product of a number and its reciprocal is 1. Thus we are motivated to make the following definition.

$$9. \quad a^{-n} = \frac{1}{a^n}.$$

For example,

$$3^{-4} = \frac{1}{3^4} = \frac{1}{81}, \qquad x^{-6} = \frac{1}{x^6}, \quad \text{and} \quad y^{-1} = \frac{1}{y}.$$

If we apply Rule 9 to $\dfrac{1}{a^{-n}}$, we get

$$\frac{1}{a^{-n}} = \frac{1}{\dfrac{1}{a^n}} = 1 \cdot \frac{a^n}{1} = a^n.$$

Thus we have the following rule:

$$10. \quad \frac{1}{a^{-n}} = a^n.$$

For example,

$$\frac{1}{3^{-2}} = 3^2 = 9 \quad \text{and} \quad \frac{1}{x^{-5}} = x^5.$$

Example 2

a. $-x^{-6} = -(x^{-6}) = -\dfrac{1}{x^6}.$

b. $\dfrac{1}{x^2} = x^{-2}$, by Rule 9.

c. $2x^{-3} = 2\left(\dfrac{1}{x^3}\right) = \dfrac{2}{x^3}$, but $(2x)^{-3} = \dfrac{1}{(2x)^3} = \dfrac{1}{8x^3}.$

d. $\dfrac{1}{(-2)^{-3}} = (-2)^3 = -8$. Note that we changed only the sign of the exponent; we should not change the sign of the base. That is,

$$\frac{1}{(-2)^{-3}} \neq (2)^3.$$

e. $x^{-2} + y^{-2} = \dfrac{1}{x^2} + \dfrac{1}{y^2}$. Note: $x^{-2} + y^{-2} \neq \dfrac{1}{x^2 + y^2}$.

We can obtain a rule for a negative power of a fraction. Observe that

$$\left(\frac{a}{b}\right)^{-2} = \frac{1}{\left(\dfrac{a}{b}\right)^2} = \frac{1}{\dfrac{a^2}{b^2}} = 1 \cdot \frac{b^2}{a^2} = \frac{b^2}{a^2} = \left(\frac{b}{a}\right)^2.$$

Generalizing this result, we have Rule 11.

$$\boxed{11. \quad \left(\frac{a}{b}\right)^{-n} = \left(\frac{b}{a}\right)^n.}$$

For example, $\left(\frac{2}{3}\right)^{-4} = \left(\frac{3}{2}\right)^4 = \frac{81}{16}$.

Using Rules 8–10, we can replace Rules 3–5 by the following rule, which holds regardless of the values of m and n.

$$\boxed{12. \quad \frac{a^m}{a^n} = a^{m-n} = \frac{1}{a^{n-m}}.}$$

For example,

$$\frac{x^3}{x^7} = x^{3-7} = x^{-4} = \frac{1}{x^4}.$$

But, more directly, we have

$$\frac{x^3}{x^7} = \frac{1}{x^{7-3}} = \frac{1}{x^4}.$$

Using exponents we can manipulate factors in a fraction. Observe that

$$\frac{x^2 y^{-3}}{z^{-4}} = x^2 \cdot y^{-3} \cdot \frac{1}{z^{-4}} = x^2 \cdot \frac{1}{y^3} \cdot z^4 = \frac{x^2 z^4}{y^3}.$$

Compare the factors in the first and last fractions. You can see that a nonzero *factor* of the numerator (or denominator) of a fraction may be equivalently expressed as a factor of the denominator (or numerator) if the sign of its exponent is *changed*. For example,

$$\frac{x^{-2}y^3}{2z^{-2}} = \frac{y^3z^2}{2x^2}.$$

The word *factor* in the preceding statement is crucial! For example,

$$\frac{1}{2^2 + 3^2} \neq 2^{-2} + 3^{-2} \quad \text{but} \quad \frac{1}{2^2 3^2} = 2^{-2} 3^{-2}.$$

Example 3

Simplify and give all answers with positive exponents.

a. $x^{-2}y^{-2} = \dfrac{1}{x^2 y^2}.$

b. $\dfrac{16x^{-4}}{32x^7} = \dfrac{1}{2x^7 x^4} = \dfrac{1}{2x^{11}}.$ Note that we canceled the common factor 16.

c. $\dfrac{-x^{-2}}{y^{-3}z^2} = -\dfrac{y^3}{x^2 z^2}.$

d. $\dfrac{x^{-7}y^6}{x^9 y^{-2}} = \dfrac{y^6 y^2}{x^9 x^7} = \dfrac{y^8}{x^{16}}.$

Although the rules of exponents stated in Sec. 2.1 assumed the exponents to be positive integers, *the rules of exponents are equally true for any exponents.* Henceforth, we assume this fact, as Example 4 shows.

Example 4

Perform the operations and write the answers with positive exponents only.

a. $x^5 x^{-2} = x^{5+(-2)} = x^{5-2} = x^3.$

b. $2x^{-2}x^{-3} = 2x^{-2-3} = 2x^{-5} = \dfrac{2}{x^5}.$

c. $(2x^3 y^{-6})^2 = 2^2(x^3)^2(y^{-6})^2 = 4x^6 y^{-12} = \dfrac{4x^6}{y^{12}}.$

d. $(x^2 y^{-4})^{-3} = (x^2)^{-3}(y^{-4})^{-3} = x^{-6}y^{12} = \dfrac{y^{12}}{x^6}.$

Alternatively,

$$(x^2 y^{-4})^{-3} = \frac{1}{(x^2 y^{-4})^3} = \frac{1}{x^6 y^{-12}} = \frac{y^{12}}{x^6}.$$

e. $\left(\dfrac{x^{-2}y^3}{z^{-4}}\right)^{-6} = \dfrac{(x^{-2}y^3)^{-6}}{(z^{-4})^{-6}} = \dfrac{x^{12}y^{-18}}{z^{24}} = \dfrac{x^{12}}{y^{18}z^{24}}.$

Another way to do this problem is

$$\left(\dfrac{x^{-2}y^3}{z^{-4}}\right)^{-6} = \left(\dfrac{y^3z^4}{x^2}\right)^{-6} = \left(\dfrac{x^2}{y^3z^4}\right)^{6} = \dfrac{x^{12}}{y^{18}z^{24}}.$$

A negative power of a number may be found with a calculator, as illustrated here.

Example	Keystroke Sequence	Display
4^{-3}	$\boxed{4}\ \boxed{y^x}\ \boxed{3}\ \boxed{+/-}\ \boxed{=}$	0.015625

Problem Set 2.2

*In Problems **1–17**, find the values of the numbers without using a calculator.*

1. 3^0.

2. $\left(\dfrac{3}{4}\right)^0$.

3. 2^{-3}.

4. 3^{-2}.

5. $\dfrac{1}{3^{-3}}$.

6. $\dfrac{3}{4^{-2}}$.

7. $2x^0 + (2x)^0$.

8. $2(-3)^0$.

9. $\dfrac{-1^0}{4^{-1}}$.

10. $-5^{-2}(25)$.

11. $\dfrac{1}{(-3)^{-3}}$.

12. $2^{-1} + 3^{-1}$.

13. $\dfrac{6^{-4}}{6^{-2}}$.

14. $3^{-2}(2^{-2})$.

15. $3^{-2} + 4(2^{-2})$.

16. $\dfrac{(3^{-2})^0}{1^{-1}}$.

17. $\left(\dfrac{2}{5}\right)^{-2}$.

*In Problems **18–32**, write each expression by using positive exponents only. Simplify.*

18. x^{-2}.

19. x^{-6}.

20. $2^{-1}x$.

21. $\dfrac{1}{x^{-3}}$.

22. $\dfrac{1}{3x^{-2}}$.

23. $3y^{-4}$.

24. $2^{-2}x^{-4}$.

25. $\dfrac{x}{4^{-2}}$.

26. $\dfrac{7^0}{x^{-1}yz^{-2}}$.

27. $x^{-5}y^{-7}$.

28. $x^{-1}y^{-2}z^4$.

29. $\dfrac{2a^2b^{-4}}{c^{-5}}$.

30. $\dfrac{a^5b^{-4}}{c^{-3}d}$.

31. $\dfrac{x^9y^{-12}}{w^2z^{-4}}$.

32. $\dfrac{(x^2 + 4x^4)^0}{x^{-2}}$.

*In Problems **33–60**, perform the operations and simplify. Give all answers with positive exponents only.*

33. x^8x^{-7}.

34. $x^{-7}x$.

35. $x^{-2}x^{-3}$.

36. $x^2x^{-4}x^9$.

37. $(2x^{-5})(4x)$.

38. $(5x^{-5})(2x^{-4})$.

39. $(xy^{-5})^{-4}$.

40. $(2x^2y^{-1})^2$.

41. $2(x^{-1}y^2)^2$.

42. $(x^{-5}y^6)^{-1}$.

43. $(3t)^{-2}$.

44. $(x^{-4}y^{-4})^4$.

45. $(x^{-5}y^5z)^{-3}$.

46. $\dfrac{10x^6}{x^{-2}}$.

47. $\dfrac{t^{-8}}{t^{-12}}$.

48. $\dfrac{-2b^{-30}}{b^{-5}}$.

49. $\dfrac{x^{-2}y^4}{x^6y^{-1}}$.

50. $\dfrac{x^{-6}x^{10}}{x^3x^{-4}}$.

51. $\dfrac{x^{-3}}{(x^{-3})^2}$.

52. $\dfrac{x^3y^3}{x^{-2}y^{-2}}$.

53. $\left(\dfrac{x}{y}\right)^{-4}$.

54. $\left(\dfrac{y}{z^{-1}}\right)^{-1}$.

55. $\left(\dfrac{8x^2}{5y^2}\right)^{-1}$.

56. $\dfrac{1}{(3x^{-1})^{-1}}$.

57. $\left(-\dfrac{z^{-1}}{x}\right)^{-1}$.

58. $\left(\dfrac{x^{-1}y^4}{z^2}\right)^{-5}$.

59. $\left(\dfrac{y^{-6}z^2}{2x}\right)^{-2}$.

60. $\left[\left(\dfrac{x}{y}\right)^{-2}\right]^{-4}$.

*In Problems **61–66**, evaluate. Round your answers to three significant digits.*

61. $(7.28)^{-2}$.

62. $(12.3)^{-4}$.

63. $\dfrac{(4.75)^{-1}}{(8.21)^{-2}}$.

64. $\left(\dfrac{4.62}{8.14}\right)^{-3}$.

65. $\dfrac{(3.24)^{-2}(1.11)^2}{(3.13-1.24)^4}$.

66. $\dfrac{(2.67+6.04)^{-3}}{(1.73)^2}$.

2.3 SCIENTIFIC NOTATION

In science and technology, we often have to deal with numbers that are either very small or very large. Writing such numbers in our usual notation may be inconvenient. For example, the longest wavelength of visible red light is about 0.00000076 meter. The speed at which that light travels through air is about 300,000,000 meters per second, and the frequency of the light is 390,000,000,000,000 hertz. To avoid having to write so many zeros to locate the decimal points in such numbers, we may express the numbers in a compact form called *scientific notation*.

Scientific notation is based on powers of 10, some of which are as follows:

$$10 = 10^1, \qquad 0.1 = \frac{1}{10} \quad = 10^{-1},$$

$$100 = 10^2, \qquad 0.01 = \frac{1}{100} \quad = 10^{-2},$$

$$1000 = 10^3, \qquad 0.001 = \frac{1}{1000} \quad = 10^{-3},$$

$$10{,}000 = 10^4, \qquad 0.0001 = \frac{1}{10{,}000} = 10^{-4}.$$

You may recall that to multiply a decimal by 10, we need only move the decimal point one position to the *right*. Thus

$$7.34 \times 10 = 73.4.$$

Similarly, we move the decimal point two and three places to the *right* to multiply by 100 (or 10^2) and 1000 (or 10^3), respectively.

$$7.34 \times 10^2 = 734 \qquad [10^2 \rightarrow 2 \text{ places to right}],$$

$$7.34 \times 10^3 = 7340 \qquad [10^3 \rightarrow 3 \text{ places to right}].$$

In each case the exponent for the power of 10 corresponds to the number of places the decimal point must be moved from its original position.

To divide a decimal by 10, 100, or 1000, we need only move the decimal point one, two, or three places to the *left*, respectively. For example,

$$\frac{7.34}{10} = 0.734.$$

Similarly, by expressing division by a positive power of 10 in terms of multiplication by a negative power of 10, we have

$$7.34 \times 10^{-1} = 0.734 \qquad [10^{-1} \rightarrow 1 \text{ place to left}],$$

$$7.34 \times 10^{-2} = 0.0734 \qquad [10^{-2} \rightarrow 2 \text{ places to left}],$$

$$7.34 \times 10^{-3} = 0.00734 \qquad [10^{-3} \rightarrow 3 \text{ places to left}].$$

We describe scientific notation as follows.

A number N is in **scientific notation** when it is expressed as the product of a decimal number between 1 and 10 and some integer power of 10. That is, $N = a \times 10^n$, where $1 \leq a < 10$ and n is an integer.

For example, the numbers 2×10^3 and 3.4×10^{-4} are in scientific notation.

Example 1

Write each number in scientific notation.

a. 2575.

Solution We want to move the decimal point three places to the left so that 2575 becomes 2.575 (which is a number between 1 and 10). To move the decimal point, we multiply by 10^{-3}. We adjust for this by multiplying by 10^3. In short, we multiply by $10^{-3} \times 10^3$, which is 1.

$$2575 = (2575 \times 10^{-3}) \times 10^3$$

$$= 2.575 \times 10^3.$$

b. 0.063.

Solution   Here we want to move the decimal point two places to the right so that 0.063 becomes 6.3, which is a number between 1 and 10. To do this we multiply by $10^2 \times 10^{-2}$:

$$0.063 = (0.063 \times 10^2) \times 10^{-2}$$

$$= 6.3 \times 10^{-2}.$$

In each answer of Example 1, observe that the power to which 10 is raised corresponds to the number of places the decimal point must be moved from its original position. Moreover:

1. If the decimal point is moved to the *left*, the exponent is *positive*;
2. If it is moved to the *right*, the exponent is *negative*.

For example,

$$18{,}000{,}000. = 1.8 \times 10^7 \quad \text{Positive 7}$$

7 places
to left

and

$$0.00027 = 2.7 \times 10^{-4}. \quad \text{Negative 4}$$

4 places
to right

Use this technique to verify the results in Example 2.

Example 2

Write each number in scientific notation.

a. $0.000624 = 6.24 \times 10^{-4}$.

b. $21{,}000{,}000 = 2.1 \times 10^7$.

c. $0.000000000000409 = 4.09 \times 10^{-13}$.

d. $6170 = 6.17 \times 10^3$.

e. $16.2 = 1.62 \times 10$.

f. $0.005930 = 5.930 \times 10^{-3}$.

g. The wavelength and frequency given at the beginning of this section can be written 7.6×10^{-7} meter and 3.9×10^{14} hertz, respectively.

To change a number from scientific notation to ordinary decimal notation, we use the exponent of the power of 10 to help us locate the decimal point. If the exponent is *positive*, the decimal point is moved to the *right*; if it is *negative*, the decimal point is moved to the *left*. This is illustrated in Example 3.

Example 3

Write each number as a decimal.

a. 6.23×10^5.

   Solution   The exponent is *positive* 5, so we move the decimal point five places to the *right*. Thus $6.23 \times 10^5 = 623,000$.

b. 4.536×10^{-4}.

   Solution   The exponent is *negative* 4, so we move the decimal point four places to the *left*. Thus $4.536 \times 10^{-4} = 0.0004536$.

c. $6.24 \times 10^{-3} = 0.00624$.

d. $2.613 \times 10^8 = 261,300,000$.

e. $-7.0030 \times 10^{-5} = -0.000070030$.

Example 4

Evaluate by using scientific notation and give the answer in scientific notation.

a. $(0.0004)^3$.

   Solution

   $$(0.0004)^3 = (4 \times 10^{-4})^3 = 4^3 \times (10^{-4})^3 = 64 \times 10^{-12}.$$

   Writing the result in scientific notation, we have

   $$64 \times 10^{-12} = (6.4 \times 10) \times 10^{-12} = 6.4 \times 10^{1-12}$$
   $$= 6.4 \times 10^{-11}.$$

b. $(30,000)(50,000,000)$.

   Solution

   $$(30,000)(50,000,000) = (3 \times 10^4)(5 \times 10^7)$$
   $$= (3 \times 5)(10^4 \times 10^7)$$
   $$= 15 \times 10^{11} = (1.5 \times 10) \times 10^{11}$$
   $$= 1.5 \times 10^{12}.$$

   Observe how we separated the operations involving powers of 10 from the operations involving the other numbers.

c. $\dfrac{0.00762}{0.00003}$.

Solution

$$\frac{0.00762}{0.00003} = \frac{7.62 \times 10^{-3}}{3 \times 10^{-5}} = \frac{7.62}{3} \times \frac{10^{-3}}{10^{-5}} = 2.54 \times 10^{2}.$$

Example 5

Coulomb's law states that the force F (in newtons) acting between two small objects having electric charges q_1 and q_2 (in coulombs) is given by

$$F = \frac{(9 \times 10^9)q_1 q_2}{r^2},$$

where r is the distance between the objects (in meters). Given that in the hydrogen atom the electron and proton each have a charge of 1.6×10^{-19} coulomb and are separated by a distance of 5.3×10^{-11} meter, compute the force between them. Give the answer in scientific notation.

Solution   Here we treat the powers of 10 and the other factors separately:

$$F = \frac{(9 \times 10^9)(1.6 \times 10^{-19})(1.6 \times 10^{-19})}{(5.3 \times 10^{-11})^2}$$

$$= \frac{(9 \times 1.6 \times 1.6)(10^9 \times 10^{-19} \times 10^{-19})}{(5.3)^2(10^{-22})}$$

$$= \left[\frac{9 \times 1.6 \times 1.6}{(5.3)^2}\right] 10^{9-19-19+22}.$$

The first factor is approximately 0.82 and the second factor is 10^{-7}. Thus

$$F = 0.82 \times 10^{-7}.$$

(The use of the equals sign here is a convenience that we shall adopt throughout the text. More precisely, we should write $F \approx 0.82 \times 10^{-7}$.) In scientific notation we have

$$F = (8.2 \times 10^{-1}) \times 10^{-7} = 8.2 \times 10^{-8} \quad \text{newton.}$$

In Example 5, the separation of powers of 10 and the other factors illustrates the mathematics of the computation. When calculators are used, such a separation may or may not be necessary. For example, a calculator that allows us to enter only numbers that are not more than 8 to 10 digits long could not directly handle this calculation without a scientific-notation capability. This is a good time for you to see if your calculator allows entry of a number in scientific notation and how such a number is displayed.

If your calculator has this capability, you enter a number in scientific notation by using the exponential (or enter-exponent) key, designated EXP, EEX, or EE (see Sec. A.2). The exponent for 10 is entered *after* the EXP key is pressed.

Enter	*Keystroke Sequence*	*Display*
2.35×10^7	2.35 EXP 7	2.35 07
6×10^{-8}	6 EXP 8 +/−	6 −08
10^{22}	1 EXP 22	1 22

To compute

$$(2.7 \times 10^{20}) \times (5.8 \times 10^{-9}),$$

we have:

Keystroke Sequence   *Display*

2.7 EXP 20 × 5.8 EXP 9 +/− =   1.566   12

Problem Set 2.3

In Problems 1–12, write each number in scientific notation.

1. 0.000006021.
2. 213,146,100,000,000.
3. 10.4.
4. 0.0624.
5. 26,245.1.
6. 2,600,000.
7. 142.
8. 0.0071.
9. 0.76.
10. 0.1.
11. 0.0348×10^2.
12. $20,300 \times 10^{-2}$.

In Problems 13–20, write each number in decimal form.

13. 2.62×10^8.
14. 1.234×10^{-8}.
15. 6.24×10^{-10}.
16. 1.006×10^4.
17. 2.020×10^{-1}.
18. 6.0411×10^{12}.
19. 7.611×10^5.
20. 2.0×10.

21. The mass of the earth, in kilograms, is usually taken to be

$$5,983,000,000,000,000,000,000,000.$$

Express this mass in scientific notation.

22. The speed of light in a vacuum is taken to be 300,000,000 meters per second. Express this speed in scientific notation.

23. A numerical value of the gravitational constant is

$$0.0000000000667.$$

Express this constant in scientific notation.

24. The Milky Way galaxy is estimated to contain about 100 billion stars. Express this number in scientific notation. (*Note:* One billion is 1,000,000,000.)

In Problems **25–32**, *perform the indicated operations and give your answer in scientific notation.*

25. $(1.3 \times 10^{-4})(2.0 \times 10^6)$.

26. $\dfrac{9.3 \times 10^{-1}}{3.1 \times 10^5}$.

27. $\dfrac{(3.0 \times 10^{11})(4.2 \times 10^{-4})}{2 \times 10^{13}}$.

28. $\dfrac{(4.8 \times 10^{-1})(5.0 \times 10^{-2})}{(3.2 \times 10^{-3})(3.0 \times 10^{-4})}$.

29. $\dfrac{(1.0 \times 10^4)^3}{2.5 \times 10^5}$.

30. $\dfrac{(1.2 \times 10^{-2})^2}{2.88 \times 10}$.

31. $\dfrac{(35{,}600{,}000)(5{,}870{,}000)}{(12{,}600)(567{,}000)(1230)}$.

32. $\dfrac{(576 \times 10^{-4})(0.00000562)}{(0.00891)(4730)(0.938)}$.

33. A current I of 10^4 amperes is produced in a conductor, with length L of 4 meters, at a point where the earth's magnetic field B is 5×10^{-5} teslas at right angles to the conductor. The force F, in newtons, on the conductor is given by

$$F = IBL$$
$$= (10^4)(5 \times 10^{-5})(4).$$

Evaluate F.

34. The electric potential at a distance of 1.73×10^{-4} meter from a point charge of 4.62×10^{-2} coulomb is

$$\frac{(9 \times 10^6)(4.62 \times 10^{-2})}{1.73 \times 10^{-4}} \quad \text{volts.}$$

Evaluate this potential.

35. The volume V of a sphere is given by $V = \frac{4}{3}\pi r^3$, where r is the radius. Find the volume of the earth (in cubic meters) if the earth's radius is 6.37×10^6 meters.

36. An equilibrium constant, K, that relates to the solubility of lead iodide is given by

$$K = [Pb^{++}][I^-]^2.$$

Calculate K if $[Pb^{++}]$ is 1×10^{-2} and $[I^-]$ is 1.2×10^{-3}.

37. Two small objects have electric charges of 3.3×10^{-6} and 4×10^{-6} coulomb, respectively. Find the force between the objects when they are placed 2 meters apart. (Refer to Example 5.)

38. In calibrating the scale of a voltmeter, the following formula is used.

$$R_x = R_T\left(\frac{I_M - I_x}{I_x}\right).$$

Find the value of R_x (in ohms) if $R_T = 2000$ ohms, $I_M = 10^{-3}$ ampere, and $I_x = 7.5 \times 10^{-4}$ ampere.

39. The resistance R (in ohms) of a wire of length L (in meters) and cross-sectional area A (in square meters) is given by $R = \rho L/A$, where ρ (the Greek letter rho) is the resistivity of the material (in ohm·meters). Find the resistance of a copper wire 5 meters long with a

cross-sectional area of 0.0000003 square meter. The resistivity of copper is 0.000000017 ohm·meter.

40. While the universe is about 10^{10} years old, humans have existed only for about 10^6 years. Determine the number of years the universe existed before humans, and express that number in scientific notation.

41. The effective capacitance C of capacitors C_1 and C_2 connected in series is given by

$$C = \frac{C_1 C_2}{C_1 + C_2}.$$

Find C (in farads) if $C_1 = 2.0 \times 10^{-6}$ farad and $C_2 = 6.0 \times 10^{-6}$ farad.

42. If the mass of a molecule of a certain substance is 2.5×10^{-25} kilogram, find the number of molecules that have a total mass of 1 kilogram.

43. If the mass (in kilograms) of a proton is 1.673×10^{-27} and that of an electron is 9.109×10^{-31}, how many times larger is the mass of a proton compared to the mass of an electron?

2.4   RADICALS

If a and b are real numbers and $b^2 = a$, then b is called a **square root** of a. For example, $3^2 = 9$, so 3 is a square root of 9. In fact, because $(-3)^2 = 9$, the negative number -3 is also a square root of 9. There is a way to distinguish between the positive and negative square roots of a number. The *positive* square root is called the **principal square root**. Thus 3 is the principal square root of 9. For convenience we usually omit the word *principal* and simply say that 3 is *the* square root of 9.

We denote the (principal) square root of a nonnegative number a by the symbol $\sqrt{a}$, where a is called the **radicand**, the symbol $\sqrt{}$ is the **radical sign**, and the symbol $\sqrt{a}$ is itself called a **radical**.

$$\boxed{\sqrt{a} = b \quad \text{means} \quad b^2 = a, \text{ where } b \geq 0.}$$

Thus $\sqrt{9} = 3$, not -3. We define $\sqrt{0}$ to be 0.

You will find the appearance of radicals in many of your other courses, for they are found in virtually every area of technology. To illustrate, the expressions

$$\frac{1}{2\pi\sqrt{LC}}, \qquad \sqrt{\frac{\beta}{d}}, \quad \text{and} \quad \sqrt{\frac{M}{2\rho_0 N_0}}$$

are, respectively, the resonant frequency of an alternating current circuit, the velocity of a compressional wave, and the distance between ions in a cubic crystal.

Example 1

a. $\sqrt{25} = 5$ (the square root of 25 is 5) because $5^2 = 25$ and 5 is positive. The radicand is 25.

b. $\sqrt{0.01} = 0.1$ (the square root of 0.01 is 0.1) because $(0.1)^2 = 0.01$ and 0.1 is positive.

c. $-\sqrt{81} = -(\sqrt{81}) = -(9) = -9$ (the *negative* of the square root of 81 is -9).

d. $\sqrt{\dfrac{49}{100}} = \dfrac{7}{10}$ because $\left(\dfrac{7}{10}\right)^2 = \dfrac{49}{100}$.

We wish to mention two things. First, troubles arise with square roots of negative numbers. For example, there is no real number whose square is -4 because the square of a real number is never negative. This means that the symbol $\sqrt{-4}$ does not represent a real number. For the present we shall not concern ourselves with square roots of negative numbers. (Such roots are called *imaginary numbers* and are discussed in Chapter 13.)

Second, a square root of a number may be irrational. One example is $\sqrt{2}$, which is often approximated by 1.414. In some problems we shall find it convenient to leave an answer in radical form instead of using a decimal approximation obtained with a calculator.

There are four properties of square roots that will be useful in our future work. To begin with, note that since $\sqrt{9}$ is a number whose square is 9, we have $(\sqrt{9})^2 = 9$. Similarly,

$$(\sqrt{6})^2 = 6.$$

A more general rule is:

$$\textbf{1.} \quad (\sqrt{a})^2 = a.$$

Moreover, because the square of a is a^2, a is a square root of a^2. If a is nonnegative, a general rule is:*

$$\textbf{2.} \quad \sqrt{a^2} = a.$$

For example, $\sqrt{3^2} = 3$.

The next two rules state that the square root of a product (or quotient) is the product (or quotient) of the square roots:

$$\textbf{3.} \quad \sqrt{ab} = \sqrt{a}\sqrt{b}.$$

$$\textbf{4.} \quad \sqrt{\dfrac{a}{b}} = \dfrac{\sqrt{a}}{\sqrt{b}}.$$

---

* For any real number a, the rule is $\sqrt{a^2} = |a|$. For example, $\sqrt{(-6)^2} = |-6| = 6$. Rules 3 and 4 assume a and b are nonnegative.

These rules allow us to simplify many radicals by writing them as a product or quotient of radicals so that one of them is easy to compute. For example, to simplify $\sqrt{24}$, we write 24 as a product of factors so that one factor is a perfect square. Then we use Rule 3.

$$\sqrt{24} = \sqrt{4 \cdot 6} = \sqrt{4}\sqrt{6} = 2\sqrt{6}.$$

We say that the factor 4 has been *removed* from the radicand, and $2\sqrt{6}$ is considered to be the *simplified form* of $\sqrt{24}$.

When using this procedure, it is best to rewrite the radicand so that one of the factors is the *largest* whose square root can be taken easily. For example, to simplify $\sqrt{48}$ we should write 48 as $16 \cdot 3$. Writing it as $4 \cdot 12$ requires more work:

$$\sqrt{48} = \sqrt{4 \cdot 12} = \sqrt{4}\sqrt{12} = 2\sqrt{4 \cdot 3}$$
$$= 2(\sqrt{4}\sqrt{3}) = 2(2)\sqrt{3} = 4\sqrt{3}.$$

More simply,

$$\sqrt{48} = \sqrt{16 \cdot 3} = \sqrt{16}\sqrt{3} = 4\sqrt{3}.$$

Example 2

Simplify each radical.

a. $\sqrt{18} = \sqrt{9 \cdot 2} = \sqrt{9}\sqrt{2} = 3\sqrt{2}.$

b. $\sqrt{20} = \sqrt{4 \cdot 5} = \sqrt{4}\sqrt{5} = 2\sqrt{5}.$

c. $\sqrt{\dfrac{11}{25}} = \dfrac{\sqrt{11}}{\sqrt{25}} = \dfrac{\sqrt{11}}{5}.$

d. $\dfrac{\sqrt{20}}{\sqrt{5}} = \sqrt{\dfrac{20}{5}} = \sqrt{4} = 2.$

When a fraction has a radical in its denominator, as with $2/\sqrt{3}$, it is sometimes convenient to rewrite the fraction in an equivalent form that does not have a radical in the denominator. This procedure is called **rationalizing the denominator**. To rationalize the denominator of

$$\frac{2}{\sqrt{3}},$$

we note that if the denominator were $(\sqrt{3})^2$, it would simplify by the rule $(\sqrt{a})^2 = a$. We can change $\sqrt{3}$ to $(\sqrt{3})^2$ with the use of the fundamental principle of fractions by multiplying both the numerator and denominator by $\sqrt{3}$.

$$\frac{2}{\sqrt{3}} = \frac{2\sqrt{3}}{\sqrt{3}\sqrt{3}} = \frac{2\sqrt{3}}{(\sqrt{3})^2} = \frac{2\sqrt{3}}{3}.$$

Actually, multiplying the numerator and denominator by $\sqrt{3}$ is equivalent to multiplying the given fraction by $\sqrt{3}/\sqrt{3}$. That is,

$$\frac{2}{\sqrt{3}} = \frac{2}{\sqrt{3}} \cdot \frac{\sqrt{3}}{\sqrt{3}} = \frac{2\sqrt{3}}{(\sqrt{3})^2} = \frac{2\sqrt{3}}{3}.$$

Example 3

Rationalize the denominator.

a. $\dfrac{1}{\sqrt{5}} = \dfrac{1}{\sqrt{5}} \cdot \dfrac{\sqrt{5}}{\sqrt{5}} = \dfrac{\sqrt{5}}{(\sqrt{5})^2} = \dfrac{\sqrt{5}}{5}.$

b. $\sqrt{\dfrac{5}{3}} = \sqrt{\dfrac{5}{3} \cdot \dfrac{3}{3}} = \sqrt{\dfrac{15}{3^2}} = \dfrac{\sqrt{15}}{\sqrt{3^2}} = \dfrac{\sqrt{15}}{3}.$

The notion of square root can be extended to other roots. Since $2^4 = 16$, we say that 2 is a *fourth root* of 16. Similarly, -2 is a fourth root of 16. More generally, if n is a positive integer,

$$\boxed{b \text{ is an } \textbf{nth} \text{ root of } a \text{ if } b^n = a.}$$

Thus 5 is a **cube** (or third) **root** of 125 because $5^3 = 125$. Whenever a number a has a *positive* nth root, we call that root the **principal nth root** of a. Thus 2 is the (principal) fourth root of 16.

If a number has a *negative* nth root but does *not* have a positive nth root, we refer to that negative root as the principal nth root. For example, the (principal) cube root of -8 is -2 because $(-2)^3 = -8$ and there are no positive cube roots of -8. We define any root of 0 to be 0.

We denote the principal nth root of a by the symbol $\sqrt[n]{a}$, where n is called the **index** of the radical. Thus $\sqrt[4]{16} = 2$ (the index is 4) and $\sqrt[3]{-8} = -2$. For a principal square root, we omit the index 2. Thus $\sqrt[2]{9} = \sqrt{9} = 3$. We define $\sqrt[n]{0}$ to be 0.

$$\boxed{\begin{array}{c} \textbf{PRINCIPAL } \textit{n} \textbf{th ROOT OF } \textit{a} \\[4pt] \sqrt[n]{a} \end{array}}$$

Example 4

a. $\sqrt[3]{-1} = -1$ because $(-1)^3 = -1$ and no positive number has -1 for its cube.

b. $\sqrt[3]{\dfrac{1}{125}} = \dfrac{1}{5}$ because $\left(\dfrac{1}{5}\right)^3 = \dfrac{1}{125}$.

c. $\sqrt[5]{0} = 0$.

Corresponding to the previous rules for square roots, we have the following rules for *n*th roots.*

1. $(\sqrt[n]{a})^n = a$.

2. $\sqrt[n]{a^n} = a$.

3. $\sqrt[n]{ab} = \sqrt[n]{a}\,\sqrt[n]{b}$.

4. $\sqrt[n]{\dfrac{a}{b}} = \dfrac{\sqrt[n]{a}}{\sqrt[n]{b}}$.

Example 5

a. $(\sqrt[5]{3})^5 = 3$ and $\sqrt[5]{3^5} = 3$.

b. $\sqrt[3]{54} = \sqrt[3]{27 \cdot 2} = \sqrt[3]{27}\,\sqrt[3]{2} = 3\sqrt[3]{2}$.

c. $\sqrt[3]{-16} = \sqrt[3]{(-8)(2)} = \sqrt[3]{-8}\,\sqrt[3]{2} = -2\sqrt[3]{2}$.

d. $\sqrt[3]{2}\,\sqrt[3]{4} = \sqrt[3]{2 \cdot 4} = \sqrt[3]{8} = 2$.

Example 6

Given the formula

$$d = \sqrt{(x_2 - x_1)^2 + (y_2 - y_1)^2},$$

find d if $x_1 = 1$, $x_2 = -7$, $y_1 = 2$, and $y_2 = 8$.

Solution

$$d = \sqrt{(x_2 - x_1)^2 + (y_2 - y_1)^2} = \sqrt{(-7 - 1)^2 + (8 - 2)^2}$$
$$= \sqrt{(-8)^2 + (6)^2} = \sqrt{64 + 36} = \sqrt{100}$$
$$= 10.$$

This formula, called the *distance formula*, is used in analytic geometry.

* In each rule we assume that each expression represents a real number.

In Example 6 we have $\sqrt{64 + 36} = \sqrt{100} = 10$. We emphasize that

$$\sqrt{64 + 36} \neq \sqrt{64} + \sqrt{36} = 8 + 6 = 14.$$

That is, in general the square root of a *sum* is **not** the sum of the square roots:

$$\sqrt{a + b} \neq \sqrt{a} + \sqrt{b}.$$

Moral: Don't try fancy tricks of your own. Stick to basics.

Example 7

The current I (in amperes) in a resistance R (in ohms) in which the power dissipation is P (in watts) is given by

$$I = \sqrt{\frac{P}{R}}.$$

Find the current if the resistance is 3×10^3 ohms and the power dissipation is 3×10^{-3} watts.

Solution

$$I = \sqrt{\frac{3 \times 10^{-3}}{3 \times 10^3}} = \sqrt{10^{-3}(10^{-3})} = 10^{-3} \text{ ampere.}$$

To find roots of numbers with a calculator, use either the square-root key $\boxed{\sqrt{x}}$ (see Sec. A.4) or the root key $\boxed{\sqrt[x]{y}}$ (see Sec. A.5), which calculates the xth root of y. (The equals key is not necessary with the square-root key.) With the root key, the radicand y is entered first, then the root key is pressed, and finally the index x is entered.

Example	Keystroke Sequence	Display
$\sqrt{75}$	$\boxed{75}$ $\boxed{\sqrt{x}}$	8.66025
$\sqrt[5]{75}$	$\boxed{75}$ $\boxed{\sqrt[x]{y}}$ $\boxed{5}$ $\boxed{=}$	2.37144

Example 8

The impedance Z of a series R-L circuit is given by

$$Z = \sqrt{X_L^2 + R^2},$$

where X_L is the inductive reactance, R is the resistance, and all quantities are measured in ohms. Find the impedance of such a circuit in which $X_L = 10$ ohms and $R = 12$ ohms.

Solution

$$Z = \sqrt{(10)^2 + (12)^2} = \sqrt{100 + 144} = \sqrt{244} = 15.6 \text{ ohms.}$$

The calculator keystroke sequence is

$$\boxed{10}\ \boxed{x^2}\ \boxed{+}\ \boxed{12}\ \boxed{x^2}\ \boxed{=}\ \boxed{\sqrt{x}}\qquad \text{Display:}\quad 15.620499$$

Problem Set 2.4

In Problems 1–38, compute the numbers without the aid of a calculator.

1. $\sqrt{36}.$ 2. $\sqrt{81}.$ 3. $\sqrt[3]{8}.$

4. $\sqrt[3]{125}.$ 5. $\sqrt{49}.$ 6. $\sqrt{100}.$

7. $\sqrt[3]{-27}.$ 8. $\sqrt[3]{-8}.$ 9. $-\sqrt[3]{-64}.$

10. $\sqrt[4]{1}.$ 11. $\sqrt[4]{16}.$ 12. $\sqrt[5]{-1}.$

13. $\sqrt[4]{0}.$ 14. $\sqrt[3]{64}.$ 15. $\sqrt[3]{-125}.$

16. $\sqrt[5]{32}.$ 17. $\sqrt[6]{64}.$ 18. $\sqrt{12 \cdot 3}.$

19. $\sqrt{0.04}.$ 20. $\sqrt{0.25}.$ 21. $\sqrt{\dfrac{1}{16}}.$

22. $\sqrt{\dfrac{1}{100}}.$ 23. $-\sqrt{25}.$ 24. $\sqrt[3]{-1}.$

25. $\sqrt{81} - \sqrt[3]{-8}.$ 26. $\sqrt{64} - \sqrt[3]{-8}.$ 27. $\dfrac{\sqrt{64} + \sqrt[3]{-64}}{\sqrt{81} + \sqrt[4]{81}}.$

28. $\sqrt[3]{|-8|}.$ 29. $\sqrt{5} \cdot \sqrt{5}.$ 30. $\sqrt[3]{7} \cdot \sqrt[3]{7} \cdot \sqrt[3]{7}.$

31. $(\sqrt[4]{4})^4 + \sqrt[4]{4^4}.$ 32. $(\sqrt{3})^2 + \sqrt[5]{2^5}.$ 33. $\sqrt{(-3)^2} - \sqrt{3^2}.$

34. $\dfrac{\sqrt{49} - \sqrt{36}}{\sqrt{100} - \sqrt[3]{1}}.$ 35. $\dfrac{(-1)^2 - (3\sqrt{9})}{|-4| + \sqrt{16}}.$ 36. $\dfrac{(-2)^2 + (-3)^4}{\sqrt[5]{1} + |2 - 3|}.$

37. $\sqrt{0.01} + \sqrt{0.0025}.$ 38. $2(\sqrt[3]{0.008}) - (0.01)^2.$

In Problems 39–63, simplify the numbers by using properties of radicals, as in Examples 2, 3, and 5.

39. $\sqrt{50}.$ 40. $\sqrt{75}.$ 41. $\sqrt{12}.$

42. $\sqrt{32}.$ 43. $\sqrt{8}.$ 44. $\sqrt{400{,}000}.$

45. $\sqrt{14{,}400}.$ 46. $\sqrt[3]{24}.$ 47. $\sqrt[4]{48}.$

48. $\sqrt[4]{162}.$ 49. $\sqrt[5]{64}.$ 50. $\sqrt[3]{-54}.$

51. $\sqrt[3]{-500}.$ 52. $\sqrt{\dfrac{5}{4}}.$ 53. $\sqrt{\dfrac{14}{9}}.$

54. $\sqrt{\dfrac{2}{25}}$.

55. $\sqrt[3]{\dfrac{10}{27}}$.

56. $\dfrac{\sqrt{90}}{\sqrt{10}}$

57. $\dfrac{\sqrt{50}}{\sqrt{2}}$.

58. $\dfrac{\sqrt[4]{64}}{\sqrt[4]{4}}$.

59. $\dfrac{\sqrt[3]{-2}}{\sqrt[3]{16}}$.

60. $\dfrac{\sqrt[5]{-4}}{\sqrt[5]{-128}}$.

61. $\sqrt{8}\cdot\sqrt{2}$.

62. $\sqrt{27}\cdot\sqrt{3}$.

63. $\sqrt[3]{4}\cdot\sqrt[3]{16}$.

In Problems **64–71**, rationalize the denominators.

64. $\dfrac{5}{\sqrt{6}}$.

65. $\dfrac{3}{\sqrt{2}}$.

66. $\dfrac{2.4}{\sqrt{6}}$.

67. $\dfrac{1}{\sqrt{7}}$.

68. $\dfrac{3}{2\sqrt{5}}$.

69. $\dfrac{4}{3\sqrt{2}}$.

70. $\sqrt{\dfrac{1}{15}}$.

71. $\sqrt{\dfrac{3}{2}}$.

In Problems **72–76**, compute the numbers. *Round your results to four significant digits.*

72. $\sqrt{84.26}$.

73. $\sqrt{0.02864}$.

74. $\sqrt[3]{0.0009264}$.

75. $\sqrt[6]{59{,}270}$.

76. $\sqrt[4]{6.083\times10^5}$.

In Problems **77** and **78**, use the distance formula in Example 6 to find d with the given information

77. $x_1 = -1,\ y_1 = 1,\ x_2 = -6,\ y_2 = 13$.

78. $x_1 = 10,\ y_1 = 2,\ x_2 = 7,\ y_2 = -2$.

79. The radius of a circle with area A is given by

$$r = \sqrt{\dfrac{A}{\pi}}.$$

Find the radius of a circle if the area is (a) 4π square meters, and (b) 15 square centimeters.

80. The relationship between the molecular velocity (v) and molecular weight (M) of two gases A and B at the same temperature is given by Graham's diffusion law:

$$\dfrac{v_A}{v_B} = \sqrt{\dfrac{M_B}{M_A}}.$$

Given that the molecular weight of hydrogen (H_2) is 2 and the molecular weight of oxygen (O_2) is 32, find the value of

$$\dfrac{v_{H_2}}{v_{O_2}}.$$

The Pythagorean theorem allows one side of a right triangle to be found when the other two sides are known. The relationships to be used are given in Fig. 2.1. In Problems **81–84**, find the length of the indicated side.

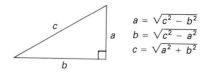

$$a = \sqrt{c^2 - b^2}$$
$$b = \sqrt{c^2 - a^2}$$
$$c = \sqrt{a^2 + b^2}$$

FIGURE 2.1

81. a if $c = 5$ meters and $b = 4$ meters.
82. a if $c = 25$ centimeters and $b = 24$ centimeters.
83. b if $c = 5 \times 10^{-2}$ meter and $a = 4 \times 10^{-2}$ meter.
84. b if $c = 7.2 \times 10^{-2}$ meter and $a = 3.1 \times 10^{-2}$ meter.
85. In an R-L series circuit, $X_L = 5$ ohms and $R = 12$ ohms. Find the circuit impedance. (Refer to Example 8.)
86. The filament of a 100-watt light bulb has a resistance of 144 ohms. Find the current through the filament. (Refer to Example 7.)

The expression $\dfrac{-b + \sqrt{b^2 - 4ac}}{2a}$ is used in solving a certain type of equation. In Problems **87–90**, evaluate this expression for the given values of a, b, and c.

87. $a = 1,\ b = 2,\ c = -15$.   88. $a = 4,\ b = 5,\ c = 0$.
89. $a = 4,\ b = -12,\ c = 9$.   90. $a = 1,\ b = -9,\ c = 14$.
91. In statistics, the standard deviation of the numbers x_1 and x_2 is given by

$$\sqrt{\frac{(\bar{x} - x_1)^2 + (\bar{x} - x_2)^2}{2}},$$

where $\bar{x} = \dfrac{x_1 + x_2}{2}$. Find the standard deviation of $x_1 = 0$ and $x_2 = 4$.

92. Compute y' (read y *prime*), where

$$y' = \frac{\dfrac{x + 5}{2\sqrt{x + 1}} - \sqrt{x + 1}}{(x + 5)^2},$$

if $x = 3$.

93. The speed v (in meters per second) for an artificial satellite to be in circular orbit around the earth is given by

$$v = \sqrt{\frac{9.8R^2}{r}},$$

where R is the radius of the earth (in meters) and r is the radius of the orbit (in meters). If $R = 6.37 \times 10^6$ meters, find the orbital speed for a satellite 3.59×10^7 meters above the surface of the earth. Round your answer to 3 significant figures.

2.5 RATIONAL EXPONENTS

In this section we will give a meaning to a rational exponent, such as in $5^{1/3}$. We want to define $5^{1/3}$ so that the rules of exponents hold. For example, by the rule $(a^m)^n = a^{mn}$, the cube of $5^{1/3}$ must be

$$(5^{1/3})^3 = 5^{3/3} = 5^1 = 5.$$

But recall that 5 is also the cube of $\sqrt[3]{5}$:

$$(\sqrt[3]{5})^3 = 5.$$

Thus we want the cube of $5^{1/3}$ to be the same as the cube of $\sqrt[3]{5}$. For this reason we define $5^{1/3}$ to be $\sqrt[3]{5}$. More generally, we define $a^{1/n}$ (where n is a positive integer*) to be the principal nth root of a:

$$\boxed{1. \quad a^{1/n} = \sqrt[n]{a}.}$$

Thus

$$2^{1/4} = \sqrt[4]{2} \quad \text{and} \quad \sqrt{x+5} = (x+5)^{1/2}.$$

With our definition of $a^{1/n}$, the other rules of exponents hold whether the exponents are integers or fractions.

Example 1

Write each of the following in radical form and find the value.

a. $4^{1/2} = \sqrt[2]{4} = \sqrt{4} = 2.$

b. $27^{1/3} = \sqrt[3]{27} = 3.$

c. $(-32)^{1/5} = \sqrt[5]{-32} = -2.$

d. $16^{-(1/2)} = \dfrac{1}{16^{1/2}} = \dfrac{1}{\sqrt{16}} = \dfrac{1}{4}.$

---

* In Rule 1 we exclude the case where a is negative and n is even. For example, $(-4)^{1/2} = \sqrt{-4}$ is not a real number. Also, in Rule 2 we assume that $\sqrt[n]{a}$ is a real number.

Example 2

Write each of the following with fractional exponents.

a. $\sqrt{7} = 7^{1/2}$.

b. $\sqrt[4]{3x^2 + 4y} = (3x^2 + 4y)^{1/4}$.

c. $x\sqrt{y} + \sqrt{xy} = xy^{1/2} + (xy)^{1/2}$.

Rule 1 can be made more general to include other rational exponents. We define $a^{m/n}$, where m and n are integers and n is positive, as follows.

$$2. \quad a^{m/n} = (\sqrt[n]{a})^m = \sqrt[n]{a^m}.$$

That is, the denominator of m/n is the index of the corresponding radical, and the numerator is the exponent for the root or the exponent in the radicand. In terms of exponents, we have $a^{m/n} = (a^{1/n})^m = (a^m)^{1/n}$. By Rule 2 we may look at $x^{4/3}$ in two ways: first, as the fourth power of the cube root of x, $(\sqrt[3]{x})^4$; and second, as the cube root of the fourth power of x, $\sqrt[3]{x^4}$. To compute $a^{m/n}$ it is often easier first to take the nth root of a and then raise the result to the mth power.

Example 3

Find the value of each of the following.

a. $8^{2/3} = (\sqrt[3]{8})^2 = 2^2 = 4$. Alternatively, we have $8^{2/3} = \sqrt[3]{8^2} = \sqrt[3]{64} = 4$.

b. $\left(\dfrac{1}{4}\right)^{3/2} = \left(\sqrt{\dfrac{1}{4}}\right)^3 = \left(\dfrac{1}{2}\right)^3 = \dfrac{1}{8}$.

c. $(-27)^{4/3} = (\sqrt[3]{-27})^4 = (-3)^4 = 81$.

d. $27^{-2/3} = \dfrac{1}{27^{2/3}} = \dfrac{1}{(\sqrt[3]{27})^2} = \dfrac{1}{(3)^2} = \dfrac{1}{9}$.

In Rule 2 we must require that $a \geq 0$ if n is even. Otherwise, the statement $\sqrt{x^2} = x^{2/2} = x$ is not always true. To see why, let $x = 6$. Then $\sqrt{x^2} = \sqrt{36} = 6 = x$. But if $x = -6$, then $\sqrt{x^2} = \sqrt{36} = 6 = -(-6) = -x$. Thus $\sqrt{x^2}$ is x if x is positive, and is $-x$ if x is negative. (Remember that if x is negative, then $-x$ is positive.) More simply this means that

$$\sqrt{x^2} = |x|.$$

Thus

$$\sqrt{6^2} = |6| = 6 \quad \text{and} \quad \sqrt{(-6)^2} = |-6| = 6.$$

Unless otherwise stated, we shall assume that all literal numbers appearing in radicals are positive. As a result, we are free to write $\sqrt{x^2} = x$.

Keep in mind that *the rules for exponents are true for all types of exponents.* These rules will be applied to fractional exponents in Examples 4 and 5.

Example 4

Write each of the following with fractional exponents and simplify if possible.

a. $\sqrt[5]{x^4} = x^{4/5}$.

b. $x\sqrt[6]{y^5} = xy^{5/6}$.

c. $(\sqrt{3})^4 = (3^{1/2})^4 = 3^{4/2} = 3^2 = 9$.

d. $\sqrt[4]{x^8} = x^{8/4} = x^2$.

e. $\sqrt[4]{a^4b^8} = (a^4b^8)^{1/4} = (a^4)^{1/4}(b^8)^{1/4} = a^{4/4}b^{8/4} = ab^2$.

Example 5

Perform the indicated operations and give the answer with positive exponents only.

a. $x^{1/2}x^{1/3} = x^{(1/2)+(1/3)} = x^{5/6}$.

b. $\dfrac{2x}{x^{1/4}} = \dfrac{2x^1}{x^{1/4}} = 2x^{1-(1/4)} = 2x^{3/4}$.

c. $\dfrac{y^{2/3}}{x^{2/3}y^{1/3}} = \dfrac{y^{1/3}}{x^{2/3}}$.

d. $(8a^3)^{2/3} = 8^{2/3}(a^3)^{2/3} = (\sqrt[3]{8})^2a^2 = 2^2a^2 = 4a^2$.

e. $(15^{-1/2})^4 = 15^{-4/2} = 15^{-2} = \dfrac{1}{15^2} = \dfrac{1}{225}$.

f. $(x^{2/9}y^{-4/3})^{1/2} = (x^{2/9})^{1/2}(y^{-4/3})^{1/2} = x^{1/9}y^{-2/3} = \dfrac{x^{1/9}}{y^{2/3}}$.

g. $\left(\dfrac{x^{12}}{y^6}\right)^{-1/3} = \dfrac{(x^{12})^{-1/3}}{(y^6)^{-1/3}} = \dfrac{x^{-4}}{y^{-2}} = \dfrac{y^2}{x^4}$.

h. $\left(\dfrac{x^{1/5}y^{6/5}}{z^{2/5}}\right)^{15} = \dfrac{(x^{1/5}y^{6/5})^{15}}{(z^{2/5})^{15}} = \dfrac{x^3y^{18}}{z^6}$.

Here are some examples of using a calculator to evaluate numbers involving fractional exponents.

Example	Keystroke Sequence	Display
$(48.5)^{1/4}$	48.5 y^x 4 $1/x$ $=$	2.63898
$27^{2/3}$	27 y^x (2 $\div$ 3) $=$	9
$(9.5)^{-2.4}$	9.5 y^x 2.4 $+/-$ $=$	0.00450

Problem Set 2.5

In Problems 1–15, find the value of each expression without the use of a calculator.

1. $25^{1/2}$.
2. $125^{1/3}$.
3. $81^{-1/2}$.
4. $4^{-1/2}$.
5. $27^{2/3}$.
6. $(\frac{1}{9})^{3/2}$.
7. $27^{-2/3}$.
8. $9^{3/2}$.
9. $(-8)^{4/3}$.
10. $(-8)^{-1/3}$.
11. $16^{3/4}$.
12. $(64^{1/2})^{1/3}$.
13. $(-\frac{1}{32})^{-1/5}$.
14. $(\frac{1}{27})^{-2/3}$.
15. $(4^3)^{2/3}$.

In Problems 16–30, rewrite each expression by using positive fractional exponents.

16. $\sqrt[3]{y}$.
17. $\sqrt{x}$.
18. $\sqrt[4]{x^9}$.
19. $\sqrt[3]{x^2}$.
20. $\dfrac{\sqrt{y}}{\sqrt[4]{x}}$.
21. $\sqrt[4]{x^3 y^5}$.
22. $\sqrt[4]{x}\,\sqrt[5]{y}$.
23. $\sqrt[6]{x^5 y^{12}}$.
24. $\dfrac{1}{\sqrt{2}}$.
25. $x^2 \sqrt[4]{x}$.
26. $\dfrac{2}{\sqrt[3]{x-5}}$.
27. $\dfrac{3}{\sqrt{x}}$.
28. $\sqrt[3]{7-x}$.
29. $\sqrt[3]{(x^2-5x)^2}$.
30. $\dfrac{1}{\sqrt[6]{(x^2-2x)^5}}$.

In Problems 31–64, perform the operations and simplify. Write your answers with positive exponents only.

31. $x^{1/2}x^{3/2}$.
32. $x^{2/3}x^{4/3}$.
33. $x^{4/3}x^{-1/3}$.
34. $x^{-1/2}x^{3/2}$.
35. $x^{-2}x^{7/2}x^{5/2}$.
36. $x^3 x^{-3/8} x^{-5/8}$.
37. $x^{1/2}x^{1/4}$.
38. $x^{3/5}x^{1/3}$.
39. $x^{1/2}(3x^{1/2}y)$.
40. $(xy^{-1/2})y^{5/2}$.
41. $(x^{1/2})^3$.
42. $(y^{-1/3})^6$.
43. $(y^6)^{1/3}$.
44. $(t^{10})^{4/5}$.
45. $2(x^{-2/3})^6$.
46. $(-y^{1/3})^6$.
47. $(2x^{-2}y^{1/3})^3$.
48. $(3x^3y^{1/2})^4$.
49. $(ab^2c^3)^{3/4}$.
50. $(x^2y)^{1/3}$.
51. $(27x^{15})^{-1/3}$.
52. $(8x^{-6})^{1/3}$.
53. $(-8x^{-6})^{1/3}$.
54. $(27^{-1}x^{15})^{-1/3}$.
55. $\dfrac{x^{4/9}y^{2/5}}{x^{1/9}y^{7/5}}$.
56. $\dfrac{x^{4/7}y^{3/20}}{x^{3/7}y^{17/20}}$.
57. $\dfrac{x^{2/3}y^{-9/4}}{x^{-4/3}y^{-1/4}}$.

58. $\dfrac{xy^{2/5}}{x^{1/2}y^{2/5}z^{-1/3}}.$

59. $\left(\dfrac{x^{-4/3}}{y^{-2/3}}\right)^{-3}.$

60. $\left(\dfrac{3^{1/4}x^{3/4}}{y^{1/2}}\right)^{4}.$

61. $\left(\dfrac{x^{-1}}{x^{1/3}}\right)^{2}.$

62. $\left(\dfrac{y^{2/3}}{x^{3/4}}\right)^{2/3}.$

63. $\left(\dfrac{2^{1/3}x^{2/3}}{x^{1/3}}\right)^{6}.$

64. $\left(\dfrac{x^{3/2}}{x^{-1}}\right)^{-4}.$

65. Experimental data indicate that the radius R of a nucleus of an isotope of mass number A is given by

$$R = r_0 A^{1/3}.$$

The value of r_0, the radius parameter, can be taken to be 1.5×10^{-13} centimeter. Determine the radius of a naturally occurring isotope of zinc whose mass number is 64.

66. In studies of the effect of an electrical current in a ferromagnetic material, the equation $W = KB^{8/5}$ occurs. If $B = 0.60$, express W in terms of K.

67. When 20 cubic meters of oxygen at a temperature of 285 K and a pressure of 10^5 newtons per square meter is adiabatically compressed to 0.5 cubic meter, its new pressure p and temperature T are given by

$$p = (10)^5\left(\frac{20}{0.5}\right)^{7/5}$$

and

$$T = (285)\left(\frac{20}{0.5}\right)^{7/5}.$$

Find p and T.

68. If the gas in Problem 67 is helium, the expressions for p and T have a fractional exponent of $\frac{5}{3}$ instead of $\frac{7}{5}$. Find the values of p and T if the gas is helium.

69. The rate R (in cubic meters per second) at which water discharges over a 90° V-notch wier is given by $R = 1.37h^{5/2}$, where h is the height (in meters) of the water above the wier crest. Find R if $h = 2.4$ meters.

2.6 REVIEW

Important Terms and Symbols

Section 2.1   Base, exponent, a^n, nth power.

Section 2.2   Zero exponent, a^0, negative exponent, a^{-n}.

Section 2.3   Scientific notation.

Section 2.4   Square root, principal square root, $\sqrt{a}$, radicand, radical sign, radical, rationalizing the denominator, nth root, principal nth root, $\sqrt[n]{a}$, index.

Section 2.5   Rational exponent, $a^{1/n}$, $a^{m/n}$.

Formula Summary

LAWS OF EXPONENTS

$$a^0 = 1. \qquad\qquad a^{-n} = \frac{1}{a^n}.$$

$$a^m a^n = a^{m+n}. \qquad\qquad (a^m)^n = a^{mn}.$$

$$\frac{a^m}{a^n} = a^{m-n} = \frac{1}{a^{n-m}}. \qquad (ab)^n = a^n b^n.$$

$$\left(\frac{a}{b}\right)^n = \frac{a^n}{b^n}. \qquad\qquad \left(\frac{a}{b}\right)^{-n} = \left(\frac{b}{a}\right)^n.$$

LAWS OF RADICALS

$$(\sqrt[n]{a})^n = \sqrt[n]{a^n} = a. \qquad\qquad \sqrt[n]{ab} = \sqrt[n]{a}\,\sqrt[n]{b}.$$

$$\sqrt[n]{\frac{a}{b}} = \frac{\sqrt[n]{a}}{\sqrt[n]{b}}. \qquad\qquad a^{1/n} = \sqrt[n]{a}.$$

$$a^{m/n} = (\sqrt[n]{a})^m = \sqrt[n]{a^m}.$$

Review Questions

1. In $(3x)^2$ the base is ___(a)___, but in $3x^2$ it is ___(b)___.

2. The cube of $2x$ is equal to _____.

3. Although $2^2 = 4$ and $(-2)^2 = 4$, by the symbol $\sqrt{4}$ we mean ___(2)(−2)___.

4. The principal square root of 16 is _____.

5. In scientific notation, the number 0.026 is written as _____.

6. $x^2(x^3) = $ ___(a)___ and $(x^3)^2 = $ ___(b)___.

7. The expression x^2/x^3 may be written as x raised to what negative power? _____

8. $7^0 = $ ___(a)___ but 0^0 is ___(b)___.

9. $64^{1/3} = \sqrt[n]{64}$, where $n = $ _____.

10. $x^{5/4} = \sqrt[n]{x^m}$, where $m = $ _____ and $n = $ _____.

11. $x^{2/3} = (\sqrt[n]{x})^m$, where $m = $ _____ and $n = $ _____.

12. $\sqrt[7]{x^5} = x^{m/n}$, where $m = $ _____ and $n = $ _____.

13. $(\sqrt[4]{x})^9 = x^{m/n}$, where $m = $ _____ and $n = $ _____.

14. $\sqrt[4]{x^4} = $ _____; $\sqrt[3]{z^6} = \sqrt[3]{(z^2)^3} = $ _____.

15. $\sqrt{32} = \sqrt{(\underline{}) \cdot 2} = \sqrt{\underline{}} \cdot \sqrt{2} = \underline{} \sqrt{2}$.

16. $\sqrt{\dfrac{5}{36}} = \dfrac{\sqrt{5}}{\sqrt{\rule{1.5cm}{0.4pt}}} = \dfrac{\sqrt{5}}{(\rule{1.5cm}{0.4pt})}.$

Answers to Review Questions

1. (a) $3x$, (b) x. **2.** $8x^3$. **3.** 2. **4.** 4. **5.** 2.6×10^{-2}. **6.** (a) x^5, (b) x^6. **7.** -1.
8. (a) 1, (b) Not defined. **9.** 3. **10.** 5, 4. **11.** 2, 3. **12.** 5, 7. **13.** 9, 4. **14.** x, z^2.
15. 16, 16, 4. **16.** 36, 6.

Review Problems

*In Problems **1–24**, determine whether each statement is, in general, true or false.*

1. $a^3 a^2 = a^6$.
2. $\left(\dfrac{1}{x}\right)^3 = \dfrac{1}{x^3}$.
3. $2^2 + 3^2 = 5^2$.

4. $(-a)^4 = a^4$.
5. $3 + 3 + 3 = 3^3$.
6. $(x^2 y)^2 = x^4 y$.

7. $(2x^2)^3 = 2^3 x^5$.
8. $\dfrac{x^5}{x^3} = x^2$.
9. $(-y)^3 = -y^3$.

10. $2^2 \cdot 2^4 = 4^8$.
11. $3x = xxx$.
12. $(-3)^3 = -3^3$.

13. $(-2)^2 = -2^2$.
14. $(2 + 5)^2 = 2^2 + 5^2$.
15. $a^3 a^7 = a^7 a^3$.

16. $(x^2)^y = x^{2y}$.
17. $(2x)^2 = 2x^2$.
18. $\sqrt{16} = -4$.

19. $\sqrt{5}\sqrt{5} = \sqrt{25}$.
20. $\sqrt{4 \cdot 2} = 4\sqrt{2}$.
21. $\sqrt[3]{-\dfrac{1}{8}} = -\dfrac{1}{2}$.

22. $\sqrt[3]{0} = 0$.
23. $\sqrt{1} + \sqrt{1} = \sqrt{2}$.
24. $(\sqrt[3]{-1})^2 = 1$.

*In Problems **25–38**, evaluate the expressions without the use of a calculator.*

25. 3^0.
26. $2\left(-\dfrac{2}{3}\right)^0$.
27. 5^{-1}.

28. $(-3)^{-1}$.
29. $4\left(-\dfrac{2}{3}\right)^{-2}$.
30. $\dfrac{2^{-1}}{4^{-2}}$.

31. $100^{1/2}$.
32. $64^{1/3}$.
33. $4^{3/2}$.

34. $(25)^{-3/2}$.
35. $(32)^{-2/5}$.
36. $\left(\dfrac{9}{100}\right)^{3/2}$.

37. $\left(\dfrac{1}{16}\right)^{5/4}$.
38. $\left(-\dfrac{27}{64}\right)^{2/3}$.

*In Problems **39–87**, simplify. Use only positive exponents in your answers.*

39. $x^6 x^4 x^3$.
40. $\dfrac{x^2 x^{20}}{y^5 y^2}$.
41. $\dfrac{(x^2)^5}{(y^5)^{10}}$.

42. $(5x^2)(2x^3)$.
43. $(-2xy^4)^5$.
44. $\left(\dfrac{2xy^3}{z^2}\right)^2$.

45. $\dfrac{(x^3)^6}{x(x^3)}$.

46. $\dfrac{(xy^2)^2}{(t^2w)^3}$.

47. $(-x)^2(-x)^3$.

48. $\dfrac{(x^2)^3(x^4)^5}{(x^3)^8}$.

49. $\dfrac{(5^7)^2}{(5^4)^4}$.

50. $\dfrac{2^6 2^{11}}{(2^5)^3}$.

51. $-\sqrt{36} + \sqrt{81}$.

52. $\dfrac{-\sqrt{144}}{|-12|}$.

53. $|-2| - \sqrt[4]{16}$.

54. $(\sqrt[3]{2})^3$.

55. $\sqrt{13} \cdot \sqrt{13}$.

56. $\dfrac{(-1)^2 - 3\sqrt{9}}{|-4| + \sqrt{16}}$.

57. $\dfrac{1}{3}\sqrt[3]{\dfrac{1}{27}}$.

58. $\dfrac{\sqrt[5]{-32}}{\sqrt{9}}$.

59. $\sqrt{0.01} + \sqrt{0.0025}$.

60. $\dfrac{\sqrt{4} - \sqrt[3]{8}}{\sqrt{6}}$.

61. $\sqrt{45}$.

62. $\sqrt{72}$.

63. $\sqrt[4]{810}$.

64. $\sqrt[3]{40}$.

65. $\sqrt{\dfrac{7}{81}}$.

66. $\sqrt{\dfrac{10}{9}}$.

67. $\dfrac{\sqrt[3]{24}}{\sqrt[3]{3}}$.

68. $\dfrac{\sqrt{72}}{\sqrt{2}}$.

69. $\sqrt{18} \cdot \sqrt{2}$.

70. $\sqrt[3]{8} \cdot \sqrt[3]{16}$.

71. $2x^{-1}x^{-3}$.

72. xy^{-1}.

73. $(4t^2)^{-2}$.

74. $(-2z)^{-3}$.

75. $(x^{-2}y^2)^3$.

76. $(ab^2c^3)^{3/4}$.

77. $\dfrac{x^{-2}y^{-6}z^2}{xy^{-1}}$.

78. $(2x^{-1}y)^{-2}$.

79. $\dfrac{x^3y^{-2}}{x^5z^2}$.

80. $2x^{1/2}y^{-3}x^{1/3}$.

81. $(-3x^{1/2}y^{2/3})^3$.

82. $\dfrac{2^0}{(2^{-2}x^{1/2}y^{-2})^3}$.

83. $\left(\dfrac{x^{-2}}{w^2y^{-2}}\right)^{-4}$.

84. $(9z^6)^{1/2}$.

85. $(16y^8)^{3/4}$.

86. $\left(\dfrac{27t^3}{8}\right)^{2/3}$.

87. $\left(\dfrac{1000}{a^9}\right)^{-2/3}$.

88. Evaluate

$$\dfrac{\sqrt[4]{(0.00593)^2(23.5)^4}}{\sqrt{(4.71 \times 10^{-4})(503)^{0.34}}}.$$

*In Problems **89–92**, rationalize the denominator.*

89. $\dfrac{6}{\sqrt{3}}$.

90. $\dfrac{2}{\sqrt{8}}$.

91. $\sqrt{\dfrac{8}{11}}$.

92. $\sqrt{\dfrac{1}{3}}$.

*In Problems **93–96**, write the numbers in scientific notation.*

93. 0.0000564.

94. 45,030,000,000.

95. 28,000,000.

96. 0.007007.

97. The elastic potential energy PE (in joules) of a spring is given by

$$PE = \tfrac{1}{2}kx^2,$$

where k is the spring constant (in newtons per meter) and x is the displacement of the spring (in meters) from the equilibrium position $(x = 0)$. Find the potential energy of a spring in each of the following cases.

   (a) $k = 3.2$, $x = 0.2$.
   (b) $k = 4$, $x = -0.2$,
   (c) $k = 2$, $x = -1.2$.

98. The area A of a circle of radius r is given by $A = \pi r^2$. By what factor does the area increase if the radius is (a) tripled, and (b) halved?

99. The frequency of vibration of a string of length L, fixed at both ends and vibrating in its fundamental mode, is given by

$$f = \frac{1}{2L}\sqrt{\frac{T}{\mu}},$$

where f is frequency, μ (mu) is mass per unit length, and T is the tension in the string. If the tension in the string is quadrupled, what happens to the frequency?

100. The resistivity of silver is 0.000000016 ohm $\cdot$ meter. A coil is to be made by using 2500 meters of wire. If the wire has a circular cross section of radius 5×10^{-2} meter, find the resistance of the coil. (Refer to Problem 39 in Sec. 2.3.)

3

Operations with Algebraic Expressions

3.1 ALGEBRAIC EXPRESSIONS

Our work with algebra involves *algebraic expressions*. An **algebraic expression** is either a number, including a literal number, or a combination of numbers obtained by one or more of the operations of addition, subtraction, multiplication, division, raising to powers, or extraction of roots. For example, some algebraic expressions are

$$7, \quad 2x^2, \quad 3 + 4\sqrt{y}, \quad \text{and} \quad \frac{x-2}{y}.$$

When an algebraic expression is a sum or difference of individual expressions, each of the individual expressions, together with the sign preceding it, is called a **term** of the algebraic expression. For example, the expression

$$5x^2 - \frac{x}{y} + 4$$

has three terms:

$$+5x^2, \quad -\frac{x}{y}, \quad +4.$$

First term   Second term   Third term

108

The term $+4$ is called a **constant term** because it has a fixed value. As another example, the expression $3x^4$ is considered to consist of one term, $+3x^4$.

If a term can be written as the product of two or more expressions, then each expression is called a **factor** of that term. For example, $6xy^2$ can be written as $(6x)(y^2)$, so $6x$ and y^2 are factors of $6xy^2$. These are not the *only* factors of $6xy^2$. We can write $6xy^2$ as $(6)(xy^2)$, so 6 and xy^2 are also factors of $6xy^2$. In fact, because $6xy^2 = (1)(6xy^2)$, certainly $6xy^2$ can be considered a factor of $6xy^2$. In general, any term is a factor of itself.

When a term is considered as the product of two factors, each factor is called the **coefficient** of the other factor. For example, in the term $6xy^2$ the factor $6x$ is the coefficient of y^2.

If a term is a constant or is the product of a constant factor and a literal factor, then the constant is called the **numerical coefficient**, or simply the **coefficient**, of the *term*. Thus the coefficient of $6xy^2$ is 6, and xy^2 is the literal factor, or **literal part**. A coefficient that is not written is taken to be 1. Thus the coefficient of x^2 is 1 (because $x^2 = 1 \cdot x^2$).

Example 1
___

The terms of the algebraic expression

$$-x^3 + \frac{x^2}{3} + 2y - 3,$$

are

$$-1x^3, \qquad \frac{1}{3}x^2, \qquad 2y, \qquad -3,$$

and their coefficients are

$$-1, \qquad \frac{1}{3}, \qquad 2, \qquad -3.$$

___

An algebraic expression may be classified according to the number of terms it contains. A **monomial** is an expression with exactly one term, such as $4x^2$. An expression with more than one term, such as $x + y + 1$, is a **multinomial**. Two special cases of a multinomial are a **binomial**, which has exactly two terms, and a **trinomial**, which has exactly three terms.

Example 2
___

$x + 5$	$x^2 - 2x + 1$	$7y$
Binomial	Trinomial	Monomial

___

The work *polynomial* refers to an algebraic expression having a special form. A **polynomial** in the variable x is an expression in which each term has the form ax^n, where a is a constant and the exponent n is a *nonnegative* integer. We call n the **degree of the term**. For example, the expression $6x^4 - 4x^3 + 5x$ can be written as $6x^4 - 4x^3 + 5x^1$. Thus

$$6x^4 - 4x^3 + 5x \qquad \text{is a polynomial in } x.$$

$$\begin{array}{c} \text{Degree} \\ \text{of term:} \end{array} \quad \begin{array}{ccc} \uparrow & \uparrow & \uparrow \\ 4 & 3 & 1 \end{array}$$

A *nonzero* constant, such as 2, is a polynomial of degree zero (because $2 = 2x^0$). The constant 0 is also a polynomial, but no degree is assigned to it.

In a polynomial, the degree of the term with the highest degree (and nonzero coefficient) is called the **degree of the polynomial**, and the coefficient of that term is called the **leading coefficient**. For example, in the polynomial

$$6x^4 - 4x^3 + 5x,$$

the term with the highest degree is $6x^4$. Thus the polynomial has degree 4 and leading coefficient 6.

Example 3

Polynomial:	$8x^2 + x$	$6 - 8x^4 - x^5$	$4 - \frac{6}{7}y$
Degree:	2	5	1
Leading coefficient:	8	-1	$-\frac{6}{7}$

Example 4

a. $\sqrt{x}$, which can be written as $x^{1/2}$, is *not* a polynomial because it cannot be put in the form ax^n where n is a nonnegative *integer*.

b. $\dfrac{3}{x^2}$, which can be written as $3x^{-2}$, is *not* a polynomial because the exponent for x in the latter form is a *negative* integer.

When using a calculator to evaluate a polynomial, you should make use of the calculator's memory feature (if your calculator has it). The store key $\boxed{\text{STO}}$ or x-to-memory key $\boxed{x \to \text{M}}$ stores the displayed number in the memory for later use. The recall memory key, $\boxed{\text{RCL}}$ or $\boxed{\text{RM}}$, displays the number in the memory, but does not affect the contents of the memory.

For example, to evaluate

$$3x^4 - 2x^3 + 7x^2 - 3x + 2$$

when $x = 6$, we can use the following keystroke sequence.

| 6 | STO | 3 | × | RCL | y^x | 4 | − | 2 | × | RCL | y^x | 3 | + | 7 | × | RCL |

| x^2 | − | 3 | × | RCL | + | 2 | = |

Display: 3692

A polynomial may involve more than one variable. An example of a polynomial in x and y is

$$x^6 + 4x^2y^3 - 3.$$

Notice that the exponents for x and y are positive integers. To find the degree of the term $4x^2y^3$, we add the exponents for x and y. Thus $4x^2y^3$ has degree $2 + 3 = 5$. The term with highest degree is x^6, so the polynomial has degree 6.

Problem Set 3.1

1. Given the term $3x^2y$, state (a) the coefficient of y, and (b) the coefficient of $3y$.
2. Given the term $-4a^2b^3c$, state (a) the coefficient of b^3, and (b) the coefficient of $-4c$.

In Problems 3–10, give the (numerical) coefficient of each term.

3. $2x^3 + 3x^2 - 4x$.   4. $3x^2 - x$.   5. $2x^2y - 2x + 1$.
6. $4x - 6x^2y^2$.   7. $4wx^2 + x - 2$.   8. x^2.

9. $2 + \dfrac{7x}{3}$.   10. $-\dfrac{x}{2} + \sqrt{2}x^2$.

In Problems 11–18, classify each expression as (a) a monomial, (b) a multinomial, (c) a binomial, or (d) a trinomial. More than one classification may apply.

11. $2x^2$.   12. $2x^3 + 6$.   13. $y^2 - 2y + 3$.
14. $4z$.   15. $3x^4 - \sqrt{3}$.   16. $2 - x^2 - x$.
17. $x^4 - x^2 + x + 1$.   18. $\sqrt{x} + \sqrt{y} - x - 5$.

In Problems 19–30, determine if the given expression is a polynomial in x. If it is, find the (a) degree and (b) leading coefficient.

19. $-4x + 2$.   20. $\dfrac{x^2}{2}$.

21. $7x^2 + 6x - 2$.   22. $3x^3 - 7x$.
23. $x + 2x^{-2}$.   24. $-x^2 + 1$.

25. $\dfrac{18 - 2x^4}{3}$.

26. $x + \dfrac{1}{x}$.

27. $3x^3 - 2x^5 + x^2$.

28. $x^6 - x^5 + x^4 + 1$.

29. 56.7.

30. $\sqrt[3]{x} + x^6$.

In Problems **31–34,** state the degree of the given polynomial in x and y.

31. $2x^3y^3 + y^5$.

32. $x^3 + y^4 + 2xy$.

33. $9 + xy^6 + (x^2)^4$.

34. $3x^2y^2 + 7x^3 + y^4 + 1$

35. Find the value of the polynomial

$$4x^3 - 5x^2 + 7x - 1.25$$

if $x = 2.48$. Round your answer to 2 decimal places.

36. Find the value of the polynomial

$$25x^4 - 20x^3 - 60x^2 + x - 230$$

if $x = 6$.

3.2 ADDITION AND SUBTRACTION OF ALGEBRAIC EXPRESSIONS

Adding or subtracting algebraic expressions involves combining *like terms*. **Like terms** are terms that have the same literal part. That is, like terms differ at most only in their numerical coefficients. For example,

$$3xy \text{ and } -5xy \text{ are like terms,}$$

as well as the pairs $3y^2$ and y^2 and 2 and 7. But

$$2x^2y \text{ and } 2xy^2 \text{ are not like terms}$$

because x^2y is not the same as xy^2. Similarly, x^2 and x are not like terms.
    We combine like terms under addition or subtraction by using the distributive law. For example,

$$3x^2 + 5x^2 = (3 + 5)x^2 = 8x^2,$$

and

$$7y - 9y = (7 - 9)y = -2y.$$

That is,

> To add (or subtract) like terms, add (or subtract) their numerical coefficients.

Example 1

Combine like terms.

a. $2x^2 + 7x^2 + x^2 = (2 + 7 + 1)x^2 = 10x^2$. (Do not forget the coefficient 1.)

b. $-5ab^2 + 2b^2a = (-5 + 2)ab^2 = -3ab^2$. Note that $-5ab^2$ and $2b^2a$ are like terms because $ab^2 = b^2a$.

To simplify

$$9x^2 + 5x + 2 - 4x^2 + x - 3,$$

we first gather together like terms (by using the commutative law):

$$9x^2 - 4x^2 + 5x + x + 2 - 3.$$

Then we combine like terms:

$$(9 - 4)x^2 + (5 + 1)x + (2 - 3) = 5x^2 + 6x - 1.$$

Example 2

Simplify by combining like terms.

a. $-x + 5y - 7x - 2y = -x - 7x + 5y - 2y = -8x + 3y$.

b. $x^2 - 3x + 2 + 3x^2 = x^2 + 3x^2 - 3x + 2 = 4x^2 - 3x + 2$.

Sometimes an expression contains grouping symbols, such as **parentheses ( )**, **brackets [ ]**, or **braces { }**, to indicate that the terms within the symbols are to be treated as a single number or quantity. There are rules for removing grouping symbols that are preceded by plus or minus signs. For a plus sign the rule is:

---
1. Grouping symbols immediately preceded by a plus sign may be removed.

Thus

$$(5x^2 + 3x) + (2x^2 - 1) = 5x^2 + 3x + 2x^2 - 1$$
$$= 7x^2 + 3x - 1.$$

[Here we assumed that a plus sign was in front of $(5x^2 + 3x)$.]
On the other hand, observe that

$$-(2x^2 - 1) = (-1)(2x^2 - 1)$$
$$= -2x^2 + 1 \qquad \text{[distributive law]}.$$

Comparing the result to the original expression, we are led to the following rule.

> **2.** If grouping symbols are immediately preceded by a minus sign, the symbols and minus sign may be removed provided that the sign of *each* of the enclosed terms is reversed.

We emphasize that

$$-(2x^2 - 1) \neq -2x^2 - 1$$

because the sign of **each** enclosed term must be reversed.

We handle subtraction of algebraic expressions by using Rule 2. For example, to subtract $2x^2 - 1$ from $5x^2 + 3x$, we have

$$(5x^2 + 3x) - (2x^2 - 1)$$

$$= 5x^2 + 3x - 2x^2 + 1 \qquad \text{[removing parentheses]}$$

$$= 3x^2 + 3x + 1 \qquad \text{[combining like terms].}$$

Example 3

Simplify.

a. $(3x^3 - x^2 - x) + (2x^3 + 7x^2 - x)$

$= 3x^3 - x^2 - x + 2x^3 + 7x^2 - x \qquad$ [removing parentheses]

$= 3x^3 + 2x^3 - x^2 + 7x^2 - x - x \qquad$ [gathering like terms]

$= 5x^3 + 6x^2 - 2x \qquad$ [combining like terms].

b. $(3x - y) - (-4y - 2x + 3)$

$= 3x - y + 4y + 2x - 3 \qquad$ [removing parentheses]

$= 3x + 2x - y + 4y - 3 \qquad$ [gathering like terms]

$= 5x + 3y - 3 \qquad$ [combining like terms].

c. $-[-2 - 4a + 3b] - [3a - 2b]$

$= 2 + 4a - 3b - 3a + 2b \qquad$ [removing brackets]

$= 2 + a - b.$

In general, grouping symbols can be removed from an expression by using the distributive law, as the next example shows.

Example 4

Remove grouping symbols and simplify.

a. $2a + 3(b - a) = 2a + 3b - 3a = -a + 3b$. Here *each* enclosed term was multiplied by 3. Note that

$$2a + 3(b - a) \neq 2a + 3b - a$$

because the $-a$ on the right side was not multiplied by 3.

b. $-4[-2 + 3x] - x^2$.

Solution   We have to multiply *each* enclosed term by -4.

$$-4[-2 + 3x] - x^2 = (-4)(-2) + (-4)(3x) - x^2$$
$$= 8 - 12x - x^2.$$

c. $3(x^2 - 5) - 4[-a^2 + 2x^2] + 5\{x^2 + 3\}$

$= 3x^2 - 15 + 4a^2 - 8x^2 + 5x^2 + 15$ [removing grouping symbols]

$= 4a^2$ [combining like terms].

Sometimes grouping symbols appear within other grouping symbols, as in

$$-2[x - (5x - 3)].$$

When removing the grouping symbols, you are less likely to make an error if you first remove the innermost symbols (which in this case are parentheses), then the next innermost, and so on. For example,

$-2[x - (5x - 3)]$

$= -2[x - 5x + 3]$ [removing parentheses]

$= -2[-4x + 3]$ [combining within brackets]

$= 8x - 6$ [removing brackets].

Example 5

Simplify $2x - 3[-2(1 - x) + x]$.

Solution

$2x - 3[-2(1 - x) + x]$

$= 2x - 3[-2 + 2x + x]$ [removing parentheses]

$= 2x - 3[-2 + 3x]$ [combining within brackets]

$= 2x + 6 - 9x$ [removing brackets]

$= -7x + 6.$

Example 6

Simplify $-2\{2 - 3[-(x + 1)]\} - x$.

Solution

$$-2\{2 - 3[-(x + 1)]\} - x$$

$$= -2\{2 - 3[-x - 1]\} - x \qquad \text{[removing parentheses]}$$

$$= -2\{2 + 3x + 3\} - x \qquad \text{[removing brackets]}$$

$$= -2\{3x + 5\} - x \qquad \text{[combining within braces]}$$

$$= -6x - 10 - x \qquad \text{[removing braces]}$$

$$= -7x - 10 \qquad \text{[combining].}$$

Problem Set 3.2

In Problems **1–44**, *simplify.*

1. $8x^2 + 3x^2 + 4x^2$.
2. $5t^3 + 4t^3 - t^3$.
3. $2y - 10y - 4y$.
4. $-5x^2 + 7x^2 - 2x^2$.
5. $5x - 8 + 2x + 3$.
6. $9x^2 - 3x - 4x^2 - 5x$.
7. $2x^{1/2} - 5x^{1/2}$.
8. $7x^{3/2} + 2x^{3/2} - 5x^{3/2}$.
9. $4x^{-1/3} - 2(x^2)^{-1/6}$.
10. $3x^{1/4} - (x^{-1/2})^{-1/2}$.
11. $4(9x + 3y + x)$.
12. $3(a - 5b) - a$.
13. $-2(7x - 8y + 2y)$.
14. $5(a - 5b - a)$.
15. $(4y + 3) + (4y - 8)$.
16. $2 - (6x + 7) - 2x^2$.
17. $(3x + 5y) - (8y - x)$.
18. $2(x^2 + 5) + (8x^2 - 4)$.
19. $(3a - 2b) + (4a - 6b + c)$.
20. $(6x + 5) - (8x + 3)$.
21. $-(a - b - c) - (-a - b)$.
22. $-(-a + b - c) + (-a + b)$.
23. $2(x^2 - 3) + 3(x^2 + 7)$.
24. $4(3 - a) - 2(a - 1)$.
25. $5(xy + z) - (3xy + 7)$.
26. $2[x + 3z] - 4(z + x) + 3\{x - 4z\}$.
27. $6(x^2 - 2x) + (3x - 5) - 4(x^2 + 2)$.
28. $2[(3x - 5) + 4]$.
29. $2[-(5 - x) + x]$.
30. $3[4x + (2 + 5x)]$.
31. $5[3x + 2(5 - x)]$.
32. $7[4 - (y + 8)]$.
33. $-\{4a - (6 + 3a)\}$.
34. $-2\{9(z^2 - 1)\}$.
35. $5x^2 + 3[8(x^2 - 1)] + 2$.
36. $[2a + 3(b - a)] + 7a$.
37. $\{9a - (3b + a + c)\} - 4\{2b + 3c\}$.
38. $4x^2 - [8x + 2(x + x^2)]$.
39. $3\{4x - 2[5 - (x + 1)]\}$.
40. $-\{2[3(2x + 5) + 6x]\}$.
41. $3x - 3[2 - 2\{x^2 - 2y\}] - x^3$.
42. $2a^2b - [ab + 3(a - \{b + c\})]$.
43. $9 - 2\{8x - 3[4(y - 2x) - 6(x - y)]\}$.
44. $3\{5 - [xy - 2(xy + x)]\} - 4x$.
45. Subtract $6x - 4$ from $4x - 8$.
46. Subtract $2(1 - x)$ from $9x^2 - 3x$.

In Problems **47–50**, *write each expression in an equivalent form in which no symbols of grouping appear and then enclose all terms that contain x within one set of grouping symbols preceded by a minus sign. For example,*

$$3x^2 - (x - 5) = 3x^2 - x + 5 = -(-3x^2 + x) + 5.$$

47. $2(-x^2 + 7x) + x$.

48. $3(2x^2 - 2) - x^2 + x$.

49. $-3(x - x^2) - 4(2 + 3x^2) - 3x$.

50. $4[1 - 2(-3 - x)] - 2[x - 2(x - 1)]$.

51. In a discussion of the oscillations of a circular disc, the expression

$$10[(10 - 100\theta) - (10 + 100\theta)]$$

occurs, where θ is the greek letter theta. Simplify the expression.

52. The moment M for a particular loaded beam is given by

$$M = 1970x - 1200(x - 3) - 1000(x - 9).$$

Find a simplified expression for M.

53. The emf of a standard cell is found to be

$$1.0187[1 - 0.0004(T - 20)] \text{ volts,}$$

where T is the surrounding temperature of the cell in °C. Find the emf of a standard cell at 22°C.

54. A rectangle has dimensions $2a$ and $3(b + c)$, whereas a smaller rectangle has dimensions a and $b + c$. Write an expression for the difference in the perimeters of the rectangles, the larger minus the smaller perimeter; perform all indicated operations and simplify.

55. If $x > 2$, find $|2 - x|$.

3.3 MULTIPLICATION OF ALGEBRAIC EXPRESSIONS

Product of Two Monomials

In this section we discuss procedures for multiplying algebraic expressions. The simplest situation is the product of two monomials, such as $(4x^2)(2xy)$. By the associative law the order of grouping is unimportant, so we can remove the parentheses:

$$(4x^2)(2xy) = 4 \cdot x^2 \cdot 2 \cdot x \cdot y.$$

By the commutative law we can rearrange the factors so that those involving the same letter are adjacent. This gives

$$4 \cdot 2 \cdot x^2 \cdot x \cdot y = 8x^3y \qquad \text{[rule of exponents].}$$

Comparing our answer to the original product leads to the following rule:

> To multiply monomials, multiply their coefficients and multiply their literal parts.

Example 1

Find the following products of monomials.

a. $(5x^2)(4x^3) = (5 \cdot 4)(x^2 \cdot x^3) = 20x^5$.

b. $(3x^2)(-2xz^2)(-z^4) = (3 \cdot -2 \cdot -1)(x^2 x)(z^2 z^4)$

$$= 6x^3 z^6.$$

Another way to work the problem is to start by multiplying the first two factors.

$$(3x^2)(-2xz^2)(-z^4) = (-6x^3 z^2)(-z^4) = 6x^3 z^6.$$

Example 2

Find $-3x^{-2}y^3(-x^4 y^{-4})$.

Solution

$$-3x^{-2}y^3(-x^4 y^{-4}) = (-3 \cdot -1)(x^{-2} x^4)(y^3 y^{-4})$$

$$= 3x^2 y^{-1} = \frac{3x^2}{y}.$$

Note that we wrote our answer with positive exponents only. We shall follow this practice.

Be careful about exponents when you multiply monomials. For example, to find $a(ab)^2$ we multiply a **not** by ab but by $(ab)^2$, which is $a^2 b^2$. The squaring *must* be done first. Thus

$$a(ab)^2 = a(a^2 b^2) = a^3 b^2.$$

Example 3

Find the following products.

a. $(2ab^2)(4a^2 c)^3$.

We *first* apply the rules of exponents to $(4a^2 c)^3$:

$$(2ab^2)(4a^2 c)^3 = 2ab^2(64a^6 c^3) = 128a^7 b^2 c^3.$$

b. $(-2x^2 y)^3(-xy^2)^2$.

$$(-2x^2 y)^3(-xy^2)^2 = (-8x^6 y^3)(x^2 y^4) = -8x^8 y^7.$$

Product of a Monomial and a Multinomial

The distributive law is the key tool that is used to multiply a monomial and a multinomial. In the product

$$2x(x + 4),$$

$2x$ is a monomial and $x + 4$ is a multinomial. By treating $2x$ as a single number and using the distributive law $a(b + c) = ab + ac$, we reduce the problem to that of multiplying monomials.

$$(2x)(x + 4) = (2x)(x) + (2x)(4) = 2x^2 + 8x.$$

$$a \ (b + c) = \quad a \quad b \ + \ a \quad c$$

More generally, we have the rule:

> To multiply a monomial and a multinomial, multiply each term in the multinomial by the monomial.

Example 4

Perform the indicated operations and simplify

a. $2xy(x^2 + 2x - 1) = (2xy)(x^2) + (2xy)(2x) - (2xy)(1)$

$$= 2x^3y + 4x^2y - 2xy.$$

b. $-3x^{1/2}(x^{3/2} - 4) = (-3x^{1/2})(x^{3/2}) - (-3x^{1/2})(4)$

$$= -3x^2 - (-12x^{1/2}) = -3x^2 + 12x^{1/2}.$$

c. $(x^2y)^2(x - y + 2) = (x^4y^2)(x - y + 2)$

$$= (x^4y^2)(x) - (x^4y^2)(y) + (x^4y^2)(2)$$

$$= x^5y^2 - x^4y^3 + 2x^4y^2.$$

d. $5(3x^2 + 3) - 2x(x - 4) = 15x^2 + 15 - 2x^2 + 8x$

$$= 13x^2 + 8x + 15 \qquad \text{[combining similar terms].}$$

e. $2a[x(x + 1) - 7] = 2a[x^2 + x - 7] = 2ax^2 + 2ax - 14a.$

Product of Two Multinomials

To find $(x + 2)(x + 3)$, which is the product of two multinomials, we use the distributive law $(a + b)c = ac + bc$, where $x + 2$ matches $a + b$ and $x + 3$ plays the role of c. That is, we take the first term in the left factor times the right factor, plus the second term in the left factor times the right factor:

$$(x + 2)(x + 3) = x(x + 3) + 2(x + 3) = x^2 + 3x + 2x + 6 = x^2 + 5x + 6.$$

$$\begin{array}{ccccccc} \uparrow\ \uparrow & \uparrow & & \uparrow\ \uparrow & & \uparrow\ \uparrow \\ (a + b) & c & = a & c & + b & c \end{array}$$

More generally, we have the rule:

> To multiply two multinomials, multiply each term of the first multinomial by the second multinomial.

Example 5

Find the product $(x^2 - 2)(x - 3)$ and simplify.

Solution

$$(x^2 - 2)(x - 3) = x^2(x - 3) - 2(x - 3) = x^3 - 3x^2 - 2x + 6.$$

Example 6

Find $x - (x + 1)(x - 2)$ and simplify.

Solution The first minus sign applies to the *product* of the multinomials and must be kept until *after* that product is found.

$$x - (x + 1)(x - 2) = x - [x(x - 2) + 1(x - 2)] = x - [x^2 - 2x + x - 2]$$
$$= x - [x^2 - x - 2] = x - x^2 + x + 2$$
$$= -x^2 + 2x + 2.$$

Example 7

Find $3x(2x + 1)(x - 2)$ and simplify.

Solution First we shall multiply $3x$ by $(2x + 1)$.

$$3x(2x + 1)(x - 2) = [3x(2x + 1)](x - 2)$$
$$= [6x^2 + 3x](x - 2)$$
$$= 6x^2(x - 2) + 3x(x - 2)$$
$$= 6x^3 - 12x^2 + 3x^2 - 6x$$
$$= 6x^3 - 9x^2 - 6x.$$

Example 8

Find $(x^2 - 2x + 1)(x^2 + x - 1)$ and simplify.

Solution

$$(x^2 - 2x + 1)(x^2 + x - 1)$$
$$= x^2(x^2 + x - 1) - 2x(x^2 + x - 1) + 1(x^2 + x - 1)$$
$$= x^4 + x^3 - x^2 - 2x^3 - 2x^2 + 2x + x^2 + x - 1$$
$$= x^4 - x^3 - 2x^2 + 3x - 1.$$

Sometimes a vertical arrangement is used to handle a multiplication problem like that in Example 8. As shown below, we take the first term in the second row (x^2) and multiply it by *each* term in the first row. The result is $x^4 - 2x^3 + x^2$. This is repeated with the second term in the second row and then with the third term. All the results are then added. Observe that to organize our work, like terms are placed in the same column.

$$
\begin{array}{r}
x^2 - 2x + 1 \\
x^2 + x - 1 \\
\hline
x^4 - 2x^3 + x^2 \\
x^3 - 2x^2 + x \\
- x^2 + 2x - 1 \\
\hline
x^4 - x^3 - 2x^2 + 3x - 1
\end{array}
$$

Add

Before concluding this section, we leave you with some important notes. Do not confuse the expression

$$x^2 - 2(x - 3) \quad \text{with} \quad (x^2 - 2)(x - 3).$$

The first expression is a difference, whereas the second one is a product. That is,

$$x^2 - 2(x - 3) = x^2 - 2x + 6,$$

but (from Example 5)

$$(x^2 - 2)(x - 3) = x^3 - 3x^2 - 2x + 6.$$

Similarly,

$$(x^2 - 2)x - 3 \neq (x^2 - 2)(x - 3).$$

Problem Set 3.3

*In Problems **1-66**, perform the indicated operations and simplify.*

1. $(2x)(5y)$.
2. $(3x)(5x^2)$.
3. $(2x)(3xy)$.
4. $-a(ac)$.
5. $3ab(a^2b)$.
6. $xy(xz)(xw)$.
7. $2xy^2(-4x^3y^2)$.
8. $-a^2b(-ac)$.
9. $x^{1/2}(3x^{1/2}y)$.
10. $(xy^{-1/2})y^{5/2}$
11. $(3x^{-4}y^5)(2xy^{-1})$.
12. $(3x^{-7}y^{-3})(x^2y^2)(4x)$.
13. $ab(a^2b)(bc^2)$.
14. $x(xy^2)(y^2z)$.
15. $2x^2yz^2(\frac{1}{2}xz^2)$.
16. $(2xy^2)^3(2y)^2$.
17. $(-3x)(4xy^2)(-2x^2y^3)$.
18. $x(x^3y^2)^2(-2xy^2)^3$.
19. $(10x^4)(3x^3)^2$.
20. $(3y^3)^2(-y)^3$.
21. $(ab^2c^3)^{3/4}(a^{1/4}b^{3/4})^5$.
22. $(x^{1/2}y^{2/3})(x^2y)^{1/3}$.
23. $a(-bc)^2(cd)^2$.
24. $(-xy)^3(-x)^2$.
25. $x(x^2 - 4x + 7)$.
26. $-2(x - 2y^2 + 7xy)$.
27. $a^2b(-3 + ab - a)$.
28. $2x(-x + 2y + 1)$.
29. $x^{1/3}(x^{2/3} + 3)$.
30. $x^{3/2}(1 - x^{1/2})$.
31. $-5xy(x^2 - y^2 + xy)$.
32. $a(b - a)(ac)$.
33. $(2x^2y)^2(x + 2y^2 - 3x^2)$.
34. $(ab)^2(a^2 - b^2c + ac^2)$.
35. $(x + 2)(x + 5)$.
36. $(x + 4)(x + 5)$.
37. $(y - 2)(3y + 2)$.
38. $(y - 2)(y + 2)$.
39. $(3x - 1)(3x - 1)$.
40. $(2x + 3)(3x + 2)$.
41. $(4x + y)^2$.
42. $(x^2 + 4)(x^4 + 4)$.
43. $(3x^2)(2xy)(x + y)$.
44. $(x + 3)(5x^2)$.
45. $(t - 2)(t^2 + 2t + 4)$.
46. $(x + 5)(x^2 - x - 1)$.
47. $(x^2 - 2)(x^2 - 5x + 1)$.
48. $(2 + x - y)(x - y)$.
49. $(x + y + 1)(x + y - 1)$.
50. $(2x - 1)(2x^3 - 3x + 1)$.
51. $4x(2x - 1)(2x + 1)$.
52. $-3x^2(x - 2)(x - 1)$.
53. $xy + y(y + x)$.
54. $(x - 3)x - 4$.
55. $(x^2 + 1)(2x) - (x^2 - 2)(2x)$.
56. $x^2y - 2x - x^2y(1 + x)$.
57. $x(x - 1) - 2(3 - x)$.
58. $2xy(x^2y) - y^2(2x^3 - 2xy)$.
59. $3x(x^2y - xy^2) - 3y(x^3 + 4x^2y)$.
60. $2(x^3y - 2x^2) - 2x^2(x^2 - 2xy)$.
61. $3(x - x^2) + (x + 1)(x - 1)$.
62. $(x + 1)(x^2 - x + 1) - 1 - x^3$.
63. $x^3 + y^3 - (x + y)(x^2 - xy + y^2)$.
64. $(-x - 2)(x + 2) - (-1 + x)x$.
65. $(1.4x + 3.2)(2.6x - 4.1)$.
66. $x + 4.54(2.32x - 1.41)$.

67. One rectangle has dimensions $2a$ and $3(b + c)$, while a smaller rectangle has dimensions $a - 1$ and $b + c$. Write an expression for the difference in the areas of the rectangles, namely, the larger area minus the smaller; perform all indicated operations and simplify.

68. In a mathematical development of the relativity of time, the following expression was encountered:

$$K\left\{(t_1 - t_2) - \frac{v}{c^2}(x_1 - x_2)\right\}.$$

Remove all symbols of grouping.

69. The condition necessary for a battery to deliver maximum power to a resistive load involves the expression

$$(R + r)^2(1) - R(2)(R + r).$$

Perform the indicated operations and simplify.

70. The approximate length of a suspension bridge cable is given by

$$a\left[1 + \frac{8}{3}\left(\frac{d}{a}\right)^3\right],$$

where a is the span and d is the sag. Perform the indicated operations.

71. The electrical resistance of a wire at temperature T is given by

$$R_0[1 + \alpha(T - T_0)],$$

where T_0 is a reference temperature and R_0 is the resistance of the wire at that temperature. The constant α (alpha) is called the *temperature coefficient of resistance*. Perform the indicated operations.

3.4 DIVISION OF ALGEBRAIC EXPRESSIONS

Division of Monomials

To divide a monomial by a monomial, we use the rules of exponents. For example,

$$\frac{6x^3y^4}{18x^2y^6} = \frac{6}{18} \cdot \frac{x^3}{x^2} \cdot \frac{y^4}{y^6} = \frac{1}{3} \cdot x \cdot \frac{1}{y^2} = \frac{x}{3y^2}.$$

You could think of this as canceling common factors, which was discussed in Sec. 1.5.

$$\frac{6x^3y^4}{18x^2y^6} = \frac{\overset{x}{\cancel{6}}\cancel{x^3}\cancel{y^4}}{\underset{3}{\cancel{18}}\cancel{x^2}\underset{y^2}{\cancel{y^6}}} = \frac{x}{3y^2}.$$

After some practice, you may be able to do some intermediate steps in your head. For example, convince yourself that

$$\frac{-36x^5y^4}{9x^8y^3} = -\frac{4y}{x^3}.$$

Example 1

Division with monomials.

a. $\dfrac{(2a^3b^2)(3a^5c)}{-3a^4b^3cd} = \dfrac{6a^8b^2c}{-3a^4b^3cd}$ [simplifying numerator]

$$= \dfrac{6}{-3} \cdot \dfrac{a^8}{a^4} \cdot \dfrac{b^2}{b^3} \cdot \dfrac{c}{c} \cdot \dfrac{1}{d}$$

$$= -2 \cdot a^4 \cdot \dfrac{1}{b} \cdot 1 \cdot \dfrac{1}{d} = -\dfrac{2a^4}{bd}.$$

b. $\left(\dfrac{x^2y^3}{2xy^2}\right)^3 = \left(\dfrac{xy}{2}\right)^3 = \dfrac{x^3y^3}{8}$. Here we chose to simplify *before* cubing. It is a general **practice** that you should follow because it often simplifies the work.

c. $\dfrac{(2a^2x)^3(-2a^4y)}{(2xy^2)^2} = \dfrac{(8a^6x^3)(-2a^4y)}{4x^2y^4} = \dfrac{-16a^{10}x^3y}{4x^2y^4} = -\dfrac{4a^{10}x}{y^3}.$

In general,

$$\dfrac{(a^2x)^2}{ay} \neq \dfrac{(ax)^2}{y} \qquad \text{[do not divide } a^2 \text{ by } a\text{]}.$$

You must square a^2x before canceling. Thus

$$\dfrac{(a^2x)^2}{ay} = \dfrac{a^4x^2}{ay} = \dfrac{a^3x^2}{y}.$$

Division of a Multinomial by a Monomial

When the numerator of a fraction is a multinomial, the fraction can be broken up into simpler fractions. This is based on the definition of division and the distributive law:

$$\dfrac{a + b + c}{d} = (a + b + c)\left(\dfrac{1}{d}\right) = a \cdot \dfrac{1}{d} + b \cdot \dfrac{1}{d} + c \cdot \dfrac{1}{d}.$$

$$= \dfrac{a}{d} + \dfrac{b}{d} + \dfrac{c}{d}.$$

Thus

$$\boxed{\dfrac{a + b + c}{d} = \dfrac{a}{d} + \dfrac{b}{d} + \dfrac{c}{d}.}$$

That is,

To divide a multinomial by a monomial, divide *each* term of the multinomial by the monomial.

For example,

$$\frac{3x^6 + x^3}{x} = \frac{3x^6}{x} + \frac{x^3}{x} = 3x^5 + x^2,$$

and

$$\frac{20x^2y^2 + 5xy^2 - 2x}{2xy} = \frac{20x^2y^2}{2xy} + \frac{5xy^2}{2xy} - \frac{2x}{2xy}$$

$$= 10xy + \frac{5y}{2} - \frac{1}{y}.$$

Students occasionally make errors when dealing with a monomial divided by a multinomial. *A monomial divided by a multinomial* **cannot**, *as a rule, be broken into simpler fractions.* That is,

$$\frac{a}{b + c} \neq \frac{a}{b} + \frac{a}{c}.$$

For example,

$$\frac{8}{6 + 2} \neq \frac{8}{6} + \frac{8}{2}$$

because $1 \neq \frac{16}{3}$.

Example 2

a. $\dfrac{5x^2y - x^3z}{-4xyz^2} = \dfrac{5x^2y}{-4xyz^2} - \dfrac{x^3z}{-4xyz^2}$

$$= -\frac{5x}{4z^2} - \left(-\frac{x^2}{4yz}\right) = -\frac{5x}{4z^2} + \frac{x^2}{4yz}.$$

b. $\dfrac{2(x + y)}{xy}$.

We first multiply 2 by $x + y$ to obtain a multinomial in the numerator.

$$\frac{2(x + y)}{xy} = \frac{2x + 2y}{xy} = \frac{2x}{xy} + \frac{2y}{xy} = \frac{2}{y} + \frac{2}{x}.$$

c. $\dfrac{3x - (2x^2y)^2 - 7x^2(x^2y^2)}{3x(2y^2)} = \dfrac{3x - 4x^4y^2 - 7x^4y^2}{6xy^2}$

$$= \dfrac{3x - 11x^4y^2}{6xy^2} = \dfrac{3x}{6xy^2} - \dfrac{11x^4y^2}{6xy^2}$$

$$= \dfrac{1}{2y^2} - \dfrac{11x^3}{6}.$$

Example 3

a. $\dfrac{3x^4 - 2x^2}{x^{1/3}} = \dfrac{3x^4}{x^{1/3}} - \dfrac{2x^2}{x^{1/3}} = 3x^{4-(1/3)} - 2x^{2-(1/3)} = 3x^{11/3} - 2x^{5/3}.$

b. $\dfrac{(4x^4y)^{-1/2}(y^4)^{-1/8}}{3xy^{-3}} = \dfrac{y^3}{3x(4x^4y)^{1/2}(y^4)^{1/8}} = \dfrac{y^3}{3x(2x^2y^{1/2})(y^{1/2})}$

$$= \dfrac{y^3}{6x^3y} = \dfrac{y^2}{6x^3}.$$

c. $\dfrac{4 + y}{y^{-6}} = \dfrac{4}{y^{-6}} + \dfrac{y}{y^{-6}} = 4y^6 + y^7.$ Another way to handle this problem is

$$\dfrac{4 + y}{y^{-6}} = (4 + y)y^6 = 4y^6 + y^7.$$

Note: $\dfrac{4 + y}{y^{-6}} \neq 4 + y(y^6) = 4 + y^7.$

We remark that

$$\dfrac{a + ax}{a} \neq \dfrac{\cancel{a} + ax}{\cancel{a}}, \quad \text{but} \quad \dfrac{a + ax}{a} = \dfrac{a}{a} + \dfrac{ax}{a} = 1 + x.$$

You may cancel only common factors, **not** terms.

Example 4

The reciprocal of the total resistance of three resistors, R_1, R_2, and R_3, connected in parallel in an electric circuit is

$$\dfrac{R_2R_3 + R_1R_3 + R_1R_2}{R_1R_2R_3}$$

Simplify this expression.

Solution

$$\frac{R_2 R_3 + R_1 R_3 + R_1 R_2}{R_1 R_2 R_3} = \frac{R_2 R_3}{R_1 R_2 R_3} + \frac{R_1 R_3}{R_1 R_2 R_3} + \frac{R_1 R_2}{R_1 R_2 R_3}$$

$$= \frac{1}{R_1} + \frac{1}{R_2} + \frac{1}{R_3}.$$

Division of a Polynomial by a Polynomial

To divide a polynomial by a polynomial, we use a long-division procedure that is similar to long division with numbers. To review that, we shall find $\frac{113}{4}$:

$$
\begin{array}{r}
28 \leftarrow \text{Quotient} \\
\text{Divisor} \rightarrow 4\overline{)113} \leftarrow \text{Dividend} \\
\underline{8} \\
33 \\
\underline{32} \\
1 \leftarrow \text{Remainder}
\end{array}
$$

The answer may be written as $28\frac{1}{4}$ or $28 + \frac{1}{4}$. Thus

$$\frac{113}{4} = 28 + \frac{1}{4} = \text{quotient} + \frac{\text{remainder}}{\text{divisor}}.$$

A way of checking a division is to verify that

$$(\text{Quotient})(\text{divisor}) + \text{remainder} = \text{dividend}.$$

Checking the last problem gives

$$(28)(4) + 1 = 113$$
$$113 = 113.$$

We now turn to a procedure for long division with polynomials, such as dividing $2x^3 - 5x^2 + 3x + 4$ by $x - 2$:

$$\frac{2x^3 - 5x^2 + 3x + 4}{x - 2} \begin{array}{l} \leftarrow \text{Dividend} \\ \leftarrow \text{Divisor} \end{array}$$

or

$$x - 2\overline{)2x^3 - 5x^2 + 3x + 4}.$$

Observe that the terms of the divisor and dividend are written in order of descending powers of x (from left to right). If this is not originally the case, you should arrange the terms in that way. Here are the steps in the long-division process:

1.

$$
\begin{array}{r}
2x^2 \leftarrow \text{Quotient} \\
x - 2 \overline{)\,2x^3 - 5x^2 + 3x + 4}
\end{array}
$$

We divide the first term of the dividend, $2x^3$, by the first term of the divisor, x. This gives $2x^2$, the first term in the quotient.

2.

$$
\begin{array}{r}
2x^2 \leftarrow \text{Quotient} \\
x - 2 \overline{)\,2x^3 - 5x^2 + 3x + 4} \\
\underline{2x^3 - 4x^2} \\
-x^2 + 3x
\end{array}
$$

We multiply the divisor, $x - 2$, by the quotient term $2x^2$ that was found in Step 1. The product is $2x^3 - 4x^2$, which is written below the corresponding like terms of the dividend and *subtracted* from them. We obtain $-x^2$ and bring down $3x$ from the original dividend (we should also bring down the 4, but it saves some work to leave it alone for now). Thus we have $-x^2 + 3x$, which we treat as a new dividend.

3.

$$
\begin{array}{r}
2x^2 - x \leftarrow \text{Quotient} \\
x - 2 \overline{)\,2x^3 - 5x^2 + 3x + 4} \\
\underline{2x^3 - 4x^2} \\
-x^2 + 3x
\end{array}
$$

We divide $-x^2$, the first term of the new dividend, by x, the first term of the divisor. This gives $-x$, the second term of the quotient.

4.

$$
\begin{array}{r}
2x^2 - x \leftarrow \text{Quotient} \\
x - 2 \overline{)\,2x^3 - 5x^2 + 3x + 4} \\
\underline{2x^3 - 4x^2} \\
-x^2 + 3x \\
\underline{-x^2 + 2x} \\
x + 4
\end{array}
$$

We multiply the divisor, $x - 2$, by the quotient term $-x$ that was found in Step 3. The product is $-x^2 + 2x$, which is written below the corresponding like terms of the new dividend and subtracted from them. We obtain x and bring down 4, the last term of the original dividend.

5.

$$2x^2 - x + 1 \quad \leftarrow \text{Quotient}$$

$$x - 2 \overline{)2x^3 - 5x^2 + 3x + 4}$$
$$\underline{2x^3 - 4x^2}$$
$$-x^2 + 3x$$
$$\underline{-x^2 + 2x}$$
$$x + 4$$
$$\underline{x - 2}$$
$$6 \leftarrow \text{Remainder}$$

We repeat the long-division process and arrive at the remainder, 6. Notice that the divisor, $x - 2$, has degree one and the remainder, 6, has degree zero. In general, *we end the division process when the remainder is zero or has degree less than the degree of the divisor.*

6. The quotient is $2x^2 - x + 1$ and the remainder is 6. Writing the answer in the form

$$\text{Quotient} + \frac{\text{remainder}}{\text{divisor}},$$

we have

$$\frac{2x^3 - 5x^2 + 3x + 4}{x - 2} = 2x^2 - x + 1 + \frac{6}{x - 2}.$$

We can check the division by making sure that

$$(\text{Quotient})(\text{divisor}) + (\text{remainder}) = \text{dividend}.$$

Doing this we have

$$(2x^2 - x + 1)(x - 2) + (6) = 2x^2(x - 2) - x(x - 2) + 1(x - 2) + 6$$
$$= 2x^3 - 4x^2 - x^2 + 2x + x - 2 + 6$$
$$= 2x^3 - 5x^2 + 3x + 4,$$

which is equal to the dividend.

Example 5

Divide $2x^4 - 5x^3 - x - 4$ by $2x - 1$.

Solution   The terms in both polynomials are in order of descending powers of x. However, in the dividend $(2x^4 - 5x^3 - x - 4)$, a power of x does not appear. The x^2-term is missing. Because long division with polynomials involves subtraction of like terms, you will make your work more orderly and hence less confusing if any missing term is inserted with a coefficient of zero. That is, the dividend is written as

$$2x^4 - 5x^3 + 0x^2 - x - 4.$$

Thus we have

$$
\begin{array}{r}
x^3 - 2x^2 - x - 1 \qquad \leftarrow \text{Quotient} \\
\text{Divisor} \rightarrow 2x - 1 \overline{)2x^4 - 5x^3 + 0x^2 - \ x - 4} \leftarrow \text{Dividend} \\
\underline{2x^4 - \ x^3} \qquad\qquad\qquad \\
- 4x^3 + 0x^2 \qquad\qquad \\
\underline{- 4x^3 + 2x^2} \qquad\qquad \\
- 2x^2 - \ x \qquad \\
\underline{- 2x^2 + \ x} \qquad \\
- 2x - 4 \\
\underline{- 2x + 1} \\
-5 \leftarrow \text{Remainder}
\end{array}
$$

The answer is

$$\frac{2x^4 - 5x^3 - x - 4}{2x - 1} = x^3 - 2x^2 - x - 1 - \frac{5}{2x - 1}.$$

Example 6

Divide $4 + 7y^2 - 8y^3 + 6y^4$ by $3y^2 - y + 1$.

Solution   First we arrange the terms of the dividend in order of descending powers of y:

$$6y^4 - 8y^3 + 7y^2 + 4.$$

Noting that the y-term is missing, we insert the term $0y$ and have

$$
\begin{array}{r}
2y^2 - 2y \ + 1 \qquad\qquad \leftarrow \text{Quotient} \\
\text{Divisor} \rightarrow 3y^2 - y + 1 \overline{)6y^4 - 8y^3 + 7y^2 + 0y + 4} \leftarrow \text{Dividend} \\
\underline{6y^4 - 2y^3 + 2y^2} \qquad\qquad\qquad \\
- 6y^3 + 5y^2 + 0y \qquad \\
\underline{- 6y^3 + 2y^2 - 2y} \qquad \\
3y^2 + 2y + 4 \\
\underline{3y^2 - \ y + 1} \\
3y + 5 \leftarrow \text{Remainder}
\end{array}
$$

We do not continue the division because the remainder has lower degree (1) than does the divisor (2). Thus

$$\frac{6y^4 - 8y^3 + 7y^2 + 4}{3y^2 - y + 1} = 2y^2 - 2y + 1 + \frac{3y + 5}{3y^2 - y + 1}.$$

Problem Set 3.4

In Problems **1–48**, perform the divisions. Give all answers in terms of positive exponents only.

1. $\dfrac{2ab}{4a}$.

2. $\dfrac{-3x^2y}{xy}$.

3. $\dfrac{-14ab^2}{7a^2}$.

4. $\dfrac{35x^3y^2}{-7xy}$.

5. $\dfrac{6abc}{-6ab}$.

6. $\dfrac{-ax^2y^5}{-x^3y^3}$.

7. $\dfrac{-16x^2yz}{-32xy^2z^3}$.

8. $\dfrac{-25ab^3c^2}{5ab^2}$.

9. $\dfrac{-a^2b(abc^2)}{a^2b^4c}$.

10. $\dfrac{(abc^2)^2(ab)}{2(ab)^2}$.

11. $\dfrac{(3xy)^2}{y}$.

12. $\dfrac{2xy^2}{(2xy)^2}$.

13. $\left(\dfrac{8x^2y}{24xy^2}\right)^2$.

14. $\left(\dfrac{3xy}{6y}\right)^2$.

15. $\dfrac{-(-2xy)^2}{14x^2y}$.

16. $\dfrac{(-3x^2y)^3}{(-2x)^2(9x^3y^2)^2}$.

17. $\dfrac{4x^2 - 6x + 8}{2}$.

18. $\dfrac{3x^2 - 8y + xy}{xy}$.

19. $\dfrac{5 - x^2 + 2x}{x}$.

20. $\dfrac{9x^2 - 15}{-3x}$.

21. $\dfrac{10x^3 - 15x + 2}{5x}$.

22. $\dfrac{3xy - 2y}{6}$.

23. $\dfrac{4x + 2y}{-2x}$.

24. $\dfrac{2x^2 - y^2x}{x^2}$.

25. $\dfrac{xy + x^2}{xy}$.

26. $\dfrac{x - xy}{x}$.

27. $\dfrac{3x^2y - x^3y^2 + 1}{x^2y^2}$.

28. $\dfrac{2 + 2y}{2y}$.

29. $\dfrac{6x^2y - 2y^2 + 7x - 4}{-3x}$.

30. $\dfrac{2x^2y^2 - 6xz^2 + 4x}{2xy^2z}$.

31. $\dfrac{-20x^2y^2 + 5xy^2 - 2x}{2xy}$.

32. $\dfrac{-6a^4b^2c^3 + 3a^5b^2c - 12a^6bc^2}{-6a^4b^2c^2}$.

33. $\dfrac{2xy - (2xy^2)^2 - 3x^3y^3}{(xy)^2}$.

34. $\dfrac{x(xy) + y(-x^2) + x^3y^3}{-x(xy)}$.

35. $\dfrac{2x(x^3y^2)^2 + 3x^2(2y^2) - x}{-xy^4}$.

36. $\dfrac{x(x - y) - y(2x)}{xy}$.

37. $\dfrac{x(x^2 - 2x + 1)}{x^2}$.

38. $\dfrac{(x - 2)^2}{x}$.

39. $\dfrac{x^{-2}(yz)^2w}{x^{-3}y^5}$.

40. $\dfrac{x^6}{(x^{-4}y^8)y^{-8}}$.

41. $\dfrac{2(2x^2y)^2}{3y^{-13}z^{-2}}$.

42. $\dfrac{x^2y^{-5}}{(x^{-8}y^6)^{-3}}$.

43. $\left(\dfrac{x^{-3}y^{-6}z^2}{2xy^{-1}}\right)^{-2}$.

44. $\left[\dfrac{x^{-1}y^4}{(z^2)^{-2}}\right]^{-5}$.

45. $\dfrac{x^{1/2} - 3x^{1/3}}{x^{1/4}}$.

46. $\dfrac{2x^{1/3} - y^3 - y^{1/4}}{y^{1/4}}$.

47. $\dfrac{(2x^{1/2}y)^3(4x)^{-1/2}}{x}$.

48. $\dfrac{(x^{1/3}y^{1/6})^3(2x^{-2})}{xy^{1/2}}$.

In Problems **49–68**, perform the long divisions.

49. $\dfrac{2x^2 + 3x - 4}{x - 2}.$

50. $\dfrac{9x^2 - 6x - 6}{3x - 1}.$

51. $\dfrac{x + 3}{x + 2}.$

52. $\dfrac{x}{x + 1}.$

53. $\dfrac{4x^2 - 7x - 5}{4x + 1}.$

54. $\dfrac{5x^2 + 26x + 8}{x + 5}.$

55. $\dfrac{x^3 + 2x^2 - 5x + 2}{x - 1}.$

56. $\dfrac{3x^3 + 2x^2 + 3x + 1}{3x + 2}.$

57. $\dfrac{1 + 2x - x^2 - 2x^3 - 3x^4}{3x + 5}.$

58. $\dfrac{8x^4 - 4x^2 + x}{2x + 1}.$

59. $\dfrac{8x^3 - 2x^2 + 4x - 3}{4x - 1}.$

60. $\dfrac{3x^3 + x^2 - 6x + 1}{3x + 1}.$

61. $\dfrac{x^4 - 2x^2 + 1}{x - 1}.$

62. $\dfrac{x^4 - 8x^2 + 16}{x + 2}.$

63. $\dfrac{x^3 + x^2 + x + 3}{x^2 - x + 1}.$

64. $\dfrac{x^3 + 3x^2 - 2x - 12}{x^2 - 4}.$

65. $\dfrac{x^4 - y^4}{x + y}.$

66. $\dfrac{x^5 + y^5}{x + y}.$

67. $\dfrac{5x^4 - 18x^3 + 7x^2 - 3x + 4}{x^2 - x + 6}.$

68. $\dfrac{x^3 + x^2 + x + 3}{x^2 - x - 1}.$

69. The efficiency ε of a reversible heat engine operating between a high-temperature reservoir at an absolute temperature T_1 and a low-temperature reservoir at temperature T_2 is given by

$$\varepsilon = \frac{T_1 - T_2}{T_1}.$$

Find another form for ε. The absolute temperature scale has a range from zero degrees through all positive values. From purely mathematical considerations, is there any restriction imposed on the value of T_1?

70. The effective resistance of two resistors R_1 and R_2 connected in parallel is

$$\frac{R_1 R_2}{R_1 + R_2}.$$

(a) Simplify this expression, if possible.
(b) Simplify the reciprocal of this expression, if possible.

71. When three capacitors C_1, C_2, and C_3 are connected in series, the total capacitance of the combination, C_t, is given by

$$C_t = \frac{C_1 C_2 C_3}{C_2 C_3 + C_1 C_3 + C_1 C_2}.$$

Find the reciprocal of C_t and perform the indicated division.

72. When a weight W is suspended from a system of three springs having elastic constants k_1, k_2, and k_3, respectively, the displacement of the weight is found to be

$$W\left(\frac{4k_1k_2 + k_1k_3 + k_2k_3}{4k_1k_2k_3}\right).$$

Perform the indicated operations.

73. The average speed $\bar{v}$ of a certain particle, in meters per second, is given by

$$\bar{v} = \frac{2t^2 + 7t + 3}{t + 3},$$

where t is in seconds.

(a) Perform the indicated division to find a simplified expression for $\bar{v}$.

(b) Show that the given expression and the simplified expression give the same value for $\bar{v}$ when $t = 2$.

3.5 SI UNITS AND CONVERSION OF UNITS

Throughout the engineering technologies we find physical quantities that are expressed as quotients. For example, pressure (p) is equal to force (F) divided by area (A), so we can write $p = F/A$. In general terms, the quotient a/b is also called the **ratio** of a to b.* Thus pressure is the ratio of force to area. Similarly, average power ($\bar{P}$) is the ratio of work to time ($\bar{P} = W/t$), and acceleration (a) is the ratio of force to mass ($a = F/m$).

In all aspects of science and technology, we must be able to measure physical quantities. The concept of ratio is basic to the process of measurement. In fact, to measure a quantity *means* to determine the ratio of that quantity to some chosen unit. For example, to measure the length of a cable with a meter stick, we determine how many times the length of the meter stick is contained in the length of the cable. If the meter stick can be fitted along the length of the cable six times, the measurement is $6/1 = 6$ meters. In this example the **unit of measurement** is the *meter*.

By selecting units for some fundamental physical quantities, we form a *system* of units. In science, technology, and industry, the universally accepted system of units is a metric system called the **International System of Units**, abbreviated SI (for Système International). The seven SI basic units are listed in Table 3.1.

TABLE 3.1 SI basic units

Quantity	Unit	Symbol
Length	meter	m
Mass	kilogram	kg
Time	second	s
Electric current	ampere	A
Thermodynamic temperature	kelvin	K
Amount of substance	mole	mol
Luminous intensity	candela	cd

* The ratio of a to b is sometimes denoted by $a:b$.

To express large multiples or small submultiples of basic units (or other units), we can use the prefixes listed in Table 3.2. The indicated symbol μ is the Greek letter *mu*.

TABLE 3.2 SI prefixes

Factor	Prefix	Symbol	Factor	Prefix	Symbol
10^{18}	exa-	E	10^{-1}	deci-	d
10^{15}	peta-	P	10^{-2}	centi-	c
10^{12}	tera-	T	10^{-3}	milli-	m
10^{9}	giga-	G	10^{-6}	micro-	μ
10^{6}	mega-	M	10^{-9}	nano-	n
10^{3}	kilo-	k	10^{-12}	pico-	p
10^{2}	hecto-	h	10^{-15}	femto-	f
10^{1}	deka-	da	10^{-18}	atto-	a

The use of prefixes makes this metric system especially convenient because it is essentially a *decimal*, or *power-of-ten*, system. You must understand how the prefixes are used. To illustrate, note that the factors represented by the prefixes *centi, milli,* and *kilo* are

$$\text{centi} \rightarrow 10^{-2} = \frac{1}{100},$$

$$\text{milli} \rightarrow 10^{-3} = \frac{1}{1000},$$

$$\text{kilo} \rightarrow 10^{3} \;\; = 1000.$$

Using the meter as our basic unit (we could use *any* unit whatever), we have

$$1 \text{ centimeter} = \frac{1}{100}(1 \text{ meter}) \quad \text{or} \quad 1 \text{ cm} = \frac{1}{100} \text{ m} = 10^{-2} \text{ m},$$

$$1 \text{ millimeter} = \frac{1}{1000}(1 \text{ meter}) \quad \text{or} \quad 1 \text{ mm} = \frac{1}{1000} \text{ m} = 10^{-3} \text{ m},$$

$$1 \text{ kilometer} = 1000 (1 \text{ meter}) \quad \text{or} \quad 1 \text{ km} = 1000 \text{ m} = 10^{3} \text{ m}.$$

Another way of stating these relationships is

$$1 \text{ m} = 100 \text{ cm} = 1000 \text{ mm} = \frac{1}{1000} \text{ km},$$

or

$$1 \text{ m} = 10^{2} \text{ cm} = 10^{3} \text{ mm} = 10^{-3} \text{ km}.$$

Example 1

a. 0.001 ampere $= 1 \times 10^{-3}$ ampere $= 1$ milliampere (mA). Note that the use of the prefix avoids the need for inserting the power of 10 in the number.

b. $50{,}000$ meters $= 50 \times 10^3$ meters $= 50$ kilometers (km).

Most physical quantities require both a *number* and a *unit* if they are to be completely specified or measured. It is meaningless to say that the length of a wire is 5. We must state whether we mean 5 m, 5 cm, 5 mm, or 5 km, for example. That is, the unit must be given. It often happens, though, that the given unit is not the desired unit. When units are treated as algebraic quantities, we can use ratios to obtain a method of converting from one unit to another.

For example, suppose we convert 2000 cm to meters. Because $1\text{ m} = 10^2$ cm, the ratio $(1\text{ m})/(10^2\text{ cm})$ is 1:

$$\frac{1\text{ m}}{10^2\text{ cm}} = 1.$$

Thus we can multiply 2000 cm by this ratio without changing its value.

$$2000\text{ cm} = (2000\text{ cm})(1) = (2000\ \cancel{\text{cm}})\left(\frac{1\text{ m}}{10^2\ \cancel{\text{cm}}}\right)$$

$$= \frac{2000}{10^2}\text{ m} = 20\text{ m}.$$

Canceling the cm units is just like canceling the common factor x in

$$2000\cancel{x} \cdot \frac{1y}{10^2\cancel{x}} = 20y.$$

That is, the units are treated like algebraic quantities. We point out, however, that the ratio (or *conversion factor*) was set up so that the unit of measurement to be canceled was placed in the denominator of the ratio and the desired unit was in the numerator. [Thus it would not have helped us to use the ratio $(10^2\text{ cm})/(1\text{ m})$.]

Example 2

Convert 2 A to microamperes.

Solution

$$2\text{ A} = (2\text{ A})(1) = (2\ \cancel{\text{A}})\left(\frac{1\ \mu\text{A}}{10^{-6}\ \cancel{\text{A}}}\right) = \frac{2}{10^{-6}}\ \mu\text{A} = 2 \times 10^6\ \mu\text{A}.$$

Example 3

Convert 3.5 g to kilograms.

Solution The gram, g, is a submultiple of the kilogram, the basic SI unit of mass. Clearly $1 \text{ kg} = 10^3 \text{ g}$, so

$$3.5 \text{ g} = (3.5 \text{ g})(1) = (3.5 \text{ g})\left(\frac{1 \text{ kg}}{10^3 \text{ g}}\right) = 3.5 \times 10^{-3} \text{ kg}.$$

Example 4

Convert 5 km to millimeters.

Solution First we shall convert kilometers to meters.

$$5 \text{ km} = (5 \text{ km})(1) = (5 \text{ km})\left(\frac{10^3 \text{ m}}{1 \text{ km}}\right) = 5 \times 10^3 \text{ m}.$$

Now we convert meters to millimeters.

$$(5 \times 10^3 \text{ m})(1) = (5 \times 10^3 \text{ m})\left(\frac{1 \text{ mm}}{10^{-3} \text{ m}}\right) = 5 \times 10^6 \text{ mm}.$$

Conversions like this are not usually written as separate steps. We write

$$5 \text{ km} = (5 \text{ km})\left(\frac{10^3 \text{ m}}{1 \text{ km}}\right)\left(\frac{1 \text{ mm}}{10^{-3} \text{ m}}\right) = 5 \times 10^6 \text{ mm}.$$

Example 5

Astronomers deal with such large distances that they find it convenient to use the *light year* as a unit of distance. One **light year** is the distance traveled by light in a vacuum in 1 year. Given that the speed of light is 3×10^8 m/s in a vacuum, determine the number of meters in a light year to three significant figures.

Solution

$$\text{Distance} = [\text{rate}][\text{time}]$$

$$= \left[\frac{3 \times 10^8 \text{ m}}{\text{s}}\right][1 \text{ yr}]\left(\frac{365 \text{ d}}{1 \text{ yr}}\right)\left(\frac{24 \text{ h}}{1 \text{ d}}\right)\left(\frac{60 \text{ min}}{1 \text{ h}}\right)\left(\frac{60 \text{ s}}{1 \text{ min}}\right)$$

$$= 9.46 \times 10^{15} \text{ m}.$$

Note the abbreviations for day (d), hour (h), minute (min), and second (s).

Example 6

Convert 5 square centimeters (cm²) to square meters (m²).

Solution

$$5 \text{ cm}^2 = (5 \text{ cm}^2)\left(\frac{1 \text{ m}}{10^2 \text{ cm}}\right)^2 = (5 \text{ cm}^2)\left(\frac{1^2 \text{ m}^2}{10^4 \text{ cm}^2}\right)$$

$$= (5 \text{ cm}^2)\left(\frac{1 \text{ m}^2}{10^4 \text{ cm}^2}\right) = 5 \times 10^{-4} \text{ m}^2.$$

Notice here that we had to *square* the ratio $(1 \text{ m})/(10^2 \text{ cm})$ so that the square centimeters unit, cm^2, could be canceled. Observe that we squared the numbers *and* the units. Students often make mistakes in doing this, so you should be careful! As the next example shows, we must *cube* the basic ratios when converting units of volume.

Example 7

The density (mass per unit volume) of copper is $8.93 \times 10^3 \text{ kg/m}^3$. Express this density in grams per cubic centimeter.

Solution

$$8.93 \times 10^3 \frac{\text{kg}}{\text{m}^3} = \left[\left(8.93 \times 10^3 \frac{\text{kg}}{\text{m}^3}\right)\left(\frac{10^3 \text{ g}}{1 \text{ kg}}\right)\right]\left(\frac{1 \text{ m}}{10^2 \text{ cm}}\right)^3$$

$$= \left[8.93 \times 10^6 \frac{\text{g}}{\text{m}^3}\right]\left(\frac{1 \text{ m}^3}{10^6 \text{ cm}^3}\right)$$

$$= 8.93 \text{ g/cm}^3.$$

The following are some useful relationships involving nonmetric units.

$$1 \text{ m} = 39.37 \text{ inches (in.)},$$

$$1 \text{ in.} = 2.54 \text{ cm},$$

$$1 \text{ mile (mi)} = 5280 \text{ feet (ft)} = 1609.344 \text{ m},$$

$$1 \text{ km} = 3280.8 \text{ ft} = 0.6214 \text{ mi},$$

$$1 \text{ yard (yd)} = 3 \text{ ft}.$$

Example 8

a. Convert a speed of 60 mi/h to feet per second.

Solution

$$\frac{60 \text{ mi}}{\text{h}} = \frac{60 \text{ mi}}{\text{h}} \times \frac{5280 \text{ ft}}{1 \text{ mi}} \times \frac{1 \text{ h}}{60 \text{ min}} \times \frac{1 \text{ min}}{60 \text{ s}} = 88 \text{ ft/s}.$$

b. Convert a height of 6 ft to meters.

Solution

$$6 \text{ ft} = 6\,\cancel{\text{ft}} \times \frac{12\,\cancel{\text{in.}}}{1\,\cancel{\text{ft}}} \times \frac{2.54\,\cancel{\text{cm}}}{1\,\cancel{\text{in.}}} \times \frac{1\ \text{m}}{10^2\,\cancel{\text{cm}}} = 1.83 \text{ m}.$$

Here is another way to do the conversion:

$$6 \text{ ft} = 6\,\cancel{\text{ft}} \times \frac{1\,\cancel{\text{km}}}{3280.8\,\cancel{\text{ft}}} \times \frac{10^3 \text{ m}}{1\,\cancel{\text{km}}} = 1.83 \text{ m}.$$

In many technical areas, we find examples of SI **derived units**, which are units that are combinations of the seven basic units given in Table 3.1. For example, *velocity* is the rate at which position changes. The SI unit of velocity, meters per second (m/s), is derived from the SI units of length and time. *Acceleration* is the rate at which velocity changes, so its unit is the ratio of units of velocity to time. Thus the SI unit of acceleration is

$$\frac{\text{m/s}}{\text{s}} = \frac{\text{m}}{\text{s}}\left(\frac{1}{\text{s}}\right) = \frac{\text{m}}{\text{s}^2},$$

where the last fraction is read *meters per second squared.*

Many SI derived units are so common that they have been given special names. For example, the unit of force is the product of the units of mass and acceleration. Thus

$$\text{Unit of force} = \text{kg}\left(\frac{\text{m}}{\text{s}^2}\right) = \frac{\text{kg m}}{\text{s}^2},$$

which is called a **newton**, N. Some of the derived units that you will encounter in your technical studies are shown in Table 3.3, where Ω is the Greek letter *omega.*

TABLE 3.3   Some SI derived units

Quantity	Name	Symbol	In terms of other units	In terms of SI base units
Force	newton	N		$\text{m} \cdot \text{kg/s}^2$
Pressure	pascal	Pa	N/m^2	$\text{kg/(m} \cdot \text{s}^2)$
Energy	joule	J	$\text{N} \cdot \text{m}$	$\text{kg} \cdot \text{m}^2/\text{s}^2$
Power	watt	W	J/s	$\text{kg} \cdot \text{m}^2/\text{s}^3$
Electric charge	coulomb	C		$\text{A} \cdot \text{s}$
Electric potential	volt	V	W/A	$\text{kg} \cdot \text{m}^2/(\text{A} \cdot \text{s}^3)$
Capacitance	farad	F	C/V	$\text{A}^2 \cdot \text{s}^4/(\text{kg} \cdot \text{m}^2)$
Resistance	ohm	Ω	V/A	$\text{kg} \cdot \text{m}^2/(\text{A}^2 \cdot \text{s}^3)$

Of course, the prefixes in Table 3.2 can be aplied to these units. To illustrate:

$$1,000,000 \text{ ohms} = 1 \times 10^6 \text{ ohms} = 1 \text{ megohm (M}\Omega\text{)},$$

$$0.000001 \text{ farad} = 1 \times 10^{-6} \text{ farad} = 1 \text{ microfarad } (\mu F).$$

Example 9

As shown in Table 3.3, the farad is a unit of capacitance, the property of a device often found in electrical circuits. When capacitors are connected in parallel, the effective capacitance of the combination is the sum of the individual capacitances. If three capacitors with capacitances of $0.5 \ \mu F$, $1.0 \ \mu F$, and $2.0 \ \mu F$ are connected in parallel, what is the effective capacitance of the combination, in farads?

Solution   Let C be the effective capacitance. We have

$$C = C_1 + C_2 + C_3$$

$$= 0.5 \ \mu F + 1.0 \ \mu F + 2.0 \ \mu F = 3.5 \ \mu F.$$

Then

$$C = 3.5 \ \mu F = (3.5 \ \mu F)\left(\frac{10^{-6} \ F}{1 \ \mu F}\right) = 3.5 \times 10^{-6} \ F.$$

Problem Set 3.5

In Problems **1–40**, *make the indicated conversions.*

1. 2.5 m to millimeters.
2. 4 m to centimeters.
3. 8.2 m to kilometers.
4. 12 cm to kilometers.
5. 15 cm to meters.
6. 100 cm to millimeters.
7. 50 mm to centimeters.
8. 10 mm to kilometers.
9. 25 mm to meters.
10. 1 μm to centimeters.
11. 6.0 μF to farads.
12. 8.2 F to microfarads.
13. 2.1 pF to microfarads.
14. 1.0 μm to centimeters.
15. 4.5×10^{-8} s to nanoseconds.
16. 0.005 A to milliamperes.
17. 6.2 μA to milliamperes.
18. 10 mA to microamperes.
19. 40 kg to grams.
20. 350 g to kilograms.
21. 300 kW to watts.
22. 0.2 W to microwatts.
23. 0.2 W to milliwatts.
24. 12,000,000 Ω to megohms.
25. 22,000 Ω to megohms.
26. 0.016 mV to volts.
27. 0.2 mV to microvolts.
28. 0.006 h to milliseconds.
29. 50 km/h to meters per second.
30. 1 mi to centimeters.
31. 1 m to miles.
32. 55 mi/h to kilometers per hour.
33. 9 yd to meters.
34. 10,000 ft to kilometers.

35. 2.4 cm^2 to square millimeters.

36. 6 m^2 to square centimeters.

37. 5.1 m^3 to cubic centimeters.

38. 800 mm^2 to square meters.

39. 500 cm^3 to cubic meters.

40. 60 mm^3 to cubic centimeters.

41. The *escape velocity* on earth is 11.2 km/s. This is the minimum initial velocity a rocket must have to escape from the earth and not return. What is the escape velocity in miles per hour? Round your answer to 2 significant digits.

42. Near the earth's surface, the acceleration due to gravity is about 9.8 m/s^2. Convert this acceleration to feet per second squared.

43. Five 1.5-μF capacitors are connected in parallel. Find the effective capacitance in farads. (Refer to Example 9.)

44. A light pulse would take about 100,000 years to travel from one edge of the Milky Way to the other. What is the distance across the Milky Way in kilometers? (Refer to Example 5.)

45. The speed of sound in dry air at 0°C is 1086 ft/s. Convert this speed to meters per second.

46. The density of gold is $1.93 \times 10^4 \text{ kg/m}^3$. Find the density of gold in grams per cubic centimeter. (Refer to Example 7.)

47. Aluminum has a density of 2.70 g/cm^3. Find the density of aluminum in kilograms per cubic meter. (Refer to Example 7.)

48. The heat of fusion of water is 3.33×10^5 J/kg. Given that 4.186 J = 1 calorie, find the heat of fusion of water in calories per gram.

49. Einstein's famous mass-energy equivalence formula is $E = mc^2$, where m is a mass and c is the speed of light. Given that $c = 3 \times 10^8$ m/s and that $1 \text{ kg} \cdot \text{m}^2/\text{s}^2 = 1$ J, find the energy in joules associated with a mass of 1 g.

50. The green line of the mercury spectrum has a wavelength of 0.00005461 cm. Given that 1 angstrom (Å) equals 10^{-7} mm, find this wavelength in angstrom units.

51. Assuming there are 365 days in each year, determine the number of seconds there are in a century.

52. The charge to mass ratio of an electron is 1.76×10^{11} C/kg. Express this ratio in microcoulombs per gram.

53. The dimensions of a room are 22 ft × 11 ft × 8 ft (in the order of length, width, and height). Find (a) the area of the room in square meters and (b) the volume of the room in cubic meters.

54. Two cities are 350 mi apart. What is the distance in kilometers?

55. The liter (L) is a commonly used unit for volume; $1 \text{ L} = 10^3 \text{ cm}^3$. If a cylindrical container has a radius of 2.0 m and a height of 3.0 m, find the volume of the tank in liters.

56. In making one revolution around the earth, the moon travels about 2.4×10^6 km in 28 days. Find the speed of the moon in meters per second.

57. The length of each edge of a cubic block is 8 cm. Find the surface area of the block in square meters.

58. Find the volume of the block in Problem 57 in liters. (See Problem 55.)

59. In addition to the pascal (Pa), other units used to measure pressure are the atmosphere (atm) and the torr. The relationships between these units are

$$1 \text{ atm} = 1.01 \times 10^5 \text{ Pa},$$

$$1 \text{ torr} = 133.3 \text{ Pa}.$$

If a gas is compressed in a tank until its pressure is 5.2 atm, find the pressure of the gas in torrs.

60. While applying the voltage-divider rule to find the voltage across a resistor in a series circuit, a student encountered the expression

$$(30 \text{ V})\left(\frac{400 \ \Omega}{400 \ \Omega + \ 2 \ k\Omega + 1.6 \ k\Omega}\right).$$

Simplify the expression and give your answer in units of volts only.

61. A *kilowatt-hour* (kWh) is the common unit that electric power companies use to measure energy consumption. For example, you use 1 kWh when you light ten 100-W lamp bulbs for 1 h or one 100-W lamp bulb for 10 h:

$$(100 \text{ W})(10 \text{ h}) = (1000 \text{ W})(1 \text{ h}) = 1 \text{ kWh}.$$

How much does it cost to use a 75-W light bulb for 30 d if it is used 24 h a day at the rate of $0.065 per kilowatt-hour?

62. Convert 1 kWh (see Problem 61) to joules (1 W $= 1$ J/s).

3.6   REVIEW

Important Terms and Symbols

Section 3.1   Algebraic expression, term of expression, constant term, factor, coefficient of term, literal part of term, monomial, multinomial, binomial, trinomial, polynomial, degree of polynomial, leading coefficient of polynomial.

Section 3.2   Like terms, grouping symbols, parentheses, ( ), brackets, [ ], braces, { }.

Section 3.4   Long division of polynomials.

Section 3.5   SI units, conversion of units, conversion factor.

Review Questions

1. An algebraic expression composed of more than one term can *always* be called a _____(a)_____; however, expressions containing exactly two and three terms are usually referred to as _____(b)_____ and _____(c)_____, respectively.

2. An algebraic expression in which every term is either a constant or the product of a constant and a positive integral power of x is called a _____.

3. For the polynomial $3x^2 - 4x + 7$, the degree is equal to _____(a)_____, the leading coefficient is _____(b)_____, and the constant term is _____(c)_____.

4. In the division of one polynomial by another, the process is ended when the degree of the _____(a)_____ is less than the degree of the _____(b)_____.

5. The coefficient of $4x^2$ is _____(a)_____; the coefficient of $-x$ is _____(b)_____; the coefficient of x is _____(c)_____.

6. Two terms are like terms if they have the same _____ part.

7. The product $(x + 3)(y - 5)$ is equal to $x(y - 5) +$ _____.

In Problems **8–10**, *insert* T (*true*) *or* F (*false*).

8. $(2x)(3y)(4z) = 8xz(3y).$ _____

9. $a(ac)^2 = a^3c^2.$ _____

10. $x(yx) = x^2y.$ _____

11. In the division $2x + 5\overline{)-8x^2 + x + 1}$, the first term of the quotient is _____ .

In Problems **12** *and* **13**, *fill in either* $+$ *or* $-$.

12. $4(x - 3y + 5) = 4x$ _____ $12y$ _____ 20.

13. $-3(1 + 4x - 6y) = -3$ _____ $12x$ _____ $18y$.

14. The expression $-\{1 - [(-1)^2(-1)] - 1\}$ is equal to _____ .

Answers to Review Questions

1. (a) Multinomial, (b) Binomials, (c) Trinomials.   2. Polynomial (in x).
3. (a) 2, (b) 3, (c) 7.   4. (a) Remainder, (b) Divisor.   5. (a) 4, (b) -1, (c) 1.
6. Literal.   7. $3(y - 5)$.   8. T.   9. T.   10. T.   11. $-4x$.   12. $-$, $+$.   13. $-$, $+$.
14. -1.

Review Problems

In Problems **1–52**, *perform the operations and simplify.*

1. $(3x + 2y - 5) + (8x - 4y + 2).$

2. $7x + 5(4x + 3).$

3. $6(a + 3b) - (8a - b - 4).$

4. $2(a + 4b) - 3(3b - 2a).$

5. $2[3(xy - 5) + 7(4 - xy)].$

6. $(4xy + 7) - 5[xy - (4 - 3xy)].$

7. $4x + [-5(2 - x) - 8].$

8. $\{1 - 2[x - (x - 1)]\} + 1.$

9. $-3\{x^2 - [3(2x - 4) + 2x^2]\}.$

10. $2\{x^2 + 3[x - (x^2 + 4)]\} + 7.$

11. $(2x^2yz)(xy^3z^6).$

12. $3xy(-2xy^2).$

13. $8ab^2(3a^2b)^2.$

14. $-xy^3(-3xz).$

15. $(2x^2y)^2(xy)^3(xy^2).$

16. $(xy)^2(xz)^2(yz)^2.$

17. $2x^{1/2}y^{-3}x^{1/3}.$

18. $(4xy^3)^{1/2}(-2x^{3/2}y)^4.$

19. $(2x^{3/4}y^{1/2})(xy^{3/2}).$

20. $x^{-3}(2x^4y^{-2}).$

21. $x(x^2 - 2x + 4).$

22. $x^2y(xy - xz + yz).$

23. $a^2b(-2a^2b + 2ab - 3).$

24. $(2xy)^2(x - y + 2xy).$

25. $(x + 3)(x - 4).$

26. $(y - 4(y^2 - 3y + 5).$

27. $(x^{1/2} + y^{1/2})(x^{1/2} - y^{1/2}).$

28. $(3 - x)(3 + x).$

29. $(x - 3)(x - 2).$

30. $(x - 3)^2.$

31. $(x + 2y)^2.$

32. $(x^2 + 1)(x^2 - 2).$

33. $(x^2 + 3)(x - 4).$

34. $(x - 3y)^2.$

35. $(3x - 1)(2x^3 - 3x^2 + 5).$

36. $(y - 4)(y^2 - 3y + 5).$

37. $(x + y + 2)(2x - 3y + 1).$

38. $(1 + x + y)(1 - x - y).$

39. $3x(x - 4) - 2(x^2 - 9).$

40. $2x^2(x^2 - xy) + 2(x^4 - x^3y).$

41. $\dfrac{ax^2y^5}{x^3y^3}.$

42. $\dfrac{2xy^2}{4y^3}.$

43. $\dfrac{(x^2y^{-1}z)^{-2}}{(xy^{1/2})^{-4}}.$

44. $\dfrac{(ab)^2(2ab^2)}{2ab^3}.$

45. $\dfrac{(-3x^2y)(2xy)}{(4x^2y)^2}.$

46. $\dfrac{(xy^2)^2(-3x)^3}{9x(xy)}.$

47. $\dfrac{x^2 - 5x + 7}{x}.$

48. $\dfrac{-3x^3 - 5x^2 + 6}{30x}.$

49. $\dfrac{x^2y - 5xy^3 + 7xy}{xy^2}.$

50. $\dfrac{6x^2y - 3y^2 + 2x - 2}{-4x}.$

51. $\dfrac{2x^2y + (2xy^2w)^2 - 4x^3y^3w}{-2xy}.$

52. $\dfrac{3xy^2 - xy^2 + 3}{xy^2}.$

In Problems 53–58, perform the long divisions.

53. $\dfrac{6x^3 + 3x^2 - 5x - 1}{2x - 1}.$

54. $\dfrac{3x^3 - 5x^2 + x + 1}{3x + 1}.$

55. $\dfrac{x^4 + x^3 + 8x - 30}{x + 3}.$

56. $\dfrac{5 - x + 4x^2 - 3x^3 - 4x^4}{4x + 3}.$

57. $\dfrac{3x + 3x^3 - 2x^4}{2x + 1}.$

58. $\dfrac{9 - x^3}{x - 2}.$

59. Convert 2.5×10^{10} W to gigawatts.

60. Convert 2.31×10^{-5} A to microamperes.

61. Convert 90 km/h to meters per second.

62. The threshold of hearing corresponds to a sound intensity of about 10^{-12} W/m². Express this intensity in microwatts per square centimeter.

Equations

4.1 TYPES OF EQUATIONS

Even a beginning student in engineering technology is soon faced with solving elementary equations. In this chapter we develop techniques to accomplish this task and apply these methods to practical situations.

An **equation** is a statement that two expressions are equal. The two expressions that make up an equation are called its **sides** or **members**. They are separated by the equality sign, $=$.

Example 1

The following are equations.

a. $3x + 1 = 16$.

b. $x^2 + 2x - 8 = 0$.

c. $\dfrac{y}{y - 4} = 6$.

d. $s = 7 - t$.

In Example 1 each equation contains at least one variable. Recall that a **variable** is a symbol, such as x, that can be replaced by any one of a set of different numbers. The most popular symbols for variables are letters from the latter part of the alphabet, such as x, y, z, s, and t. For example, the equation $3x + 1 = 16$ is said to be in the variable x, and the equation $y/(y - 4) = 6$ is in the variable y. The equation $s = 7 - t$ is in two variables, s and t.

We *never* allow a variable in an equation to have a value for which any expression in that equation is undefined. Thus in

$$\frac{y}{y - 4} = 6,$$

y cannot be 4, because this would make the denominator zero (we cannot divide by zero). In some equations the allowable values of a variable are restricted for physical reasons. For example, if the variable t represents time, negative values of t may not make sense. Hence we should assume that $t \geq 0$.

Equations in which some of the constants are not specified but are represented by letters, such as a, b, c, or d, are called **literal equations**, and the letters are called **literal constants** or **arbitrary constants**. For example, in the literal equation $x + a = b$, we may consider a and b to be arbitrary constants. A formula, such as $C = 2\pi r$, which expresses a relationship between certain quantities, may be regarded as a literal equation.

In the equation

$$3x + 1 = 16,$$

if x takes on a specific value, then the resulting equation may be either true or false. For example, substituting 5 for x gives

$$3(5) + 1 = 16$$
$$16 = 16,$$

which is true. On the other hand, if x is zero, the equation is

$$3(0) + 1 = 16$$
$$1 = 16,$$

which is *false*. Based on our results, the original equation is called a *conditional equation* in the following sense:

A **conditional equation** is an equation that is true for at least one, but not all, of the allowable values of its variables.

If we apply the distributive law to the left side of the equation

$$2(x + 1) = 2x + 2,$$

the result is identically equal to the right side $2x + 2$. Thus the equation is always a true statement, regardless of the value of x. We call this equation an *identity*:

An **identity** is an equation that is true for all allowable values of its variables.

To **solve** an equation means to find *all* values of its variables for which the equation is true. These values are called **solutions** of the equation and are said to **satisfy** the equation. When only one variable is involved, a solution is also called a **root**. The set of all solutions of an equation is called the **solution set** of that equation. Often a letter representing an unknown quantity in an equation is called an **unknown**. Example 2 illustrates these terms.

Example 2

a. In the equation $x + 3 = 4$, the variable x is the unknown. The only value of x that satisfies the equation is obviously 1. Hence 1 is a root and the solution set is $\{1\}$.

b. The equation $W_1 + W_2 = 7$ is an equation in two unknowns, W_1 and W_2. One solution is the *pair* of values $W_1 = 1$ and $W_2 = 6$. However, there are obviously infinitely many solutions. Can you think of another?

c. One root of the equation $x^2 + 2x - 8 = 0$ is -4, because substituting -4 for x makes the equation true:

$$(-4)^2 + 2(-4) - 8 = 0$$
$$16 - 8 - 8 = 0$$
$$0 = 0.$$

Thus -4 satisfies the equation.

d. The equation $2 + 2x^2 = x^2 + 2 + x^2$ is an identity, because both sides are identically equal to $2 + 2x^2$.

Some equations are false regardless of the value of the variable involved. An example is the equation $x + 2 = x$ because there is no number x which when added to 2 gives the original number x. Such an equation is called an **impossible equation**. Since there is no solution, the solution set is a set with no elements in it, $\{\ \ \}$. It is called the **empty set** (or **null set**) and is denoted by $\varnothing$. Do not confuse the set $\varnothing$ with $\{0\}$, which is not empty; $\{0\}$ has the number zero in it.

When solving an equation, we want any operation on it to result in another equation having exactly the same solutions as the given equation. Equations with the same solutions are called **equivalent equations**. For example, you can easily see by inspection that the equations

$$2x = 4 \quad \text{and} \quad 3x = 6$$

have exactly the same solution, 2. Thus these equations are equivalent.

An equation will be transformed into an equivalent equation by performing any of the following two operations.

1. **Adding (or subtracting) the same expression to (or from) both sides of an equation.**
   For example, if

$$5x = 3 + 4x,$$

   subtracting $4x$ from both sides gives the equivalent equation

$$5x - 4x = 3 + 4x - 4x,$$

   or simply

$$x = 3.$$

2. **Multiplying or dividing both sides of an equation by the same constant, except zero.**
   For example, if

$$8x = 4,$$

   dividing both sides by 8 gives the equivalent equation

$$\frac{8x}{8} = \frac{4}{8},$$

   or simply

$$x = \frac{1}{2}.$$

Performing Operations 1 or 2 on an equation *guarantees* that the resulting equation is equivalent to the original equation.

When solving an equation, we sometimes have to perform operations other than 1 and 2. These operations do not always result in equivalent equations. They include the following.

3. **Multiplying both sides of an equation by an expression involving the variable.**
   For example, by inspection the only root of $x - 1 = 0$ is 1. Multiplying each side by x gives $x^2 - x = 0$, which is satisfied if x is zero or 1 (check this by substitution). But zero does *not* satisfy the *original* equation. Thus the equations are not equivalent. The "root" zero, which was introduced in $x^2 - x = 0$, is referred to as an **extraneous root**.

4. **Dividing both sides of an equation by an expression involving the variable.**
   For example, you may check that the equation $x^2 = x$ is satisfied when x is 0 or 1. Dividing both sides by x gives $x = 1$, whose only root is 1. We do not have equivalence because a root has been "lost." Note that when x is 0, which is a root of the given equation, division by x implies division by zero, which is not defined.

5. **Raising both sides of an equation to equal powers.**

For example, squaring each side of the equation $x = 2$ gives $x^2 = 4$, which is true if x is 2 or -2. But -2 does not satisfy the *original* equation. Thus the equations are *not* equivalent.

From our discussion it is clear that when Operations 3–5 are performed, we must be careful about drawing conclusions concerning the roots of a given equation. Operations 3 and 5 *can* produce an equation with more roots. Thus you should check whether or not each "solution" obtained by these operations satisfies the *original* equation. Operation 4 *can* produce an equation with fewer roots. In this case, any "lost" roots may never be determined. Avoid Operation 4 whenever possible. In the next section, we will use Operations 1 and 2 to solve equations.

Problem Set 4.1

In Problems 1–6, use substitution to determine which of the given numbers, if any, satisfy the given equation.

1. $9x - x^2 = 0$;   1, 0.

2. $20 - 9x = -x^2$;   5, 4.

3. $y + 2(y - 3) = 4$; $\dfrac{10}{3}$, 1.

4. $2x + x^2 - 8 = 0$;   2, -4.

5. $x(7 + x) - 2(x + 1) = -7$; -1, -7.

6. $x(x + 1)^2(x + 2) = 0$;   0, -1, 2.

In Problems 7–10, determine if the given equation is an identity. See Example 2(d).

7. $x - 5 = 2(x - 10) - x + 15$.

8. $x - 3(x + 2) + 4x - 7 = -13 + 2x$.

9. $3x(7) - 5 + x = 21x$.

10. $2\left(\dfrac{x}{2} + 3\right) - x = 6$.

*In Problems 11–20, determine what operation was applied to the first equation to obtain the second equation. State whether or not the operation **guarantees** that the equations are equivalent.*

11. $x - 5 = 4x + 10$; $x = 4x + 15$.

12. $8x - 4 = 16$; $x - \frac{1}{2} = 2$.

13. $x = 4$; $x^2 = 16$.

14. $\frac{1}{2}x^2 + 3 = x - 9$; $x^2 + 6 = 2x - 18$.

15. $x^2 = 2x$; $x = 2$.

16. $4x + 9 = 2x$; $4x = 2x - 9$.

17. $x^2 = x$; $x^3 = x^2$.

18. $x^2 = 6x$; $x = 6$.

19. $3x + 1 = x - 5$; $2x + 1 = -5$.

20. $9 - 4x = 6x - 1$; $9 = 10x - 1$.

4.2   LINEAR EQUATIONS

The principles presented thus far will now be applied to solve the equation

$$3x + 2 = 17.$$

Essentially, we perform Operations 1 and 2 of the preceding section until the solution is *obvious*. More specifically, our goal is to obtain an equivalent equation in which the variable is isolated on one side of the equation. To achieve our goal, we make use of opposites and reciprocals.

Here are the steps to solve

$$3x + 2 = 17.$$

The only term involving x is $3x$, which we first isolate. To remove the 2 that is *added* to $3x$, we *subtract* 2 from *both* sides (Operation 1) and simplify. This is equivalent to adding the opposite of 2, which is -2, to both sides.

$$3x + 2 - 2 = 17 - 2$$
$$3x = 15. \tag{1}$$

Now, because x is *multiplied* by 3, we *divide* **both** sides of Eq. 1 by 3 (Operation 2) and simplify. This is equivalent to multiplying both sides by the reciprocal of 3, which is $\frac{1}{3}$.

$$\frac{3x}{3} = \frac{15}{3}$$
$$x = 5.$$

Here x is isolated and the solution is obviously 5. Since this equation was obtained from the original equation by applying Operations 1 and 2, both equations are equivalent. Thus the solution of $3x + 2 = 17$ is 5. We may write our solution as $x = 5$. To check that 5 is a solution, we use substitution. In the original equation we replace x by 5 to see that we get a true statement.

$$3(5) + 2 \overset{?}{=} 17$$
$$15 + 2 \overset{?}{=} 17$$
$$17 = 17.$$

This confirms that 5 is a solution. It is the only solution.

We emphasize that *you must perform the* **same** *operation to* **both** *sides of an equation.* If

$$3x = 15, \quad \text{then} \quad x \neq 15 - 3.$$

Here the left side was *divided* by 3, but on the right side 3 was *subtracted* from 15. You should divide *both* sides of $3x = 15$ by 3, as we did earlier.

The equation that we solved, $3x + 2 = 17$, is an example of a *linear equation*:

A **linear equation** in the variable x is an equation that can be written in the form

$$ax + b = 0, \tag{2}$$

where a and b are constants and $a \neq 0$.

A linear equation is also called a **first-degree equation** or an *equation of degree 1*, because the highest power of the variable that occurs in Eq. 2 is the first. The equation $3x + 2 = 17$ is a linear equation because it can be expressed as $3x + (-15) = 0$. This matches Eq. 2 where a is 3 and b is -15. However, $2x^2 + 5 = 0$ is *not* a linear equation. For the time being, we will be concerned with solving linear equations. *Every linear equation in one variable has one and only one root.*

Example 1

Solve $-10x - 9 = 0$.

Solution We want to isolate the variable x. Here 9 is *subtracted* from $-10x$, so we *add* 9 to both sides (Operation 1).

$$-10x - 9 = 0$$

$$-10x - 9 + 9 = 0 + 9 \qquad \text{[adding 9 to both sides]}$$

$$-10x = 9 \qquad \text{[simplifying]}$$

$$\frac{(-10)x}{-10} = \frac{9}{-10} \qquad \text{[dividing both sides by } -10\text{]}$$

$$x = -\frac{9}{10}.$$

Check: $\quad -10\left(-\frac{9}{10}\right) - 9 \overset{?}{=} 0$

$$9 - 9 \overset{?}{=} 0$$

$$0 = 0.$$

Example 2

Solve $3.264 = -8.124 + 5.740t$.

Solution Here the unknown is t.

$$3.264 = -8.124 + 5.740t$$

$$3.264 + 8.124 = -8.124 + 5.740t + 8.124 \qquad \text{[adding 8.124 to both sides]}$$

$$11.388 = 5.740t \qquad \text{[simplifying]}$$

$$\frac{11.388}{5.740} = \frac{5.740t}{5.740} \qquad \text{[dividing both sides by 5.740]}$$

$$1.984 = t \quad \text{(approximately)}$$

or

$$t = 1.984.$$

Check: $3.264 \overset{?}{=} -8.124 + 5.740(1.984)$

$3.264 \overset{?}{=} -8.124 + 11.388$

$3.264 = 3.264.$

Example 3

Solve $6x - 3(4x - 5) = 5(9 - x)$.

Solution   First we get the terms involving x on one side of the equation and the constant terms on the other side. We start by using the distributive law to remove the parentheses.

$$6x - 12x + 15 = 45 - 5x \qquad \text{[removing parentheses]}$$

$$-6x + 15 = 45 - 5x \qquad \text{[simplifying left side]}$$

$$-6x + 15 + 5x = 45 - 5x + 5x \qquad \text{[adding } 5x \text{ to both sides]}$$

$$-x + 15 = 45$$

$$-x + 15 - 15 = 45 - 15 \qquad \text{[subtracting 15 from both sides]}$$

$$-x = 30$$

$$(-1)(-x) = (-1)30 \qquad \text{[multiplying both sides by } -1]$$

$$x = -30.$$

The next three examples involve equations in which fractions appear.

Example 4

Solve $-\dfrac{3}{8}u + 1 = -2$.

Solution

$$-\frac{3}{8}u + 1 = -2$$

$$-\frac{3}{8}u + 1 - 1 = -2 - 1 \qquad \text{[subtracting 1 from both sides]}$$

$$-\frac{3}{8}u = -3.$$

We clear fractions by multiplying both sides by the denominator 8 (Operation 2).

$$8\left(-\frac{3}{8}u\right) = 8(-3) \qquad \text{[multiplying both sides by 8]}$$

$$-3u = -24 \qquad \text{[because } 8(-\tfrac{3}{8}u) = -(8\cdot\tfrac{3}{8})u = -3u\text{]}$$

$$\frac{(\cancel{-3})u}{\cancel{-3}} = \frac{-24}{-3} \qquad \text{[dividing both sides by } -3\text{]}$$

$$u = 8.$$

Example 5

Solve $3\left(\dfrac{2x}{5} - \dfrac{1}{2}\right) = 2x.$

Solution

$$3\cdot\frac{2x}{5} - 3\cdot\frac{1}{2} = 2x \qquad \text{[removing parentheses]}$$

$$\frac{6x}{5} - \frac{3}{2} = 2x.$$

When two or more terms of an equation have fractions, to clear fractions you may multiply both sides by the least common denominator (L.C.D.).* Here the L.C.D. is 10.

$$10\left(\frac{6x}{5} - \frac{3}{2}\right) = 10(2x) \qquad \text{[multiplying both sides by 10]}$$

$$10\cdot\frac{6x}{5} - 10\cdot\frac{3}{2} = 20x \qquad \text{[removing parentheses]}$$

$$12x - 15 = 20x$$

$$12x - 15 - 12x = 20x - 12x \qquad \text{[subtracting } 12x \text{ from both sides]}$$

$$-15 = 8x$$

$$-\frac{15}{8} = x \qquad \text{[dividing both sides by 8]}$$

or

$$x = -\frac{15}{8}.$$

* The *least common denominator* of two or more fractions is the smallest number with all the denominators as factors. That is, the L.C.D. is the least common multiple of all the denominators. See Sec. 0.4.

Example 6

Solve $\dfrac{8y}{3} = \dfrac{7y + 5}{6} + 8$.

Solution We first clear fractions by multiplying *both* sides by the L.C.D. 6.

$$6\!\left(\frac{8y}{3}\right) = 6\!\left(\frac{7y + 5}{6} + 8\right) \qquad \text{[multiplying both sides by L.C.D.]}$$

$$6\!\left(\frac{8y}{3}\right) = 6\!\left(\frac{7y + 5}{6}\right) + 6(8) \qquad \text{[distributive law]}$$

$$2(8y) = (7y + 5) + 6(8)$$

$$16y = 7y + 5 + 48$$

$$16y = 7y + 53$$

$$9y = 53 \qquad \text{[subtracting } 7y \text{ from both sides]}$$

$$y = \frac{53}{9} \qquad \text{[dividing both sides by 9].}$$

Now let us solve some literal equations. If we want to express a particular letter in a formula in terms of the other letters, this letter is considered the unknown. The procedures used are the same as those in the previous examples.

Example 7

The equation $P_1 V_1 = P_2 V_2$ is a statement of Boyle's law of gases, where P_1 and P_2 are variables denoting pressures and the variables V_1 and V_2 denote volumes. Express V_1 in terms of P_1, P_2, and V_2.

Solution Here V_1 is considered to be the unknown and we must isolate it.

$$P_1 V_1 = P_2 V_2$$

$$\frac{P_1 V_1}{P_1} = \frac{P_2 V_2}{P_1} \qquad \text{[dividing both sides by } P_1]$$

$$V_1 = \frac{P_2 V_2}{P_1} \qquad \text{[cancellation].}$$

When we divided both sides by P_1, we assumed that $P_1 \neq 0$ because we cannot divide by zero. Similar assumptions will be made in solving other literal equations.

Example 8

The relationship between Fahrenheit and Celsius temperature readings can be expressed by the equation

$$\frac{F - 32}{180} = \frac{C}{100},$$

where F and C represent the corresponding temperatures. Solve for F.

Solution   Here F is the unknown. Multiplying both sides by 180 (this is more efficient than clearing fractions), we have

$$180\left(\frac{F - 32}{180}\right) = 180\left(\frac{C}{100}\right)$$

$$F - 32 = \frac{9}{5}C$$

$$F = \frac{9}{5}C + 32.$$

Example 9

The formula for the perimeter* P of a rectangle is $P = 2l + 2w$, where l is the length and w is the width. Solve for w.

Solution   We think of w as the unknown and must isolate it.

$$P = 2l + 2w$$

$$P - 2l = 2l + 2w - 2l \qquad \text{[subtracting } 2l \text{ from both sides]}$$

$$P - 2l = 2w$$

$$\frac{P - 2l}{2} = \frac{2w}{2} \qquad \text{[dividing both sides by 2]}$$

$$\frac{P - 2l}{2} = w.$$

Example 10

In a heat-measurement experiment, a mass m_1 of water at a high temperature T_h is added to a mass m_2 of water at a lower temperature T_l. At thermal equilibrium, the final temperature T_f of the mixture satisfies the equation

$$m_1(1000)(T_h - T_f) = m_2(1000)(T_f - T_l).$$

If 4 kg of water at 60°C are mixed with 3 kg of water at 30°C, find the final temperature of the mixture.

* The perimeter of a geometric figure is the length of its boundary.

Solution Here $T_h = 60°C$ and $T_l = 30°C$. The mass of water at the higher temperature is $m_1 = 4$ kg, and at the lower temperature it is $m_2 = 3$ kg. Substituting these values into the given equation and solving for T_f, we have

$$(4)(1000)(60 - T_f) = 3(1000)(T_f - 30)$$

$$4(60 - T_f) = 3(T_f - 30) \qquad \text{[dividing both sides by 1000]}$$

$$240 - 4T_f = 3T_f - 90 \qquad \text{[removing parentheses]}$$

$$240 = 7T_f - 90 \qquad \text{[adding } 4T_f \text{ to both sides]}$$

$$330 = 7T_f \qquad \text{[adding 90 to both sides]}$$

$$T_f = \frac{330}{7} = 47.1°C \qquad \text{[dividing both sides by 7].}$$

Problem Set 4.2

In Problems 1–66, solve the equations.

1. $x + 3 = 0$.
2. $x - 6 = 0$.
3. $4 - x = 0$.
4. $8 = y + 6$.
5. $8x = 36$.
6. $0.2x = 7$.
7. $\dfrac{x}{8} = 3$.
8. $\dfrac{x}{3} = \dfrac{7}{2}$.
9. $6x = 4x$.
10. $2x - 3 = 4$.
11. $3 - 5x = 9$.
12. $9x - 12x = 0$.
13. $6x + 4x = 20$.
14. $6y + 5y - 3 = 41$.
15. $\dfrac{x}{-7} = 2$.
16. $-2x - 3 = -4$.
17. $1 - 3y = -8$.
18. $-1 = 4 + 2u$.
19. $-5 = 7 - 8u$.
20. $5x - 10x = 15$.
21. $-x = -15$.
22. $-4x = 8$.
23. $3x - 8 = 7$.
24. $6x + 7 = 7$.
25. $\dfrac{2x}{5} = -\dfrac{3}{2}$.
26. $\dfrac{9}{8}x = \dfrac{3}{2}$.
27. $-\dfrac{2y}{3} = \dfrac{5}{2}$.
28. $\dfrac{z}{2} = \dfrac{z}{3}$.
29. $3x - \dfrac{1}{5} = 4$.
30. $3x - \dfrac{9}{4} = 2$.
31. $8x = 4x - 12$.
32. $3x + 2x = 4x + 6$.
33. $3x + 6 = 7x - 2$.
34. $9x - 4 = 9 - 4x$.
35. $2(y - 5) = y + 1$.
36. $-3(y - 1) = 4y + 17$.
37. $7(3 - 2z) = 3 - 5z$.
38. $8 - 6z = 4(3z + 5)$.
39. $(4x + 3) - (7 - x) = 7x$.
40. $2(x - 1) - (3x + 7) = x$.
41. $2(x - 1) - 3(x - 4) = 4x$.
42. $x = 2 - 2[2x - 3(1 - x)]$.
43. $\dfrac{x}{5} = 2x - 6$.
44. $\dfrac{5y}{7} - \dfrac{6}{7} = 2 - 4y$.
45. $t + \dfrac{t}{2} = 6$.
46. $\dfrac{x}{3} - 4 = \dfrac{x}{5}$.

47. $y = \dfrac{3}{2}y - 4.$

48. $\dfrac{x}{2} + \dfrac{x}{3} = 7.$

49. $-\dfrac{1}{12} = -2 - \dfrac{3}{4}x.$

50. $\dfrac{x-3}{4} = 5.$

51. $\dfrac{3}{4} = \dfrac{2-x}{3}.$

52. $\dfrac{3}{8} = \dfrac{2x-2}{6}.$

53. $\dfrac{3}{4}(x-1) = 7 + x.$

54. $2\left(x - \dfrac{1}{5}\right) = 3x.$

55. $3x + \dfrac{x}{5} - 5 = \dfrac{1}{5} + 5x.$

56. $y - \dfrac{y}{2} + \dfrac{y}{3} - \dfrac{y}{4} = \dfrac{y}{5}.$

57. $w + \dfrac{w}{2} - \dfrac{w}{3} + \dfrac{w}{4} = 5.$

58. $\dfrac{z}{3} + \dfrac{3}{4}z = \dfrac{9}{2}(z-1).$

59. $\dfrac{x+2}{3} - \dfrac{2-x}{6} = x - 2.$

60. $\dfrac{x}{5} + \dfrac{2(x-4)}{10} = 7.$

61. $\dfrac{3}{4}(z-3) = \dfrac{9}{5}(3-z).$

62. $\dfrac{2y-7}{3} + \dfrac{8y-9}{14} = \dfrac{3y-5}{21}.$

63. $\dfrac{2}{3}(x-5) = 4x + \dfrac{1}{2}.$

64. $\dfrac{x}{3} + 1 = 4\left(x - \dfrac{1}{2}\right).$

65. $2[3z + 4(z-1)] = -(7-z).$

66. $2\{4x - [8 + 2(5x-4)]\} = 0.$

The relationships in Problems **67–86** *occur in physics, chemistry, and various branches of engineering technology. Express the indicated symbol(s) in terms of the remaining symbols.*

67. $P_1 V_1 = P_2 V_2; \quad V_2.$

68. $PV = nRT; \quad R.$

69. $P = i^2 R; \quad R.$

70. $E = mc^2; \quad m.$

71. $V = \pi r^2 h; \quad h.$

72. $T^2 = 4\pi^2\left(\dfrac{L}{g}\right); \quad L.$

73. $K = \tfrac{1}{2}mv^2; \quad m.$

74. $F = \dfrac{1}{2\pi}\left(\dfrac{e}{m}\right)b; \quad b.$

75. $F = k\dfrac{QQ'}{r^2}; \quad Q'.$

76. $I = \dfrac{n\mathcal{E}}{r + nR}; \quad \mathcal{E}.$

77. $V = V_0\left(\dfrac{P_1}{P_2}\right)\left(\dfrac{T_2}{T_1}\right); \quad T_2.$

78. $y = mx + b; \quad x, m.$

79. $v = v_0 - at; \quad a.$

80. $S = v_0 t + \tfrac{1}{2}at^2; \quad v_0.$

81. $V = 2\pi r^2 + 2\pi rh; \quad h.$

82. $F = \dfrac{9}{5}C + 32; \quad C.$

83. $Q = mc(t_2 - t_1) + mL; \quad t_2.$

84. $Q = kA\left(\dfrac{t_2 - t_1}{d}\right)\tau; \quad t_1, t_2.$

85. $I = \dfrac{E}{R}(1 - e^{-Rt/L}); \quad E.$

86. $V = E(x_b - x_a); \quad x_b.$

87. Find the final temperature when 2 kg of water at 79°C is mixed with 4 kg of water at 40°C. (See Example 10.)

88. When 3 kg of water at temperature T_h is mixed with 2 kg of water at 10°C, the final temperature of the water is 15°C. Find T_h to one decimal place. (See Example 10).

89. When 5 kg of water at 80°C is mixed with a certain mass of water at 10°C, the final temperature is 40°C. Find, to one decimal place, the mass of the water that was initially at 10°C. (See Example 10.)

90. In analyzing a circuit by nodal analysis, the following equation occurs:

$$\frac{V - 100}{40} + \frac{V}{80} + \frac{V - 150}{60} = 0.$$

Solve for V (to one decimal place).

91. Use the formula $P = 2l + 2w$ to find the width w of a rectangle whose perimeter P is 960 m and whose length l is 360 m.

92. Use the formula $A = \frac{1}{2}bh$ to find the height h of a triangle whose area A is 75 cm² and whose base b is 15 cm.

93. When radar is used on a highway to determine the speed of a car, a radar beam is sent out and reflected from the moving car. The difference F in frequency (in cycles per second) between the original and reflected beams is given by

$$F = \frac{vf}{334.8},$$

where v is the speed of the car (in miles per hour) and f is the frequency of the original beam (in megacycles per second). Suppose you are driving along a highway with a speed limit of 55 mi/h. A police officer aims a radar beam with a frequency of 2450 (megacycles per second) at your car, and the officer observes the difference in frequency to be 420 (cycles per second). Can the officer claim that you were speeding?

94. When solid objects are heated, they expand in length—which is why expansion joints are placed in bridges and pavements. Generally, when the temperature of a solid body of length l_0 is increased from T_0 to T, the body's length l is given by

$$l = l_0[1 + \alpha(T - T_0)],$$

where α (α is the Greek letter *alpha*) is called the *coefficient of linear expansion*. Suppose a metal rod 1 m long at 0°C expands 0.001 m when it is heated from 0°C to 100°C. Find the coefficient of linear expansion.

95. At time t, the height s of an object thrown up from the ground with a velocity v_0 is given by

$$s = v_0 t - 16t^2.$$

If an object is to have a height of 16 ft after 2 s, find v_0 (in feet per second).

96. In a certain chemical solution, the hydrogen ion concentration, written $[H^+]$, in moles per liter is given by

$$1 \times 10^{-4} = [H^+] \times 1 \times 10^{-2}.$$

Solve for $[H^+]$ and give your answer in scientific notation.

97. Applying the conservation of energy principle to analyze the motion of a cylinder rolling down an incline of height h results in the equation

$$mgh = \frac{1}{2}mv^2 + \frac{1}{2}\left(\frac{1}{2}mR^2\right)\left(\frac{v}{R}\right)^2.$$

Here m and R are the mass and radius of the cylinder, g is the acceleration due to gravity, and v is the speed of the cylinder at the bottom of the incline. Solve for v^2.

4.3 PROPORTION

In physics and chemistry, a type of equation that occurs frequently is that of a *proportion*.

> A **proportion** is a statement that two ratios are equal. That is, it is an equation of the form
>
> $$\frac{a}{b} = \frac{c}{d} \quad \text{or} \quad a:b = c:d.$$

The numbers a and d above are called the **extremes** of the proportion, whereas b and c are called the **means**. Either of the above proportions can be read *a is to b as c is to d.* To say that the three numbers a, b, and c are in the ratio $2:3:5$ means that $a/b = 2/3$, $b/c = 3/5$, and $a/c = 2/5$. Thus 8, 12, and 20 are in the ratio $2:3:5$.

If we multiply both sides of the proportion

$$\frac{a}{b} = \frac{c}{d}$$

by bd, we get

$$bd\left(\frac{a}{b}\right) = bd\left(\frac{c}{d}\right)$$

$$ad = bc \quad \text{[cancellation]}.$$

That is, *the product of the extremes is equal to the product of the means.* This fact is often useful in solving a proportion, as the next example shows.

Example 1

Solve the following proportions for x.

a. $\dfrac{7}{5} = \dfrac{2}{x}$.

Solution We set the product of the extremes equal to the product of the means. (This is called **cross multiplication**.)

$$7 \cdot x = 2 \cdot 5$$
$$7x = 10$$
$$x = \frac{10}{7}.$$

b. $(3 - x):4 = (x + 2):5$.

Solution

$$\frac{3-x}{4} = \frac{x+2}{5}$$

$$5(3-x) = 4(x+2)$$

$$15 - 5x = 4x + 8$$

$$15 = 9x + 8$$

$$7 = 9x$$

$$x = \frac{7}{9}.$$

Example 2

A triangle whose shortest side has a length of 14 units is similar to a triangle having sides of lengths 10, 15, and 20 units (see Fig. 4.1). Find the lengths of the other two sides of the triangle.

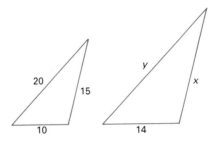

FIGURE 4.1

Solution From geometry, recall that if two triangles are similar, then the lengths of their corresponding sides are proportional. Let x be the the side corresponding to the side 15. Then

$$\frac{x}{15} = \frac{14}{10}.$$

We can find x immediately by simply multiplying both sides by 15.

$$x = 15\left(\frac{14}{10}\right) = 21.$$

Let y be the side corresponding to the side 20. Then

$$\frac{y}{20} = \frac{14}{10}$$

$$y = 20\left(\frac{14}{10}\right) = 28.$$

Thus the other two sides of the triangle have lengths of 21 and 28 units.

Example 3

In an architect's drawing, the scale used is $\frac{1}{4}$ cm = 2 m. Find the scale length of an object if its true length is 56 m.

Solution Let x be the scale length of the object in centimeters. Then x has the same relation to 56 as $\frac{1}{4}$ has to 2. Thus

$$\frac{x}{56} = \frac{\frac{1}{4}}{2}$$

$$\frac{x}{56} = \frac{1}{8}$$

$$x = 56\left(\frac{1}{8}\right) = 7 \text{ cm.}$$

Example 4

Given the proportion

$$\frac{a+b}{a-b} = \frac{5}{2},$$

find the ratio of a to b.

Solution Cross multiplying and solving for a/b, we have

$$2(a+b) = 5(a-b)$$

$2a + 2b = 5a - 5b$ [removing parentheses]

$2b = 3a - 5b$ [subtracting $2a$ from both sides]

$7b = 3a$ [adding $5b$ to both sides]

$7 = \dfrac{3a}{b}$ [dividing both sides by b]

$\dfrac{7}{3} = \dfrac{a}{b}$ [dividing both sides by 3].

Thus the ratio of a to b is $\frac{7}{3}$. This does *not* mean that a must be 7 and b must be 3. Why?*

Example 5

On the Celsius temperature scale, the normal freezing point of water is 0°C and the normal boiling point is 100°C. On the Fahrenheit scale, the corresponding temperatures are 32°F and 212°F. By using a proportion, determine the temperature on the Celsius scale that corresponds to 100°F.

* Because $\frac{7}{3}$ might be the reduced form of the ratio a/b. For instance, we could have $a = 14$ and $b = 6$.

Solution   The temperature scales are shown in Fig. 4.2, where t is the Celsius temperature that we want to find. Because these temperature scales each have uniform divisions and use the

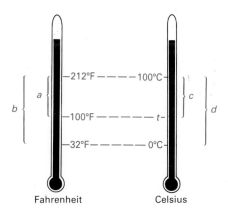

FIGURE 4.2

same fixed points (the freezing and boiling points of water), the proportion

$$\frac{a}{b} = \frac{c}{d}$$

holds, where a, b, c, and d are the lengths of the temperature intervals shown. Thus

$$\frac{212 - 100}{212 - 32} = \frac{100 - t}{100 - 0}$$

$$\frac{112}{180} = \frac{100 - t}{100}$$

$$100(112) = 180(100 - t)$$

$$11{,}200 = 18{,}000 - 180t$$

$$180t = 6800$$

$$t = \frac{6800}{180} = 37.8°C.$$

Problem Set 4.3

In Problems **1–10,** *solve the given proportion.*

1. $\dfrac{x + 1}{5} = \dfrac{3}{7}.$

2. $\dfrac{x - 3}{6} = \dfrac{x}{5}.$

3. $\dfrac{4}{8 - x} = \dfrac{3}{4}.$

4. $\dfrac{2y - 3}{4} = \dfrac{6y + 7}{3}.$

5. $x:\frac{3}{2} = 2\frac{1}{3}:3\frac{1}{6}$.

6. $3\frac{1}{2}:2\frac{1}{2} = 6:x$.

7. $\dfrac{x+3}{4} = \dfrac{x+1}{2}$.

8. $\dfrac{7+2(x+1)}{3} = \dfrac{8x}{5}$.

9. $\dfrac{\frac{3}{2}}{2} = \dfrac{\frac{1}{4}}{x}$.

10. $\dfrac{\frac{1}{2}}{x} = \dfrac{8}{\frac{2}{3}}$.

In Problems **11–14**, *find the ratio of a to b.*

11. $\dfrac{a-b}{a+b} = \dfrac{4}{13}$.

12. $\dfrac{a+2b}{b-a} = \dfrac{7}{2}$.

13. $\dfrac{2(a+2b)}{13b-a} = \dfrac{1}{2}$.

14. $\dfrac{2(2b-a)}{3(a-b)} = \dfrac{1}{3}$.

15. A triangle in which the longest side has a length of 14 m is similar to a triangle with sides of lengths 3, 5, and 6 m. Find the lengths of the other two sides of the triangle.

16. The length and width of a sheet metal plate must be in the ratio of 7:5, respectively. If the length of the plate is to be 56 cm, what must its width be?

17. A Wheatstone bridge is a device that can be used to measure an unknown resistance. In a balanced condition it is governed by the proportion

$$\frac{R_1}{R_2} = \frac{R_3}{R_4},$$

where the R's are the resistances in ohms in each arm of the bridge. If $R_1 = 3\,\Omega$, $R_2 = 5\,\Omega$, $R_3 = 7\,\Omega$, and $R_4 = 1 + x$ ohms, find x.

18. If a block is on a frictionless inclined plane as shown in Fig. 4.3, the following proportion applies:

$$\frac{F}{W} = \frac{h}{l},$$

where F is the force to just start the block moving up the inclined plane, W is the weight of the block, and h and l are the height and length, respectively, of the inclined plane. Find l if $W = 50\,\text{N}$, $F = 5\,\text{N}$, and $h = 2\,\text{m}$.

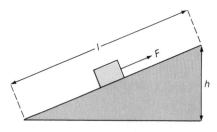

FIGURE 4.3

19. Graham's law of diffusion of two gases is given by the proportion

$$\frac{\text{Rate of diffusion of gas A}}{\text{Rate of diffusion of gas B}} = \frac{\sqrt{\text{Density of B}}}{\sqrt{\text{Density of A}}}.$$

Suppose gas A is nine times as dense as B (that is, take the density of B to be x and that of A to be $9x$). In a certain diffusion apparatus, and under constant temperature and pressure, gas A diffuses 8 cm/s. Under the same conditions, what is the rate of diffusion of B?

20. For the proportion in Problem 19, suppose gas A is 25 times as dense as gas B. How do their rates of diffusion compare?

21. Under conditions of constant pressure, Charles's law of gases is given by the proportion

$$\frac{V_1}{V_2} = \frac{T_1}{T_2},$$

where V_1 and V_2 are the volumes of the gas at temperatures T_1 and T_2, respectively. Suppose that at a particular pressure gas A has a volume of 200 cm^3 at a temperature of 273 K (the Kelvin scale). What volume will it occupy at 373 K? Give your answer to the nearest cubic centimeter.

22. Under conditions of constant temperature, Boyle's law for gases is given by the proportion

$$\frac{P_1}{V_2} = \frac{P_2}{V_1},$$

where P_1 and P_2 are the pressures of the gas at volumes V_1 and V_2, respectively. If 10 L of hydrogen gas at a pressure of 1 atm is allowed to expand at constant temperature to a new volume of 18 L, find the resulting pressure of the gas.

23. A step-up transformer having 100 turns on the input winding is connected to a 10-V generator. The desired output voltage is to be 40 V. To find the number of turns, N_2, that the output winding must have, it is necessary to solve the proportion

$$\frac{V_1}{V_2} = \frac{N_1}{N_2},$$

where $V_1 = 10$, $V_2 = 40$, and $N_1 = 100$. Find N_2.

24. Two circles have radii of r_1 and r_2 and corresponding areas of A_1 and A_2. Show that $A_1 : A_2 = r_1^2 : r_2^2$.

25. Use a proportion to find (to one decimal place) the temperature on the Celsius temperature scale that corresponds to 200°F. (See Example 5.)

26. Use a proportion to find the temperature on the Fahrenheit scale that corresponds to 50°C. (See Example 5.)

27. Suppose we define a new temperature scale on which the normal freezing and boiling points of water are $-20°$ and $200°$, respectively. What temperature on this scale corresponds to (a) 50°C and (b) 50°F? (See Example 5.)

28. On the Kelvin temperature scale, the freezing and boiling points of water are 273 K and 373 K, respectively. What temperature on the Kelvin scale corresponds to 50°C? (See Example 5.)

29. At a temperature of 0°C, the frequency f_s emitted by a source and the frequency f_0 heard by a stationary observer are related by the proportion

$$\frac{f_0}{f_s} = \frac{331}{331 \pm v_s}.$$

In this special case of the Doppler effect, v_s is the speed of the source in meters per second. The plus sign is used if the source recedes from the observer, and the minus sign if it approaches the observer. What frequency would you hear in a laboratory if a tuning fork with a frequency of 256 Hz were moved *toward* you at a speed of 4 m/s? Give your answer to the nearest hertz.

30. In Problem 29, what frequency would you hear if the same tuning fork were moved *away* from you at 4 m/s?

31. For the frequencies in Problem 29, if the *source* is stationary and the observer is moving, then under the same conditions the Doppler effect gives

$$\frac{f_0}{f_s} = \frac{331 \pm v_0}{331},$$

where v_0 is the speed of the observer in meters per second. Here the plus sign is used if the observer moves toward the source. If you were approaching a stationary fire siren emitting a frequency of 512 hertz at a speed of 20 m/s, what frequency would you hear?

32. In Problem 31, if you were moving *away* from the given source at the given speed, what frequency would you hear?

33. When a galvanometer of resistance R_G (in ohms) is used as an ammeter (Fig. 4.4), a shunt

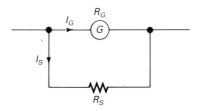

FIGURE 4.4

resistance R_S is used. The relationship between the currents I_G (through the galvanometer) and I_S (through the shunt resistance) is given by

$$\frac{I_G}{I_S} = \frac{R_S}{R_G}.$$

Given that $I_G = 0.005$ A, $I_S = 4.995$ A, and $R_G = 50\,\Omega$, find R_S to two decimal places.

4.4 TRANSLATING ENGLISH TO ALGEBRA

In the next section you will begin to solve word problems. This involves taking a situation stated in words and expressing it in terms of algebraic statements. To prepare you for this, we shall now show you how some "translations" are handled.

Example 1

Translate each of the following word statements into an algebraic statement.

a. Twice a given number is 46.

Solution Let the given number be x. Then twice this number translates into $2x$. Because this expression is (or equals) 46, the complete translation is

$$2x = 46.$$

b. m_1 is 4 more than m_2.

Solution We translate 4 more than m_2 as either $m_2 + 4$ or $4 + m_2$. Because m_1 is equal to this expression, the complete translation is

$$m_1 = m_2 + 4.$$

This situation can also be stated as "m_2 is 4 *less* than m_1." We translate 4 less than m_1 as $m_1 - 4$, *not* as $4 - m_1$. Thus

$$m_2 = m_1 - 4.$$

c. The sum of 4 times x and y is 10^4.

Solution

$$4x + y = 10^4.$$

d. The reciprocal of the difference of x and y.

Solution

$$\frac{1}{x - y}.$$

e. 12% of x.

Solution We write 12% as $\frac{12}{100}$ or as 0.12. The translation in decimal form is

$$0.12x.$$

f. 12% more than x.

Solution Do not confuse this with Part e. The translation is

$$x + 0.12x \quad \text{or} \quad 1.12x.$$

Example 2

Translate the following statement into an algebraic statement.

A transformer has six times as many turns on the input winding (N_1) as there are turns on the output winding (N_2).

Solution Because the order of the key words in the preceding statement is

six, times, input, as, output,

you might think that the translation is

$$6N_1 = N_2.$$

This is **incorrect**. It is crucial that you carefully analyze what the given statement means. Essentially, it asserts that whatever the number of turns on the output winding, the number of turns on the input winding is six times as many. Because there are N_2 turns on the output winding, there are $6N_2$ turns on the input winding. Thus

$$N_1 = 6N_2.$$

Example 3

When a battery of emf $\mathcal{E}$ (in volts) is connected to a group of resistors in parallel, the current I (in amperes) in the battery is equal to the product of the emf and the sum of the reciprocals of the resistances. Write an equation for the current I in a battery that is connected to two parallel resistors R_1 and R_2. Evaluate the current if $\mathcal{E} = 12$ V, $R_1 = 6\,\Omega$, and $R_2 = 3\,\Omega$.

Solution   Because I is the product of $\mathcal{E}$ and $1/R_1 + 1/R_2$, the equation is

$$I = \mathcal{E}\left(\frac{1}{R_1} + \frac{1}{R_2}\right).$$

By substituting 12 for $\mathcal{E}$, 6 for R_1, and 3 for R_2, we obtain

$$I = 12\left(\frac{1}{6} + \frac{1}{3}\right) = 2 + 4 = 6 \text{ A}.$$

Problem Set 4.4

*In Problems **1–28**, translate the word statement into an algebraic statement.*

1. The sum of x and y is 4.
2. The difference of s and 16 is 10.
3. 16 is 5 more than m_1.
4. y is 4 less than x.
5. w is 5 less than the difference of x and y.
6. x is 4 more than the product of y and z.
7. Four times a number x is y.
8. The sum of 4 and the product of 6 and w is 80.
9. There are twice as many resistors (r) as capacitors (c).
10. There is five times as much alcohol (a) as water (w).

11. A rectangle has a length l that is 4 cm less than 5 times its width w (in centimeters).

12. The power P of a lens (in diopters) is the reciprocal of its focal length f (in meters).

13. Mass m_2 is 8% of mass m_1.

14. Mass m_2 is 8% more than mass m_1.

15. Mass m_2 is 8% less than mass m_1.

16. When five times a number n is decreased by 3, the result is 14.

17. The product of x and the sum of a and b is equal to the sum of b times x and a times x. Show that the resulting equation is an identity.

18. If 3 is decreased by 15 times a number x, the result is 3, plus 4 times the number.

19. Twice the sum of x and 3 is equal to five times the result of subtracting 9 from x.

20. The sum of a number n and 2 is multiplied by the value of the number decreased by 2. The result is equal to -4 plus the square of the number.

21. The reciprocal of t_1 plus the reciprocal of t_2 is equal to the reciprocal of T.

22. The sum of three consecutive integers is 36. Assume that the first integer is n.

23. The distance s traveled by a particle along a path is equal to the product of its velocity v and the time t it travels.

24. The number z exceeds $y - 4z$ by $5(z - \frac{1}{5}y)$.

25. Fahrenheit temperature F is equal to $\frac{9}{5}$ Celsius temperature C, plus 32.

26. If a mass were converted into energy, the amount E of energy liberated is equal to the product of the mass m and the square of the speed of light c.

27. An interpretation of Newton's second law leads to the result that W, the weight of an object, is equal to the mass m of the object times the acceleration g due to gravity.

28. The product of the pressure P of a gas and the volume V of the gas is equal to a constant k.

In Problems 29–34, answer the question by using an algebraic expression.

29. A solution is 25% acid, by volume. How much acid is there in x milliliters of the solution?

30. A tank contains 100 gal of an alcohol and water solution. If x gallons are alcohol, how many gallons are water?

31. The average temperature today is 22°C. The average temperature tomorrow is expected to be x°C lower. What is the expected average temperature tomorrow?

32. A light bulb uses 75 W of power per hour. How many kilowatts are used in t hours?

33. A p-watt light bulb burns 24 h a day for 30 d. How many kilowatt-hours are consumed? (Refer to Problem 61 of Problem Set 3.5.)

34. The width of a rectangle is w and the length is $w/2$. What is the area of the rectangle?

35. A spring stretches 5.3 cm for each newton of weight that is suspended from it. If the initial length of the spring is 20 cm, write an equation that relates the length L (in centimeters) of the stretched spring to the load W (in newtons) suspended from it. From this equation, find the length of the spring when 2.0 N are suspended from it.

36. The weight of a large lump of clay decreases by 9.8 N for each kilogram of clay that is removed from it.
    (a) If the initial weight is 45 N, write an equation that gives the weight W (in newtons) of the clay after a mass of m kilograms of clay is removed from it.
    (b) Find the weight of clay after a mass of 2 kg has been removed.

37. A 12-V battery is connected to three resistors in parallel. If $R_1 = 12\,\Omega$, $R_2 = 4\,\Omega$, and $R_3 = 3\,\Omega$, find the current in the battery. (See Example 3.)

4.5 WORD PROBLEMS

We now turn to *word* problems. With a word problem the equation to be solved is not handed to you. You have to set it up by taking the facts and relationships stated verbally in the problem and translating them into an equation. The following suggestions may be used as a guide.

GUIDE TO SOLVING WORD PROBLEMS

1. Read the problem more than once so that you clearly understand all of the given facts and relationships and what you are asked to find.
2. Choose a letter to represent an unknown quantity that you want to find.
3. Take the relationships and facts given in the problem and translate them into an equation involving the letter. Drawing a diagram is often useful.
4. Solve the equation and check your solution to see if it answers the question that the problem posed. (Watch out for answers that seem unreasonable!) Sometimes the solution to the *equation* is not the answer to the *question* asked, but it may be of use in obtaining that answer. In some cases a solution may have to be rejected for physical reasons.

The following model problems illustrate basic techniques and concepts. Study them carefully before you proceed to the exercises. We begin with a percentage problem.

Example 1

63 is 35% of what number?

Solution Let n represent the unknown number. In decimal form, 35% is 0.35, so the equation to solve is

$$63 = 0.35n.$$

Dividing both sides by 0.35 gives

$$\frac{63}{0.35} = \frac{0.35n}{0.35}$$

$$180 = n.$$

Example 2

To produce a certain compound, chemicals A and B must be combined in the ratio of 2:3, respectively (by mass). If 350 kg of the compound must be produced, find the amount of each chemical that must be used.

Solution   The ratio of 2:3 means that the compound consists of two parts A and three parts B. Let m be the mass of each part, in kilograms. Figure 4.5 shows the situation.

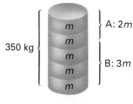

FIGURE 4.5

From the diagram we have

$$2m + 3m = 350$$

$$5m = 350$$

$$m = \frac{350}{5} = 70.$$

But $m = 70$ is *not* the answer to the original problem. Each *part* has a mass of 70 kg. The mass of A is $2m = 2(70) = 140$, and the mass of B is $3m = 3(70) = 210$. Thus 140 kg of chemical A and 210 kg of chemical B must be used.

We now turn to a *rate* problem. You may recall that

Distance = (rate)(time).

Two other forms of this are

$$\text{Time} = \frac{\text{distance}}{\text{rate}}, \qquad \text{rate} = \frac{\text{distance}}{\text{time}}.$$

Example 3

Suppose a person can row about 11 km/h in still water. The speed of the current in a stream is 3 km/h. How far upstream can the person row so as to be back at the starting point in 2 h?

Solution   Upstream, against the current, the rate of the boat is $11 - 3 = 8$ km/h. Downstream, with the current, the rate of the boat is $11 + 3 = 14$ km/h. Let d be the distance (in kilometers) that the person can row upstream (see Fig. 4.6). Then the time (in hours) to row

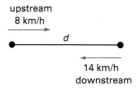

FIGURE 4.6

upstream at 8 km/h is distance/rate, or $d/8$. Downstream, the distance is also d, but the time at 14 km/h is distance/rate, or $d/14$. Thus

$$\left(\begin{array}{c} \text{Time} \\ \text{upstream} \end{array}\right) + \left(\begin{array}{c} \text{time} \\ \text{downstream} \end{array}\right) = \text{total time}$$

$$\frac{d}{8} + \frac{d}{14} = 2$$

$$56\left(\frac{d}{8} + \frac{d}{14}\right) = 56(2) \qquad \text{[multiplying both sides by 56, the L.C.D.]}$$

$$56 \cdot \frac{d}{8} + 56 \cdot \frac{d}{14} = 112$$

$$7d + 4d = 112$$

$$11d = 112$$

$$d = \frac{112}{11} = 10\frac{2}{11}.$$

Thus the person can row upstream a distance of $10\frac{2}{11}$ km.

There is another way of arriving at the answer. Instead of finding the distance right away, we let the unknown be the *time* it takes to row upstream. Let t be the time, in hours, it takes to row upstream at 8 km/h. Because the total time traveled is 2 h, then $2 - t$ is the time it takes to row downstream at 14 km/h (see Fig. 4.7).

Time upstream $= t$
8 km/h

14 km/h
Time downstream $= 2 - t$

FIGURE 4.7

Now, distance $=$ (rate)(time) and

distance upstream $=$ distance downstream.

Therefore, we have

$$(\text{Rate})(\text{time}) = (\text{rate})(\text{time})$$

$$8t = 14(2 - t)$$

$$8t = 28 - 14t$$

$$22t = 28$$

$$t = \frac{28}{22} = \frac{14}{11}\,\text{h}.$$

Thus the distance upstream is $(\text{rate})(\text{time}) = (8)(\frac{14}{11}) = \frac{112}{11} = 10\frac{2}{11}\,\text{km}$.

Example 4

Entering a certain storage tank are three pipes: A, B, and C. Pipe A can fill the tank in 2 h, pipe B in 4 h, and pipe C in 5 h. How long will it take to fill the tank if all three pipes are used?

Solution Let t be the time, in hours, it takes to fill the tank when all three pipes are used. Now, pipe A fills $\frac{1}{2}$ of the tank in 1 hr, so in t hours it fills $t \cdot \frac{1}{2}$ (or $t/2$) of the tank. Similarly, in t hours pipes B and C fill $t/4$ and $t/5$ of the tank, respectively. When the three pipes are used together for t hours, the whole tank is filled (that is, the portion filled is 1). Thus

$$\frac{t}{2} + \frac{t}{4} + \frac{t}{5} = 1.$$

Multiplying both sides by 20, the L.C.D., we have

$$20\left(\frac{t}{2} + \frac{t}{4} + \frac{t}{5}\right) = 20(1)$$

$$10t + 5t + 4t = 20$$

$$19t = 20$$

$$t = \frac{20}{19} = 1\frac{1}{19}.$$

The time required is $1\frac{1}{19}$ hours.

We now give some examples of *mixture* problems.

Example 5

How many liters of antifreeze that is 70% alcohol (70 percent by volume is alcohol) must be added to 10 L of a 35% solution to yield a 50% solution?

Solution Drawings like Fig. 4.8 are very helpful in problems like this. Refer to that drawing as you follow the discussion. Let x be the number of liters of the 70% solution to be added to

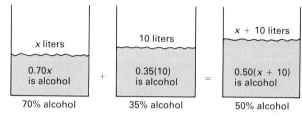

FIGURE 4.8

the 10 L of the 35% solution. Then we end up with $x + 10$ liters, of which 50% must be alcohol. That is, $0.50(x + 10)$ is alcohol. This alcohol comes from two sources: $0.70x$ comes from the 70% solution and $0.35(10)$ comes from the 35% solution. Thus

$$0.70x + 0.35(10) = 0.50(x + 10)$$

$$0.70x + 3.5 = 0.50x + 5$$

$$0.20x = 1.5$$

$$x = \frac{1.5}{0.20} = 7.5.$$

Thus 7.5 L of the 70% solution must be added.

Example 6

A chemical manufacturer mixes a 30% acid solution (30% by volume is acid) with an 18% acid solution. How much of each solution should be used to obtain 500 L of a 25% acid solution?

Solution Let x be the number of liters of the 30% solution to be used. Then to get a total of 500 L of a 25% solution, there must be $500 - x$ liters of the 18% solution (see Fig. 4.9). The total amount of acid in the 500 L of the 25% solution is $0.25(500) = 125$ L. This acid comes

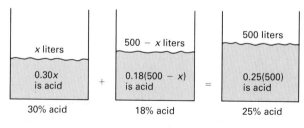

FIGURE 4.9

from two sources: $0.30x$ comes from the 30% solution and $0.18(500 - x)$ comes from the 18% solution. Thus

$$0.30x + 0.18(500 - x) = 125$$

$$0.30x + 90 - 0.18x = 125$$

$$0.12x = 35$$

$$x = \frac{35}{0.12} = \frac{3500}{12}$$

$$= 291\tfrac{2}{3}.$$

Thus $500 - x = 500 - 291\tfrac{2}{3} = 208\tfrac{1}{3}$. The manufacturer should mix $291\tfrac{2}{3}$ L of the 30% solution with $208\tfrac{1}{3}$ L of the 18% solution.

Our last example deals with forces acting on a beam. This topic is typically found in technical physics courses.

Example 7

If two downward forces F_1 and F_2 act on a very light beam (Fig. 4.10), the beam will balance on the pivot when $F_1 d_1 = F_2 d_2$. Here d_1 and d_2 are the distances of F_1 and F_2, respectively, from the pivot. The distances d_1 and d_2 are called **lever arms**, and the product of a force and its lever arm is called a **torque**.

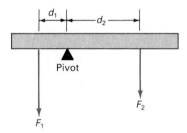

FIGURE 4.10

a. If $F_1 = 20$ N, $F_2 = 13$ N, and $d_1 = 0.5$ m, find d_2 so that the beam balances.

Solution

$$F_1 d_1 = F_2 d_2$$

$$(20)(0.5) = (13)d_2$$

$$\frac{20(0.5)}{13} = d_2$$

$$d_2 = \frac{10}{13} \text{ m.}$$

Thus the force of 13 N should be applied $\tfrac{10}{13}$ m to the right of the pivot.

b. If the beam in Fig. 4.11 is in balance on the pivot shown, how large is the force F?

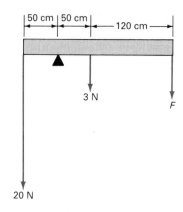

FIGURE 4.11

Solution For the beam to balance, the sum of the clockwise torques—that is, the torques associated with forces tending to cause clockwise rotation about the pivot—must equal the sum of the counterclockwise torques.

Sum of clockwise torques = sum of counterclockwise torques.

$$3(50) + F(120 + 50) = 20(50)$$

$$150 + 170F = 1000$$

$$170F = 850$$

$$F = \frac{850}{170} = 5 \text{ N}.$$

Problem Set 4.5

1. 90 is 30% of what number?
2. 72 is 45% of what number?
3. 75 is 12% of what number?
4. 34 is 85% of what number?
5. What percentage of 200 is 8?
6. What percentage of 60 is 9?
7. What percentage of 50 is 55?
8. What percentage of 1400 is 70?
9. Approximately 21% of the air we breathe is oxygen. To the nearest milliliter, how many milliliters of air contain 1 mL of oxygen?
10. A college dormitory houses 210 students. This fall, rooms are available for 76 freshmen. On the average, 95% of those freshmen who request room applications actually reserve a room. How many room applications should the college send out if it wants to receive 76 reservations?
11. In order to produce a certain compound, chemicals A and B must be combined, by mass, in

the ratio of $3:11$, respectively. If 175 grams of the compound are needed, how many grams of each of A and B must be used?

12. A lab assistant is to prepare 84 mL of a chemical solution. It is to be made up of 2 parts alcohol and 3 parts acid. How much of each should be used?

13. A certain alloy is made up of 8 parts of metal A, 3 parts of metal B, and 1 part of metal C by mass. How much of each metal is needed to make 168 kg of the alloy?

14. How many kilograms each of chemicals A, B, and C must be combined to obtain 93 kg of a compound that consists of A, B, and C in the ratio $1:2:3$, respectively, by mass?

15. A builder makes a certain type of concrete by mixing together 1 part cement, 3 parts sand, and 5 parts stone (by volume). If he wants 585 ft^3 of concrete, how many cubic feet of each ingredient does he need?

16. According to *The Consumer's Handbook* [Paul Fargis, ed., (New York: Hawthorn, 1974)], a good oiled furniture finish contains two parts boiled linseed oil and one part turpentine. If you need a pint (16 fluid oz) of this furniture finish, how many fluid ounces of turpentine are needed?

17. How many kiloliters of a 60% acid solution (60% is acid by volume) must be added to 12 kL of a 35% acid solution so that the resulting solution is 50% acid?

18. A 6-L car radiator is two-thirds full of water. How much of a 90% antifreeze solution (90% is alcohol by volume) must be added to it to make a 10% antifreeze solution in the radiator?

19. A chemical manufacturer plans to mix a 20% acid solution (20% is acid by volume) with a 30% acid solution to get 700 L of a 24% acid solution. How many liters of each must be used?

20. To produce iron for miscellaneous castings, how many kilograms of ferrosilicon (86% silicon) should be added to 2000 kg of the base iron (2.17% silicon) to give the iron a 2.25% silicon content, making it easier to machine?

21. How many milliliters of water must be evaporated from 80 mL of a 12% salt solution (12% is salt by volume) so that what remains is a 20% salt solution?

22. A company manufactures a drain cleaner. The cleaner consists of a chemical compound and metal shavings. The chemical compound not only loosens grease, but when dissolved in water it gives off heat, which speeds up the reaction, and it reacts with the metal to generate hydrogen, which also loosens dirt and grease. The company markets two forms of the cleaner: industrial strength, of which 9% is metal shavings (by weight); and household strength, of which 6% is metal shavings. A motel chain has placed an order with the company to supply them with 12,000 kg of a new form of the cleaner, which is 8% metal shavings. To fill the order, the company will mix the industrial and household forms. How many kilograms of each should go into the mixture?

23. On the moon, a lunar rover traveled from point B to point B at the rate of 5 km/h and returned to A along the same path at the rate of 15 km/h. The *total* traveling time was 2 h. Find the distance from A to B.

24. Suppose that the lunar rover in Problem 23 traveled from A to B at 6 km/h and returned at 10 km/h. If the total traveling time was 3 h, find the distance from A to B.

25. Suppose that the total time for the trip in Problem 23 was 3 h. Based on the rates given there, find the distance from A to B.

26. If you travel 120 km from A to B at an average speed of 60 km/h and then you return to A at an average speed of 40 km/h, what is your average speed for the entire trip? *Note*: The answer is not 50 km/h.

27. A pilot, flying against a headwind, traveled from A to B at 250 km/h. She flew back, with the wind, at 300 km/h. Her trip from B to A took 1 h less than the time for the trip from A to B. Find the distance from A to B.

28. From two airports that are 300 km apart, two airplanes leave at the same time and fly toward each other. One flies at 275 km/h and the other at 325 km/h. How long will it take for the planes to pass each other? *Hint*: When they pass, the sum of the distances traveled by the planes is 300 km.

29. A chemical company can fill a tank car with an industrial solvent with a regular pump in 20 min. Another pump, one that the company keeps in reserve, can fill the tank car in 30 min. How many minutes would it take to fill the tank car if both pumps were used together?

30. Water is flowing into a tank by means of pipes A and B. Pipes A and B can fill the tank, individually, in 2 h and 5 h, respectively. However, water is also flowing out of the tank into another tank by pipe C, which can completely empty the original tank in 4 h. How long would it take to fill the original tank if it were initially empty and pipes A, B, and C were all opened?

31. For an airplane flight, the *point of no return* is the point on the flight where it will take as much *time* to fly on to the destination as to fly back to the starting point. Suppose an airplane is to fly from Honolulu to San Francisco, for which the air distance is 2397 statute miles. If the airplane has an average speed of 350 mi/h in still air and there is an average tail wind of 50 mi/h, find the point of no return. Give your answer to the nearest mile from Honolulu. (*Hint*: The speed of the plane to San Francisco is its speed in still air *plus* the speed of the wind. To Honolulu, it is its speed in still air *minus* the speed of the wind.)

32. One of the most important defoliating insects is the gypsy moth caterpillar, which feeds on foliage of shade, forest, and fruit trees. A homeowner lives in an area in which the gypsy moth has become a problem. She wishes to spray the trees on her property before more defoliation occurs. She needs 128 oz of a solution made up of 3 parts of insecticide A and 5 parts of insecticide B. The solution is then mixed with water. How many ounces of each insecticide should be used?

33. If the beam shown in Fig. 4.12 is balanced on the pivot, how large is the force F? (See Example 7.)

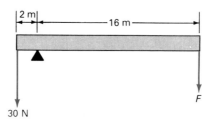

FIGURE 4.12

34. If the beam shown in Fig. 4.13 is balanced on the pivot, how large is the force F? (See Example 7.)

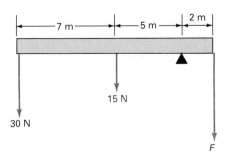

FIGURE 4.13

35. If the beam shown in Fig. 4.14 balances on the pivot, determine d. (See Example 7.)

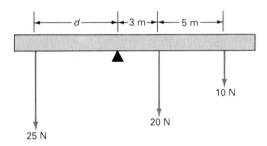

FIGURE 4.14

36. A space shuttle orbits the earth in a circular path of radius 6600 km at constant speed. If the period of its motion, the time to complete one orbit, is 98 min, what is the speed of the shuttle to the nearest kilometer per hour? (Approximate π by 3.1416.)

37. A sharpshooter heard her bullet strike a target 2 s after firing her rifle. If the bullet travels 580 m/s and sound travels 331 m/s, find the distance to the target. Give your answer to the nearest meter.

38. A student scored an 82 and an 88 on his first two math exams. What score does he have to get on his next exam so that the average of the three exams will be 90? (*Hint:* To find the average of three numbers, add the numbers and divide this sum by 3.)

39. On a certain day, temperature readings were taken on the hour, and the temperature rose 2° hourly from 3 A.M. until noon. If the average hourly temperature reading for this period was 26°C, what was the temperature at 8 A.M?

40. Assuming that the velocity of sound in air increases 60 cm/s for each degree Celsius of increase in temperature, estimate the velocity of sound in meters per second at 0°C from the following data. When the temperature is -3°C, a sound produced at A is heard at B after an interval of $5\frac{1}{5}$ s; when the temperature is 19°C, the interval is 5 s.

41. The water level in a certain reservoir is 6 m deep, but the level is sinking at the rate of 4 cm a day. The water in another reservoir is 4.92 m deep and is rising 5 cm a day. After how many days will the depths of the two reservoirs be the same, and what will this depth be?

42. A person rowed downstream for 10 km and then rowed upstream for the same period of time. However, the rower covered only 5 km going back. If the rate of the current of the stream was $1\frac{1}{4}$ km/h, find how fast the person can row in still water.

43. The top section of a transmitting antenna tower was blown over by the wind. This top section, still attached to the bottom section, touched the ground at a point 20 m from the base of the tower. If the top section was 5 m longer than the bottom section, how high was the original tower?

44. Two airplanes leave two airports that are 300 km apart at the same instant and fly toward each other. One plane flies 50 km/h faster than the other and the planes pass each other in 30 min. Find the speed of each plane.

45. A beam has weights of 50 N and 100 N attached to its ends. A pivot is located 10 cm from the center of the beam in the direction of the 100-N weight. If the beam balances on the pivot, how long is the beam?

46. In a manufacturing process, a solution which is 50% chemical A and 50% chemical B is to be added to 3000 gal of chemical A at a constant rate so that after 30 min chemical B is 20% of the mixture. At what constant rate, in gallons per minute, should the solution be added?

47. Figure 4.15 shows an arrangement of gears with gears B and C on a common shaft.* (The gears are shown as cylinders and the teeth on the gears are not indicated.) Gears A and B

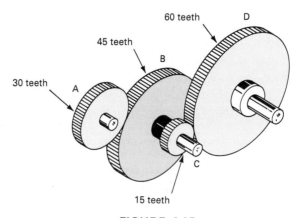

FIGURE 4.15

are meshed as are the pair C and D. Gear A has 30 teeth, B has 45 teeth, C has 15 teeth, and D has 60 teeth. For the pair of gears A and B, the following formula holds:

$$T_A \cdot \omega_A = T_B \cdot \omega_B,$$

where T_A and T_B are the number of teeth on A and B, respectively, and ω_A and ω_B are the number of revolutions per minute (rpm) made by A and B, respectively. (Here ω is the Greek lower case letter *omega* and is called the *angular velocity* of the gear.) A similar formula holds for gears C and D. Suppose the gear system (or *gear train*) is in an automobile transmission and gear A makes 3000 rpm. How many revolutions per minute does gear D make?

    * Dale Ewen and LeRoy Heaton, *Physics for Technical Education*, © 1981, p. 245. Reprinted by permission of Prentice Hall.

4.6 REVIEW

Important Terms and Symbols

Section 4.1   Equation, side of equation, variable, literal equation, arbitrary constant, conditional equation, identity, solution of equation, root of equation, solution set, impossible equation, empty set, ∅, equivalent equations, extraneous root.

Section 4.2   Linear (or first-degree) equation.

Section 4.3   Proportion, extremes, means, cross multiplication.

Useful Information

RULES FOR OBTAINING EQUIVALENT EQUATIONS

1. If the same number is *added* to (or *subtracted* from) both sides of an equation, then the resulting equation is equivalent to the original equation.

2. If both sides of an equation are *multiplied* (or *divided*) by the same **nonzero** number, then the resulting equation is equivalent to the original equation.

PROPORTION PROPERTY

$$\frac{a}{b} = \frac{c}{d} \quad \text{if and only if} \quad ad = bc$$

Review Questions

1. An equation that can be written in the form $ax + b = 0$, where $a \neq 0$, is called a(n) _____ equation.

2. In the equation $x + 3 = 3(x - 5)$ we call $x + 3$ the __(a)__ side and $3(x - 5)$ the __(b)__ side.

3. The equation $x + 4 = 5$ __(is)(is not)__ equivalent to $x + 6 = 7$. The equation $x + 5 = 7$ (a)

   __(is)(is not)__ equivalent to $2x + 10 = 9$. (b)

4. The equation $7 - 4x = 9$ is of the __(a)__ degree and its solution set is __(b)__.

5. With regard to being an identity or a conditional equation, we would classify the equation $2(x + 3) + 1 = 7 + 2x$ as a(n) _____.

6. If $p_1/p_2 = t_1/t_2$, then $t_2 = $ _____.

7. The number of roots of a linear equation is _____.

8. If the solution of $ax + b = 0$ is $x = -a$, then b must be equal to _____.

9. If the solution of $ax + b = 0$ is $x = 1/a$, then b must be equal to _____.

10. The statement "2 is to x as 3 is to 4" can be written algebraically as _____.

11. The statement $a:b = c:d$ is referred to as a(n) _____.

Answers to Review Questions

1. Linear, or first-degree.   **2.** (a) Left,   (b) Right.   **3.** (a) Is,   (b) Is not.

4. (a) First,   (b) $\{-\frac{1}{2}\}$.   **5.** Identity.   **6.** $\dfrac{t_1 p_2}{p_1}$.   **7.** One.   **8.** a^2.   **9.** -1.

10. $\dfrac{2}{x} = \dfrac{3}{4}$, or $2:x = 3:4$.   **11.** Proportion.

Review Problems

In Problems 1–18, solve the equations.

1. $4x + 1 = 3$.

2. $9x - 7 = 11$.

3. $5 = 8 - 2y$.

4. $6y = 0$.

5. $8 - \dfrac{4x}{3} = 10$.

6. $6x - \dfrac{1}{3} = 5$.

7. $\dfrac{3}{4} z + 2 = \dfrac{1}{3}$.

8. $\dfrac{1}{10} - \dfrac{2z}{5} = 4$.

9. $\dfrac{4x + 3}{4} = 7$.

10. $\dfrac{2}{5} = \dfrac{5 - 3x}{10}$.

11. $9(3u + 2) = 3 - (u + 7)$.

12. $4\left(u - \dfrac{5}{7}\right) = -3u$.

13. $\dfrac{3}{2}(x - 8) = 2x + 4$.

14. $3\{2x + 4[(7 - 2x) - 5x]\} = 0$.

15. $5[3 - 2(3x - 4)] = 18 - 9x$.

16. $7x - [8x + 4(x + 2)] = -(1 + x)$.

17. $\dfrac{2y - 7}{3} + \dfrac{8y - 9}{14} = \dfrac{3y - 5}{21}$.

18. $\dfrac{x}{2} + 1 = \dfrac{x}{3}$.

In Problems 19–22, use the given formulas to express the given symbols in terms of the remaining symbols.

19. $E = 4\pi k \dfrac{Q}{A}$; Q.

20. $\varepsilon_1 = i_2 R_1 + i_3 R_1 + i_2 R_2$; i_3.

21. $n - 1 = C + \dfrac{C'}{\lambda^2}$; C'.

22. $\sigma = \dfrac{n_0 - n_e}{\lambda} L$; n_0, n_e.

23. If 2 is to $x + 5$ as 4 is to $x - 5$, find x.

24. 28 is 35% of what number?

25. 96 is 80% of what number?

26. 162 is what percentage of 900?

27. 9 is what percentage of 75?

28. Symbolize the statement: In a circuit there are four times as many resistors (r) as capacitors (c).

29. A builder has a client who wants an L-shaped living and dining area in her new house. See Fig. 4.16. The area is to be a total of 385 square feet. What should the length l of the living area be?

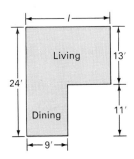

FIGURE 4.16

30. Into a graduated container two-thirds full, 10 L of fluid were poured and it was found to be five-sixths full. How many liters does the container hold?

31. A new insect spray is in the experimental stages. It contains the remarkable new "killer" ingredient K-57. A lab assistant has two spray formulas available: Formula A, of which 10% is K-57; and Formula B, of which 16% is K-57. So far, Formula A has proved to be too weak. On the other hand, Formula B seems too strong to be used near house pets. The lab assistant is told to mix Formula A with 400 mL of Formula B so that the result is 14% K-57. How many milliliters of Formula A should be used?

32. Suppose that the lab assistant in Problem 31 had needed exactly 500 mL of a 14% K-57 solution. How much of each formula would be used?

33. A construction firm has a government contract to build a swimming pool for the use of certain public officials. According to the contract, the pool must be completed within the next 21 d. The construction supervisor knows that the regular crew would take 45 d to build it. To meet the deadline, the supervisor decides to use a second crew, who can build the pool by themselves in 30 d. How long will it take both crews to construct the pool if they work together?

34. Suppose that a lunar rover traveled from point A to point B at 8 km/h and returned along the same path at 12 km/h. If the total time were 4 h, find the distance from A to B.

35. Suppose that 108 kg of a certain chemical compound must be made up, by mass, of one part chemical A, three parts chemical B, and five parts chemical C. How many kilograms of each chemical must be used?

36. After you stain furniture, *The Consumer's Handbook** urges that you protect the color by sealing the stain. "For a homemade sealer, mix one part shellac and five parts denatured alcohol." If you need a pint (16 fluid ounces) of sealer, how many fluid ounces of shellac and how many ounces of denatured alcohol are needed?

37. When the temperature of a rod of length l_0 is increased from T_0 to T, its length l is given by

$$l = l_0\{1 + \alpha(T - T_0)\},$$

where α is a constant. Solve for T_0.

* Paul Fargis, ed. (New York: Hawthorn, 1974), p. 90.

5

Functions and Graphs

5.1 FUNCTIONS

At the end of the seventeenth century, Gottfried Wilhelm Leibniz, one of the inventors of calculus, introduced the term *function* into the mathematical vocabulary. In addition to being essential to the study of calculus, the concept of a function has a role in science and technology in describing relationships between physical quantities.

Briefly, a function is a special type of input-output relationship in which one quantity (the *output*) depends on another quantity (the *input*). For example, when an object is dropped from a rest position, the distance s it falls (output) depends on time t (input). To express this dependence, we say that s is a *function of t*.

A functional relationship like this is often specified by a formula that shows what must be done to the input to find the output. For example, it can be shown that s is given in terms of t by the formula

$$s = 16t^2,$$

where s is in feet and t is in seconds. Thus

$$\text{If } t = 2, \quad \text{then} \quad s = 16(2)^2 = 64, \tag{1}$$

so corresponding to the input 2 (seconds) is the output 64 (feet).

We can think of the formula $s = 16t^2$ as defining a **rule**: Square t and multiply the result by 16. This rule assigns to each input number t exactly one output number s, and we call this rule a *function* in the following sense:

A **function** is a rule that assigns to each input number exactly one output number. The set of all input numbers to which the rule applies is called the **domain** of the function. The set of all output numbers is called the **range**.

For the distance function defined by the formula $s = 16t^2$, t cannot be negative because negative time makes no sense. Thus the domain consists of all nonnegative numbers, that is, all real numbers t such that $t \geq 0$. In (1) we found that when the input is 2, the output is 64. Thus 64 is in the range, which consists precisely of all output numbers.

A variable that represents input numbers for a function is called an **independent variable**. A variable that represents output numbers is called a **dependent variable** because its value *depends* on the value of the independent (input) variable. We say that the dependent variable is a *function of* the independent variable. In other words, output is a function of input. Thus with the formula $s = 16t^2$, the independent variable is t, the dependent variable is s, and s is a function of t.

As another example, the formula (or equation)

$$y = x + 2 \tag{2}$$

defines y as a function of x. For each value of x, the corresponding value of y is obtained by the following rule: Add 2 to x. That is, to the input x is assigned the output $x + 2$, which is y. Thus

$$\text{If } x = 1, \text{ then } y = 1 + 2 = 3;$$
$$\text{If } x = -6, \text{ then } y = -6 + 2 = -4.$$

The independent variable is x and the dependent variable is y.

Usually, letters such as f, g, h, F, and G are used to name functions. For example, we just saw that the formula $y = x + 2$ defines y as a function of x, where the rule is add 2 to the input. Suppose we give this rule the name f. Then we say that f is the function. To indicate that f assigns to the input 1 the output 3, we write

$$f(1) = 3,$$

where the symbol $f(1)$ is read f *of* 1. We call $f(1)$ the **value** of the function f at 1. Similarly, $f(-6) = -4$. More generally, if x is any input, we have the following notation:

The symbol $f(x)$, which is read f *of* x, denotes the output number that the function f assigns to the input number x in the domain of f.

Thus the output $f(x)$ is the same as y. That is, $y = f(x)$. However, because $y = x + 2$, we may simply write the formula

$$f(x) = x + 2.$$

To show you how to use this formula, we shall find $f(3)$, which is the output corresponding to the input 3. We simply replace each x in $f(x) = x + 2$ by 3:

$$f(3) = 3 + 2 = 5.$$

Likewise, we have the following **function values**:

$$f(8) = 8 + 2 = 10;$$
$$f(-4) = -4 + 2 = -2.$$

It is essential that you clearly understand the notation $f(x)$. The name of the rule (or function) is f, and $f(x)$ is the *output* that corresponds to the input x. Thus $f(x)$ does *not* mean f times x.

Quite often, functions are defined by *functional notation*. For example, the formula

$$g(x) = x^3 + x^2$$

defines the function g that assigns to the input x the output $x^3 + x^2$. That is, to apply g we add the cube and the square of an input number. A few sample function values are

$$g(2) = 2^3 + 2^2 = 8 + 4 = 12$$
$$g(-1) = (-1)^3 + (-1)^2 = -1 + 1 = 0$$
$$g(t) = t^3 + t^2$$
$$g(3z) = (3z)^3 + (3z)^2 = 27z^3 + 9z^2.$$

Note that $g(3z)$ was found by replacing each x in $x^3 + x^2$ by $3z$.

When we refer to the function g defined by the formula $g(x) = x^3 + x^2$, for convenience we shall feel free to call the formula itself a function. Thus we speak of "the function $g(x) = x^3 + x^2$," and similarly we refer to "the function $y = x + 2$."

Example 1

Given the function $f(x) = 2x + 3$, find the following function values.

a. $f(4)$.

Solution

$$f(4) = 2(4) + 3 = 8 + 3 = 11.$$

b. $f(x + h)$.

Solution   We replace x in $2x + 3$ by $x + h$.

$$f(x + h) = 2(x + h) + 3 = 2x + 2h + 3.$$

Note that to find $f(x + h)$ we do *not* write the function and add h. That is, $f(x + h) \neq f(x) + h$:

$$f(x + h) \neq 2x + 3 + h.$$

(Compare this with the correct answer just given.)

c. $f[f(x)]$.

Solution   The input for f is $f(x)$, or $2x + 3$. Thus

$$f[f(x)] = f[2x + 3] = 2(2x + 3) + 3$$
$$= 4x + 6 + 3 = 4x + 9.$$

Example 2

If $g(t) = t^2 - 3t + 1$, find the following function values.

a. $g(-2)$.

Solution   Here t is the independent variable. To find $g(-2)$ we replace each t in $t^2 - 3t + 1$ by -2.

$$g(-2) = (-2)^2 - 3(-2) + 1 = 4 + 6 + 1 = 11.$$

b. $g(t^2)$.

Solution   We replace each t in $t^2 - 3t + 1$ by t^2.

$$g(t^2) = (t^2)^2 - 3(t^2) + 1 = t^4 - 3t^2 + 1.$$

c. $g(2w)$.

Solution

$$g(2w) = (2w)^2 - 3(2w) + 1 = 4w^2 - 6w + 1.$$

d. $g\left(\dfrac{1}{t}\right)$.

Solution

$$g\left(\frac{1}{t}\right) = \left(\frac{1}{t}\right)^2 - 3\left(\frac{1}{t}\right) + 1 = \frac{1}{t^2} - \frac{3}{t} + 1.$$

Example 3

The length l of a certain rectangle is twice its width w (see Fig. 5.1). Express the perimeter of the rectangle as a function of its width.

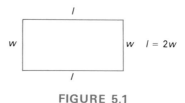

FIGURE 5.1

Solution   The perimeter P is the sum $l + w + l + w$, so

$$P = 2l + 2w.$$

We want P to be exclusively in terms of w. Since we are given that $l = 2w$, by substitution we have

$$P = 2(2w) + 2w = 6w.$$

Thus the formula $P = 6w$ expresses P as a function of w.

In Example 3 we expressed the perimeter P of a certain rectangle as a function of the width w: $P = 6w$. In technical work it is common to also use the letter P to name the function and, hence, indicate the functional dependence by writing

$$P = P(w).$$

Here the letter P is used two ways. On the right side, P represents the function. On the left side P represents the dependent variable. We could give the answer to the perimeter problem as

$$P(w) = 6w.$$

When a function is given by a formula, usually its domain is not stated. In such a case, we assume that **the domain consists of all real numbers for which that formula makes sense and gives function values that are real numbers**. For example, the domain of

$$f(x) = \frac{1}{x - 3}$$

consists of all real numbers except 3, because the expression $1/(x - 3)$ has a denominator of 0 for $x = 3$ but is a real number for all other values of x. In some cases the domain of a function is restricted for physical reasons. To illustrate, the previous distance function $s = 16t^2$ has domain $t \geq 0$ because t represents time.

Example 4

Find the domain of

$$F(t) = \sqrt{t}.$$

Solution The square root of a number is a real number only when the number is nonnegative. Thus the domain consists of all real numbers t such that $t \geq 0$.

Functions with particular forms have special names. In the next example we look at perhaps the simplest type of function there is: a *constant function*.

Example 5

The equation $g(x) = 2$ defines a *constant function*. For any input, the output is 2. The domain is all real numbers but the range consists only of 2. For example,

$$g(4.1) = 2$$

$$g(-420) = 2.$$

In general, a **constant function** has the form $g(x) = c$, where c is a constant.

A constant function belongs to a broader class of functions called *polynomial functions*. In general, a function of the form

$$f(x) = c_n x^n + c_{n-1} x^{n-1} + \cdots + c_1 x + c_0,$$

where n is a nonnegative integer and $c_n, c_{n-1}, \ldots, c_0$ are constants with $c_n \neq 0$, is called a **polynomial function** (in x). The number n is called the **degree** of the function, and c_n is the **leading coefficient**. Thus $f(x) = 3x^2 - 8x + 9$ is a polynomial function of degree 2 with leading coefficient 3. Likewise, $g(x) = 4 - 2x$ has degree 1 and leading coefficient -2. Polynomial functions of degree 1 or 2 are called **linear** or **quadratic functions**, respectively. Hence $g(x) = 4 - 2x$ is linear and $f(x) = 3x^2 - 8x + 9$ is quadratic. Note that a nonzero constant function, such as $f(x) = 5$ [which can be written as $f(x) = 5x^0$], is a polynomial function of degree 0. The constant function $f(x) = 0$ is also considered a polynomial function but has no degree assigned to it. The domain of any polynomial function is all real numbers.

Example 6

a. $f(x) = x^3 - 6x^2 + 7$ is a polynomial (function) of degree 3 with leading coefficient 1.

b. $g(x) = \dfrac{2x}{3}$ is a linear function with leading coefficient $\dfrac{2}{3}$.

c. $f(x) = \dfrac{2}{x^3}$ is *not* a polynomial function. Because $f(x) = 2x^{-3}$ and the exponent for x is not a nonnegative integer, this function does not have the proper form for a polynomial. Similarly, $g(x) = \sqrt{x}$ is not a polynomial because $g(x) = x^{1/2}$.

Example 7

The function $f(x) = |x|$ is called the **absolute value function**. Its domain consists of all real numbers. Some function values are

$$f(6) = |6| = 6$$
$$f(-2) = |-2| = 2$$
$$f(0) = |0| = 0.$$

Sometimes more than one equation is needed to define a function, as Example 8 shows.

Example 8

Let

$$f(x) = \begin{cases} x + 2, & \text{for } x \ge 1, \\ 5x, & \text{for } x < 1. \end{cases}$$

This is called a **compound function** because it is defined by more than one equation. The value of x determines which equation to use. Let us compute some function values.

$f(4)$:   Because $4 \ge 1$, we substitute 4 for x in $x + 2$.
$$f(4) = 4 + 2 = 6.$$
$f(1)$:   Because $1 \ge 1$, we substitute 1 for x in $x + 2$.
$$f(1) = 1 + 2 = 3.$$
$f(-4)$:   Because $-4 < 1$, we substitute -4 for x in $5x$.
$$f(-4) = 5(-4) = -20.$$

The function is defined for all values of x, so the domain consists of all real numbers.

Example 9

The voltage pulse V (in volts) produced by a signal generator is a function of time t (in seconds) and is given by

$$V = V(t) = \begin{cases} 2.4t, & \text{for } 0 \le t \le 1, \\ -4, & \text{for } 1 < t \le 2, \\ 2t + 6, & \text{for } 2 < t \le 4. \end{cases}$$

Find $V(\tfrac{1}{2})$, $V(2)$, and $V(3)$.

Solution

Because $0 \le \frac{1}{2} \le 1$, $V(\frac{1}{2}) = 2.4(\frac{1}{2}) = 1.2$.

Because $1 < 2 \le 2$, $V(2) = -4$.

Because $2 < 3 \le 4$, $V(3) = 2(3) + 6 = 12$.

Observe that the function is defined for all values of t between 0 and 4, inclusive, so the domain consists of all t where $0 \le t \le 4$.

If f is a function, any value of x for which $f(x) = 0$ is called a **zero** of the function. For example, we obtain zeros of $f(x) = 2x - 3$ by substituting 0 for $f(x)$ and solving for x.

$$2x - 3 = 0,$$

$$2x = 3,$$

$$x = \frac{3}{2}.$$

Thus the only zero of f is $\frac{3}{2}$. That is, $f(\frac{3}{2}) = 0$.

We have been using the term *function* in a restricted sense because, in general, the inputs or outputs do not have to be numbers. Moreover, a function does not always have to be defined by means of an equation. For example, a table of chemical elements and their abbreviations does not involve numbers and it assigns to each element its abbreviation (exactly one output). A function is implied here because there is a *correspondence*, whereby to each input there is assigned exactly one output. The next example shows another functional correspondence that is not given by an algebraic formula.

Example 10

The table in Fig. 5.2 gives the data obtained in the study of the elongation of a vertical coil spring. It gives a correspondence between the load F, in newtons, that is suspended on the

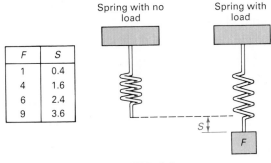

F	S
1	0.4
4	1.6
6	2.4
9	3.6

FIGURE 5.2

spring and the elongation S, in centimeters, that it produces. For example, a load of 1 N stretches the spring 0.4 cm. For each load there is exactly one elongation, and vice versa. If F is the independent variable, then S is a function of F, say

$$S = S(F),$$

where $S(F)$ is the elongation corresponding to a load of F. Thus

$$S(1) = 0.4, \qquad S(4) = 1.6, \qquad S(6) = 2.4, \quad \text{and} \quad S(9) = 3.6.$$

The domain consists of the numbers 1, 4, 6, and 9; the range consists of 0.4, 1.6, 2.4, and 3.6. Similarly, if S is the independent variable, then F is a function of S, $F = F(S)$, and $F(0.4) = 1$, and so on.

We conclude this section with some important remarks. First, the letters used to define a function are not important. For example, the function $f(z) = z^2$ is the same function as $g(x) = x^2$. In both cases the domain consists of all real numbers and the output is obtained by squaring the input.

Second, not all equations in x and y define y as a function of x. For example, let

$$y^2 = x.$$

If x is 9, then $y^2 = 9$, so y can be 3 or -3. Thus to the input 9 there are assigned not one but *two* output numbers, 3 and -3. This violates the definition of a function, so y is *not* a function of x.

Finally, some equations in two variables define either variable as a function of the other variable. For example, if

$$s = 2t,$$

then for each input t there is exactly one output, $2t$. Thus s is a function of t. However, solving the equation for t gives

$$t = \frac{s}{2}.$$

For each input s there is exactly one output, $s/2$. Thus t is a function of s. We note that Example 10 illustrated (by means of a table) how two variables simultaneously can be functions of each other.

Problem Set 5.1

In Problems 1–14, give the domain of each function.

1. $f(x) = \dfrac{3}{x}$.

2. $g(x) = \dfrac{4}{x^2}$.

3. $g(x) = \dfrac{x}{3}$.

4. $f(x) = \dfrac{x+2}{5}$.

5. $h(x) = \sqrt{6x}$.

6. $f(r) = 7r - 2$.

7. $H(z) = 10$.

8. $h(t) = (2t + 1)^2$.

9. $F(t) = 3t^2 + 5$.

10. $G(s) = \dfrac{s + 1}{s}$.

11. $H(x) = \dfrac{x}{x + 2}$.

12. $f(y) = \dfrac{8 - y}{4 - 7y}$.

13. $f(x) = \dfrac{3x - 1}{2x + 5}$.

14. $f(x) = \dfrac{1}{\sqrt{x}}$.

In Problems **15–32**, find the function values for each function.

15. $f(x) = 5x$; $f(0)$, $f(3)$, $f(-4)$.

16. $g(x) = 2x - 5$; $g(-1)$, $g(4)$, $g(-\frac{1}{2})$.

17. $h(t) = 4 - 3t$; $h(1)$, $h(-\frac{2}{3})$, $h(\frac{1}{2})$.

18. $H(s) = s^2 - 3$; $H(4)$, $H(\sqrt{2})$, $H(\frac{2}{3})$.

19. $f(x) = 7x$; $f(s)$, $f(t + 1)$, $f[f(x)]$.

20. $G(x) = 2 - x^2$; $G(-8)$, $G(u)$, $G(u^2)$.

21. $g(u) = u^2 + u$; $g(-2)$, $g(2v)$, $g(x^2)$.

22. $h(v) = \dfrac{1}{\sqrt{v}}$; $h(16)$, $h(\frac{1}{4})$, $h(1 - x)$.

23. $f(x) = 12$; $f(2)$, $f(t + 8)$, $f(-\sqrt{17})$.

24. $H(x) = (x + 4)^2$; $H(0)$, $H(2)$, $H(t - 4)$.

25. $f(x) = x^2 + 2x + 1$; $f(1)$, $f(-1)$, $f\left(\dfrac{x}{2}\right)$.

26. $f(x) = \dfrac{x^2}{x + 3}$; $f(0)$, $f(t^3)$, $f(xy)$.

27. $y = F(t) = \dfrac{t}{t - 3}$; $F(0)$, $F(-1)$, $F(t + 2)$.

28. $g(x) = |x - 3|$; $g(10)$, $g(3)$, $g(-3)$.

29. $F(t) = \begin{cases} 1, & \text{if } t > 0 \\ 0, & \text{if } t = 0; \\ -1, & \text{if } t < 0 \end{cases}$ $F(10)$, $F(-\sqrt{3})$, $F(0)$, $F(-\frac{18}{5})$.

30. $f(x) = \begin{cases} 4, & \text{if } x \geq 0 \\ 3, & \text{if } x < 0 \end{cases}$; $f(3)$, $f(-4)$, $f(0)$.

31. $G(x) = \begin{cases} x, & \text{if } x \geq 3 \\ 2 - x, & \text{if } x < 3 \end{cases}$; $G(8)$, $G(3)$, $G(-1)$, $G(1)$.

32. $h(r) = \begin{cases} 3r - 1, & \text{if } r > 2 \\ r^2 - 4r + 7, & \text{if } r < 2 \end{cases}$; $h(3)$, $h(-3)$, $h(2)$.

33. Suppose $f(x) = 2.41x^2 - 3.12x + 8.14$. (a) Find $f(2.3)$. (b) Find $2f(2.3)$. (c) Find $f[2(2.3)]$.

34. Suppose $f(x) = 0.4ax^2 + 0.3a^2x + 2.31$. (a) Find $f(3)$. (b) Find $f(3)$ if $a = -4.1$.

In Problems 35 and 36, find the function values for each function. Round your answers to three significant digits.

35. $f(x) = x^{2/3}$; $f(2), f(14), f(22.3)$.

36. $g(x) = \sqrt[8]{x - 3}$; $g(5), g(20), g(85.3)$.

In Problems 37–40, is y a function of x? Is x a function of y?

37. $y - 3x - 4 = 0$.

38. $x^2 + y = 0$.

39. $y = 7x^2$.

40. $x^2 + y^2 = 1$.

41. Express the perimeter P of a square as a function of the length of a side, l.

42. Express the area A of a square as a function of the length of a side, l.

43. Express the area A of a circle as a function of (a) its radius r; (b) its diameter d.

44. A solid cylinder has a radius of 2 cm and a height h, in centimeters. Express the total surface area A of the cylinder as a function of h.

45. Suppose a ball is thrown up from the ground and the equation $s = 14.7t - 4.9t^2$ gives the height s (in meters) of the ball after t seconds. (a) Find the heights when $t = 1$ and $t = 2$. (b) Is t a function of s? (c) Is s a function of t?

46. The distance s, in meters, that an object will fall from rest in a vacuum in t seconds is given by $s = f(t) = 4.9t^2$. Find $f(0), f(1)$, and $f(2)$. From a practical standpoint, what would you define the domain of f to be?

47. An automobile has an average fuel economy of 12.7 km/L. Express the number of liters n of gasoline used as a function of distance d traveled, in kilometers.

48. In Problem 47, express the number of liters n of gasoline used as a function of the distance d traveled, in *meters*.

49. A cylindrical storage tank with a radius of 2 m and height of 6 m is filled with an industrial solvent. As liquid is drained from the tank, the height of the liquid decreases at the rate of 1 m/h. Express the volume V of liquid in the tank as a function of the time t, in hours, that the liquid is drained.

50. The period T (in seconds) of a simple pendulum is given by

$$T(l) = 2\pi\sqrt{\frac{l}{g}},$$

where l is the length of the pendulum in meters and g is the acceleration due to gravity, which is taken to be 9.8 m/s². Find $T(2)$ to two decimal places.

51. A metal plate that is to be used in the construction of an electromechanical device is shown in Fig. 5.3. Express the area A of the metal surface as a function of a. Give your answer in simplest form.

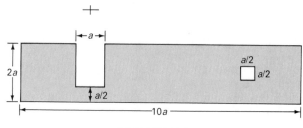

FIGURE 5.3

52. An open-top box is to be made from a 12-in. square piece of sheet metal by cutting out equal squares from each corner and then folding up the sides. (See Fig. 5.4.) Express the volume V of the box as a function of the length x of a side of the squares cut out.

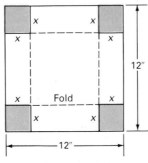

FIGURE 5.4

53. The acceleration g due to gravity, in meters per second squared, at an altitude h meters above the earth's surface is given, approximately by

$$g(h) = \frac{3.98 \times 10^{14}}{[(6.37 \times 10^6) + h]^2}.$$

(a) What is the approximate value of g at the earth's surface?
(b) What is the approximate value of g at a distance of 3.7×10^8 m from the earth's surface? (That would be near the moon!)
(c) Consider the values of g as h gets large. Based on your observation, would you ever expect a spaceship to escape *entirely* from the earth's influence?

54. In a laboratory experiment, a potential difference of 10 V is applied across an initially uncharged capacitor in series with a resistor, as shown in Fig. 5.5. The readings obtained

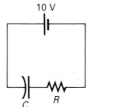

t (sec)	V (volts)
0	0
3	6.3
6	8.7
9	9.5
15	9.9

FIGURE 5.5

for the potential difference V, in volts, across the capacitor at various times t, in seconds, are given in the table. For each time there corresponds exactly one potential difference. Thus this table defines V as a function of t, $V = V(t)$.
(a) List the numbers in the domain of V.
(b) Find $V(6)$ and $V(15)$.

In Problems **55** *and* **56**, *find the zeros of the given function.*

55. $f(x) = 4x - 5$.

56. $f(t) = \frac{3}{4} - 2t$.

5.2 GRAPHS IN RECTANGULAR COORDINATES

A **rectangular** (or *Cartesian*) **coordinate system** allows us to specify and locate points in a plane. It also provides a geometric way to represent equations (in two variables) as well as functions. We obtain such a system as follows.

In a plane two real number lines, called **coordinate axes**, are constructed perpendicular to each other (one horizontal and one vertical) so that their origins coincide (see Fig. 5.6). Their point of intersection is called the **origin** of the coordinate

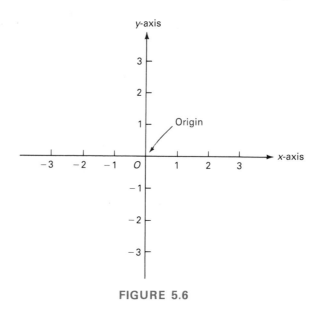

FIGURE 5.6

system and is denoted by the letter *O*. The positive numbers on the horizontal axis, commonly called the ***x*-axis**, are to the *right* of the origin. The positive numbers on the vertical axis, called the ***y*-axis**, are *above* the origin. [Depending on the situation, we use variables other than *x* and *y* to label the axes (see Example 1).] The unit distance on the *y*-axis need not necessarily be the same as that on the *x*-axis. You will see later (Example 4) that it can be convenient to use a different unit on the *y*-axis.

The plane on which the coordinate axes are placed is called a **rectangular coordinate plane** or, more simply, an ***xy*-plane**. Each point in the *xy*-plane can be assigned a pair of numbers to indicate its position. To do this for point *P* in Fig. 5.7(a), we draw perpendiculars from *P* to the *x*- and *y*-axes. They meet the *x*-axis at 3 and the *y*-axis at 2. The number 3 is called the ***x*-coordinate** or **abscissa** of *P*, and 2 is called the ***y*-coordinate** (or **ordinate**) of *P*. We say that the **rectangular coordinates** of *P* are given by the **ordered pair** (3, 2). The word *ordered* is important in the sense that the *x*-coordinate is *always* written before the *y*-coordinate. For example, note that in Fig 5.7(b) the point with coordinates (2, 3) is *not* the same as the point with coordinates (3, 2). That is,

$$(3, 2) \neq (2, 3).$$

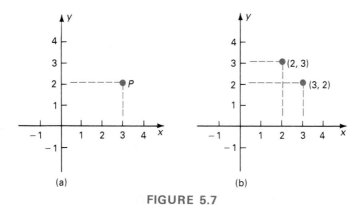

FIGURE 5.7

In general, if P is any point in the xy-plane, then its rectangular coordinates are given by an ordered pair of the form (x, y) [see Fig. 5.8].

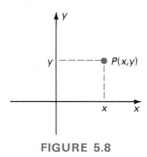

FIGURE 5.8

Thus with each point in a given coordinate plane, we can associate exactly one ordered pair (x, y) of numbers. On the other hand, it should be clear that with each ordered pair of numbers, we can associate exactly one point in that plane. Because of this *one-to-one* correspondence between the points in the plane and all ordered pairs of numbers, we shall refer to a point P with abscissa x and ordinate y simply as the point (x, y), or as $P(x, y)$. Moreover, we shall feel free to use the words *point* and *ordered pair* interchangeably.

In Fig. 5.9 we have located, or *plotted*, various points in the xy-plane. For example, to plot the point $(1, -4)$, we first move one unit to the right of the origin along the x-axis. Then we move four units downward from the x-axis. We have labeled this point with its coordinates. Notice the following:

- The origin has coordinates $(0, 0)$.
- Any point on the x-axis has a y-coordinate of 0.
- Any point on the y-axis has an x-coordinate of 0.

The coordinate axes divide the plane into four regions called **quadrants**. They are numbered in a counterclockwise order, as in Fig. 5.10. Quadrant I, or the *first*

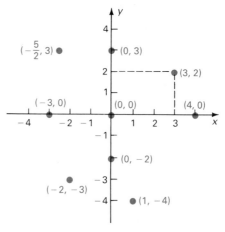

FIGURE 5.9

quadrant, consists of all points with positive coordinates. The second quadrant consists of all points with a negative x-coordinate and a positive y-coordinate, and so on. A point on an axis does not lie in any quadrant.

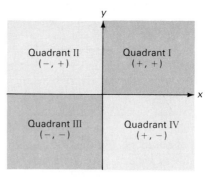

FIGURE 5.10

Using a rectangular coordinate system, we can geometrically represent the solutions of an equation in two variables. For example, let us consider the equation

$$y = 2x + 1. \qquad (1)$$

A solution is a *pair* of numbers: a value of x and a value of y that make the equation true. To find solutions we shall assign arbitrary values to x and determine corresponding values of y. For example,

$$\text{If } x = 1, \quad \text{then} \quad y = 2(1) + 1 = 3.$$

Thus one solution is $x = 1$, $y = 3$, which we represent by the ordered pair $(1, 3)$. The point corresponding to this is plotted in Fig. 5.11(a). Similarly,

$$\text{If } x = -1 \quad \text{then} \quad y = 2(-1) + 1 = -1.$$

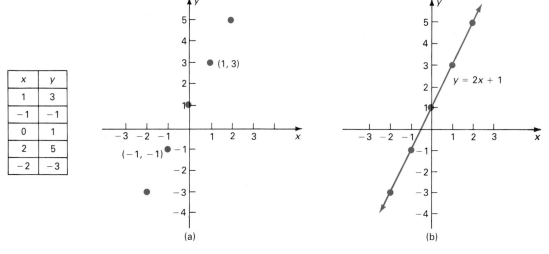

x	y
1	3
−1	−1
0	1
2	5
−2	−3

FIGURE 5.11

This solution is represented by the ordered pair $(-1, -1)$, which is plotted in Fig. 5.11(a). Figure 5.11 shows a table listing other solutions, and the corresponding points in the plane are shown in Fig. 5.11(a). If we could plot all possible solutions we would have the *graph* of the equation.

The **graph of an equation** in two variables consists of all points in the plane whose coordinates satisfy that equation.

Because Eq. 1 has infinitely many solutions, it seems impossible to determine its graph precisely. However, we are concerned only with the graph's general shape. For this reason we have to plot only enough points so that we may intelligently guess its proper shape. Then we join those points by a smooth curve wherever conditions permit. This process is called *sketching the graph of the equation,* or *graphing the equation.* We start with the point having the least x-coordinate, namely $(-2, -3)$, and progress through the points having increasingly larger x-coordinates. We finish with the point having greatest x-coordinate, namely, $(2, 5)$ [see Fig. 5.11(b)]. Observe that the points appear to lie on a straight line. Here we assume that the graph extends indefinitely, which may be indicated by arrows. At times we refer to the graph of an equation in two variables as a *curve*; thus we speak of the curve $y = 2x + 1$.

Let us now graph the equation

$$y = x^2 + 2x - 3.$$

If $x = 1$, then $y = 0$; if $x = -2$, then $y = -3$. Thus we get the points $(1, 0)$ and $(-2, -3)$, which are plotted in Fig. 5.12. At this stage you should not hastily conclude that the graph is a straight line. Without a guarantee that the graph is a

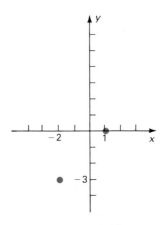

FIGURE 5.12

straight line, two points are *never* sufficient for a graph. We must plot as many points as necessary to make the general behavior of the graph reasonably apparent. Figure 5.13 gives a table of x- and y-values, and Fig. 5.13(a) shows the corresponding points in the plane. Although the more points we plot, the better our graph is, these points give us a good idea of the graph's general shape. A sketch of (a portion of) the graph appears in Fig. 5.13(b).

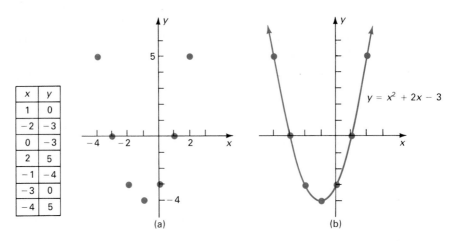

x	y
1	0
-2	-3
0	-3
2	5
-1	-4
-3	0
-4	5

$y = x^2 + 2x - 3$

(a)   (b)

FIGURE 5.13

Example 1

Graph the equation $s = t^3 + 1$.

Solution Using t for the horizontal axis and s for the vertical axis in a ts-plane, we get Fig. 5.14. Although the points $(-\frac{1}{4}, \frac{63}{64})$ and $(\frac{1}{4}, \frac{65}{64})$ in the table are difficult to show in the graph, their location reassures us that we have the proper shape of the curve.

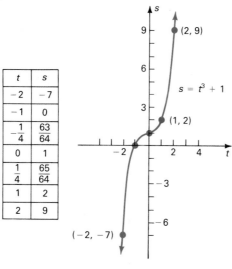

t	s
−2	−7
−1	0
−$\frac{1}{4}$	$\frac{63}{64}$
0	1
$\frac{1}{4}$	$\frac{65}{64}$
1	2
2	9

FIGURE 5.14

Example 2

Graph the equation $x = 2$.

Solution   We can think of this as an equation in x and y if we write it as $x = 2 + 0y$. Here y can be any value, but x must be 2 (see Fig. 5.15). The graph is a vertical line, that is, a line parallel to the y-axis.

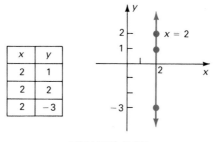

x	y
2	1
2	2
2	−3

FIGURE 5.15

Functions can be graphed in a coordinate plane. If f is a function with independent variable x and dependent variable y, then the graph of f is simply the graph of the equation $y = f(x)$.

> The **graph of a function** f consists of all points $(x, f(x))$, where x is in the domain of f.

The vertical axis can be labeled either y or $f(x)$ and is referred to as the **function-value axis**. *We always label the horizontal axis with the independent variable.*

Example 3

Sketch the graph of the absolute value function $f(x) = |x|$.

Solution   We label the vertical axis as $f(x)$ [see Fig. 5.16]. Each point on the graph has the form $(x, f(x))$. For example, the point $(-3, 3)$ means that $f(-3) = 3$.

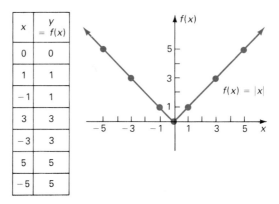

x	$y = f(x)$
0	0
1	1
-1	1
3	3
-3	3
5	5
-5	5

FIGURE 5.16

Example 4

a. Graph the function $z = f(r) = \dfrac{100}{r}$.

Solution   We use the independent variable r to label the horizontal axis. The vertical axis can be labeled either z or $f(r)$.   When $r = 0$ this function is not defined, so there is no corresponding point in the rz-plane (see Fig. 5.17). Notice that our choice of the unit distance on each axis makes the graphing easy to handle.

b. The electric potential V, in volts, at a distance r, in meters, from a small object with an electric charge of $\frac{1}{9} \times 10^{-7}$ C is given by

$$V(r) = \frac{100}{r}.$$

Sketch the graph of this function.

Solution   Observe that this function is identical in form to that in Part (a). Here, however, the variable r has a physical meaning—it represents a measurable distance. For this reason the values of r must be restricted to positive values. The graph of this electric potential function is identical to the portion of the graph in Part (a) that lies in the first quadrant.

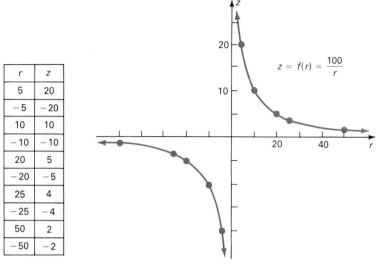

r	z
5	20
−5	−20
10	10
−10	−10
20	5
−20	−5
25	4
−25	−4
50	2
−50	−2

FIGURE 5.17

Example 5

The current i in a 1-Ω resistor as a function of the power P developed in the resistor is given by

$$i = i(P) = \sqrt{P}.$$

Sketch the graph of this function.

Solution   We use the independent variable P to label the horizontal axis. The vertical axis can be labeled either $i(P)$ or i (see Fig. 5.18). Recall (from Sec. 2.4) that $\sqrt{P}$ denotes the *principal* square root of P. Thus $i(9) = \sqrt{9} = 3$, not -3. Also, the domain is all $P \geq 0$, because the square root of a negative number is *not* a real number.

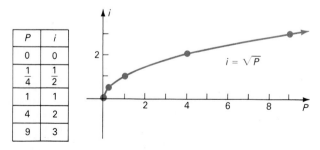

P	i
0	0
$\frac{1}{4}$	$\frac{1}{2}$
1	1
4	2
9	3

FIGURE 5.18

Example 6

One automobile in a 5-s drag race is given a handicap by being allowed to start at a point ahead of the starting line. Its distance d, in meters, from the starting line is given by $d = d(t) = 3t^2 + 5$, where t is elapsed time in seconds. Sketch the graph of this function.

Solution See Fig. 5.19. Notice that we have used only nonnegative values for time because negative time has no meaning. Also, we have $0 \leq t \leq 5$.

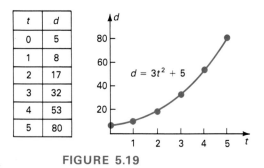

t	d
0	5
1	8
2	17
3	32
4	53
5	80

$d = 3t^2 + 5$

FIGURE 5.19

Example 7

a. For the spring in Fig. 5.2 (see page 189), the elongation S is a function of load F and is given by $S = 0.4F$. The graph of this function is given in Fig. 5.20. The independent variable F represents the load on the spring and must be positive or zero. Thus we restrict the graph to nonnegative values of F (such that the spring does not stretch out of shape).

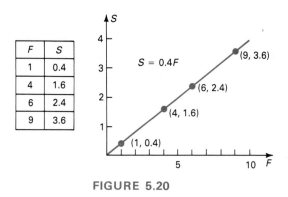

F	S
1	0.4
4	1.6
6	2.4
9	3.6

$S = 0.4F$

(9, 3.6)
(6, 2.4)
(4, 1.6)
(1, 0.4)

FIGURE 5.20

b. In Fig. 5.5 (see page 193) a table was given in which the potential difference V across a capacitor was a function of time t. That table, repeated here, gives the following time-volt

t (seconds)	0	3	6	9	15
V (volts	0	6.3	8.7	9.5	9.9

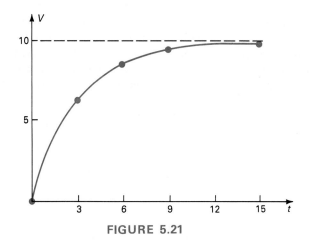

FIGURE 5.21

pairs: $(0, 0)$ $(3, 6.3)$, $(6, 8.7)$, $(9, 9.5)$, and $(15, 9.9)$. In Fig. 5.21 we have plotted each pair. We can approximate points in between the data by connecting the data points with a smooth curve. By observing the graph, we note that as time increases, the voltage increases, getting closer and closer to 10. That is, as t gets large the curve "settles down" near the dashed line. Such a line is called a **horizontal asymptote** for the curve. It is *not* part of the graph.

Example 8

Sketch the graph of the compound function

$$f(x) = \begin{cases} x, & \text{if } 0 \leq x < 3, \\ x - 1, & \text{if } 3 \leq x \leq 5, \\ 4, & \text{if } 5 < x \leq 7. \end{cases}$$

Solution The domain of f is $0 \leq x \leq 7$. The graph is given in Fig. 5.22. The *hollow dot* means that the point is *not* included in the graph.

x	f(x)
0	0
1	1
2	2
3	2
4	3
5	4
6	4
7	4

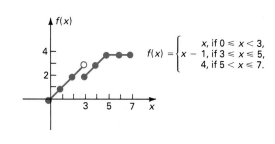

FIGURE 5.22

There is a simple technique, called the **vertical-line test**, to tell whether a curve in the xy-plane is the graph of a function. If a *vertical* line L can be drawn that intersects a curve in at least two points, then the curve is *not* the graph of a function of x. Such a condition implies the existence of an input number x with more than one output number y. However, if no such vertical line can be drawn, then the curve is the graph of a function of x. Thus the graphs in Fig. 5.23 do not represent functions of x, but those in Fig. 5.24 do.

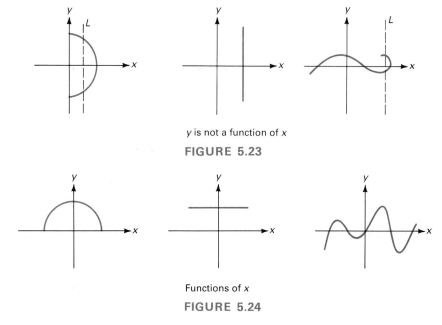

y is not a function of x

FIGURE 5.23

Functions of x

FIGURE 5.24

Example 9

Sketch the graph of $x = 2y^2$.

Solution   Here it is convenient to choose values of y and then find the corresponding values of x. See Fig. 5.25. Using the vertical-line test, you can see that the equation does *not* define a function of x.

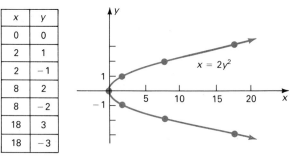

x	y
0	0
2	1
2	−1
8	2
8	−2
18	3
18	−3

$x = 2y^2$

FIGURE 5.25

There is a connection between the real zeros of a function and the graph of the function. The real zeros are the x-coordinates of the points where the graph meets the x-axis. These x-coordinates are called *x-intercepts* of the graph. To illustrate, in Fig. 5.26 the graph of $y = x^2 - 2x - 3$ meets the x-axis at the points $(-1, 0)$ and $(3, 0)$. Thus -1 and 3 are zeros of $f(x) = x^2 - 2x - 3$. That is, $f(-1) = 0$ and $f(3) = 0$.

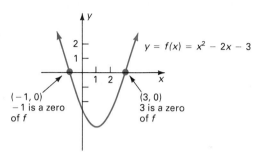

FIGURE 5.26

Problem Set 5.2

In Problems **1** *and* **2***, locate and label each of the given points and state, if possible, the quadrant in which each point lies.*

1. $(2, 7)$, $(8, -3)$, $(-\frac{1}{2}, -2)$, $(0, 0)$.   2. $(-4, 5)$, $(3, 0)$, $(1, 1)$, $(0, -6)$.

3. The following results were obtained in an experiment to measure the current i (in amperes) as a function of time t (in seconds) in an electrical circuit containing a resistor and an inductor:

t	0	2	4	6	8	10	12
i	0	6.3	8.7	9.5	9.8	9.9	10.0

Represent these data graphically and connect the points by a smooth curve.

4. In a certain experiment, a mass m was suspended from a vertical spring and allowed to come to rest. It was then pulled down 10 cm below its equilibrium position and released. The position s (in centimeters) of the mass as a function of time t (in seconds) was found to be as follows:

t	0	0.5	1	1.5	2	2.5	3	3.5	4
s	-10	-8	0	8	10	8	0	-8	-10

Represent these data graphically and connect the points by a smooth curve. Based on the data, can we say that t is a function of s?

*In Problems **5–38**, sketch the graph of the given equation or function. For any equations in x and y, indicate those for which y is **not** a function of x.*

5. $y = x$.

6. $y = 3x$.

7. $y = -2x + 1$.

8. $y = 3x - 5$.

9. $s = f(t) = t^2$.

10. $z = f(x) = 3x^2 - 1$.

11. $f(x) = -x^2 + 2$.

12. $f(x) = \frac{1}{2}x^2 + 4$.

13. $f(x) = |x| + 1$.

14. $f(x) = x^2 + x + 1$.

15. $h(w) = w^2 - 4w + 1$.

16. $g(t) = 2\sqrt{t}$.

17. $y = x^3$.

18. $y = x^3 - 2x^2 + x$.

19. $g(x) = x^3 - 3x$.

20. $x = -4$.

21. $f(z) = \dfrac{1}{z}$.

22. $f(t) = t(2 - t)$.

23. $y = \sqrt{4 - x^2}$.

24. $y = -8$.

25. $F(w) = 1 + w$, where $1 \le w \le 4$.

26. $f(x) = \dfrac{2}{x - 4}$.

27. $y = g(x) = 2$.

28. $y = |x - 3|$.

29. $x = 0$.

30. $x + y = 1$.

31. $x = -3y^2$.

32. $s = f(t) = \sqrt{t - 5}$.

33. $2x + y - 2 = 0$.

34. $f(x) = -2.1x^2 + 4.2x + 3.5$.

35. $f(x) = \begin{cases} x, & \text{if } 0 < x \le 3, \\ 4, & \text{if } 3 < x \le 5, \\ -1, & \text{if } x > 5. \end{cases}$

36. $f(x) = \begin{cases} 2x, & \text{if } x \le 2, \\ 4, & \text{if } x > 2. \end{cases}$

37. $F(x) = \begin{cases} x^2, & \text{if } 0 \le x \le 2, \\ 2x, & \text{if } 2 < x \le 3, \\ 6, & \text{if } x > 3. \end{cases}$

38. $g(x) = \begin{cases} x + 6, & \text{if } x \ge 3, \\ x^2, & \text{if } x < 3. \end{cases}$

39. Figure 5.27(a) shows the graph of $y = f(x)$.
  (a) Estimate $f(0)$, $f(2)$, $f(4)$, and $f(-2)$.
  (b) What is the domain of f?
  (c) What is the range of f?
  (d) What is a real zero of f?

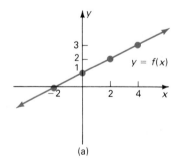

(a)

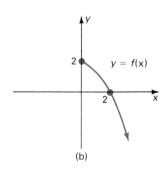

(b)

FIGURE 5.27

40. Figure 5.27(b) shows the graph of $y = f(x)$.
    (a) Estimate $f(0)$ and $f(2)$.
    (b) What is the domain of f?
    (c) What is the range of f?
    (d) What is a real zero of f?
41. Figure 5.28(a) shows the graph of $y = f(x)$.
    (a) Estimate $f(0)$, $f(1)$, and $f(-1)$.
    (b) What is the domain of f?
    (c) What is the range of f?

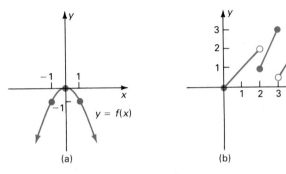

(a) (b)

FIGURE 5.28

42. Figure 5.28(b) shows the graph of $y = f(x)$.
    (a) Estimate $f(0)$, $f(2)$, $f(3)$, and $f(4)$.
    (b) What is the domain of f?
    (c) What is the range of f?
43. An automobile is moving in a straight line at a speed of 20 m/s. The brakes are applied and the vehicle is brought uniformly to rest. The speed v of the automobile is given by

$$v = f(t) = 20 - 2t,$$

where t is the time elapsed, in seconds, after the brakes are applied. Sketch the graph of this function for $0 \le t \le 10$.

44. In Problem 43, the distance d moved by the automobile, in meters, after t seconds is given by

$$d = d(t) = 20t - t^2.$$

Sketch the graph of this function for $0 \le t \le 10$.

45. For a tuned electronic circuit, the output voltage V (in volts) as a function of frequency f (in kilohertz) was observed to be as follows:

f	1300	1325	1375	1400	1425	1475	1500
V	0.01	0.03	0.25	0.95	0.35	0.10	0.05

Represent the given data graphically and connect the points by a smooth curve.

46. While monitoring a radioactive sample with a detection instrument, a technician observed the disintegration rate R (in counts per minute) as a function of time t (in hours) to be as follows:

t	R	t	R
0.5	9535	6.0	1800
1.0	8190	7.0	1330
1.5	7040	8.0	980
2.0	6050	9.0	720
3.0	4465	10.0	535
4.0	3300	11.0	395
5.0	2430	12.0	290

Represent the given data graphically and connect the points by a smooth curve.

47. If a horizontal force F acts on a 20-N block resting on a horizontal surface and if the coefficients of static and kinetic friction are 0.3 and 0.1, respectively, then the frictional force f (in newtons) acting on the block can be written as a function of F defined by

$$f = g(F) = \begin{cases} F, & \text{if } 0 \leq F \leq 6, \\ 2, & \text{if } F > 6. \end{cases}$$

Sketch the graph of g for values of F from 0 to 10 N.

48. The speed v of sound in air as a function of the temperature T (in °C) of the air is given by

$$v = f(T) = 331\sqrt{\frac{T + 273}{273}},$$

where v is in meters per second. Sketch the graph of this function for $0 \leq T \leq 100$.

49. The branch voltage V in a circuit as a function of time t is given by the function

$$V = V(t) = \begin{cases} 2t, & \text{if } 0 \leq t \leq 1, \\ 4 - 2t, & \text{if } 1 \leq t \leq 2, \\ 0, & \text{if } t > 2. \end{cases}$$

Sketch the graph of this function. Based on the graph, can we say that t is a function of V?

50. The stopping distance d of an automobile as a function of its initial speed v is given by

$$d = f(v) = \frac{v^2}{2.5},$$

where v is in meters per second and d is in meters.

(a) Sketch the graph of this function.

(b) What happens to the stopping distance if the initial speed is doubled? *Hint*: Express $f(2v)$ in terms of $f(v)$. (The result should caution all drivers!)

*In Problems **51** and **52**, find zeros of the given functions by graphical means.*

51. $f(x) = x^2 + 2x$.

52. $f(x) = x^2 - x - 2$.

53. A signal generator produces a square-wave pulse V, in volts, given by

$$V = f(t) = \begin{cases} 2.3, & \text{for } 0 \le t < 1, \\ -2.3, & \text{for } 1 \le t < 2, \\ 0, & \text{for } t \ge 2. \end{cases}$$

Sketch the graph of this function.

54. Which of the graphs in Fig. 5.29 represent functions of x?

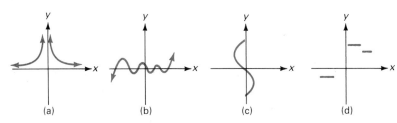

(a) (b) (c) (d)

FIGURE 5.29

5.3 REVIEW

Important Terms and Symbols

Section 5.1 Function, domain, range, independent variable, dependent variable, $f(x)$, function value, functional notation, constant function, polynomial function, compound function, zero of a function.

Section 5.2 Coordinate axes, origin, x-axis, y-axis, rectangular coordinate plane, xy-plane, x-coordinate, abscissa, y-coordinate, ordinate, rectangular coordinates, ordered pair, (x, y), quadrant, graph of an equation, graph of a function, function-value axis, horizontal asymptote, vertical-line test.

Review Questions

1. If f is a function, the set of all input numbers is called the __(a)__ of f. The set of all output numbers is the __(b)__ of f.

2. The domain of the function $f(x) = x + \dfrac{1}{x}$ consists of all real numbers except _____.

3. The sign of the abscissa of a point in the third quadrant is _____.

4. The point three units to the right of the y-axis and two units below the x-axis has coordinates _____.

5. If $f(x) = -x^2 - 1$, then $f(-1) = $ __(a)__. If $g(x) = 2$, then $g(3) = $ __(b)__.

6. The point $(2, -6)$ lies in quadrant __(a)__, whereas the point $(-2, 6)$ lies in quadrant __(b)__.

7. The abscissa of the point $(2, 3)$ is __(a)__ and its ordinate is __(b)__.

8. A variable representing input numbers of a function is called a(n) __(dependent)__ __(independent)__ variable.

9. Which of the graphs in Fig. 5.30 represent functions of x? _____ .

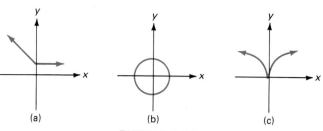

(a)   (b)   (c)

FIGURE 5.30

Answers to Review Questions

1. (a) Domain, (b) Range. 2. Zero. 3. Negative. 4. $(3, -2)$. 5. (a) -2, (b) 2.
6. (a) IV, (b) II. 7. (a) 2, (b) 3. 8. Independent. 9. a and c.

Review Problems

In Problems 1-6, give the domain of the function.

1. $f(x) = x^2 + x$.

2. $g(x) = 7 - x$.

3. $F(t) = \dfrac{t}{(t - 1)(t - 2)}$.

4. $G(x) = \dfrac{\sqrt{x}}{x - 1}$.

5. $h(x) = 4$.

6. $f(s) = \dfrac{\sqrt{s - 1}}{2}$.

In Problems 7-14, find the function values for the given function.

7. $f(x) = 3x^2 - 4x + 7$; $f(0)$, $f(-3)$, $f(5)$, $f(t)$.

8. $g(x) = 4$; $g(4)$, $g(\frac{1}{100})$, $g(-156)$, $g(x + 4)$.

9. $G(x) = \sqrt{x - 1}$; $G(1)$, $G(10)$, $G(t + 1)$, $G(x^2)$.

10. $F(x) = \dfrac{x - 3}{x + 4}$; $F(-1)$, $F(0)$, $F(5)$, $F(x + 3)$.

11. $h(u) = \dfrac{\sqrt{u + 4}}{u}$; $h(5)$, $h(-4)$, $h(x)$, $h(u - 4)$.

12. $H(s) = \dfrac{(s - 4)^2}{3}$; $H(-2)$, $H(7)$, $H(\frac{1}{2})$, $H(x^2)$.

13. $f(x) = \begin{cases} 4, & \text{if } x < 2 \\ 8 - x^2, & \text{if } x > 2 \end{cases}$; $f(4)$, $f(-2)$, $f(0)$, $f(10)$.

14. $h(q) = \begin{cases} q, & \text{if } -1 \le q < 0 \\ 3 - q, & \text{if } 0 \le q < 3 \\ 2q^2, & \text{if } 3 \le q \le 5 \end{cases}$; $h(0)$, $h(4)$, $h(-\frac{1}{2})$, $h(\frac{1}{2})$.

In Problems **15** *and* **16**, *find* $f(x + h)$ *and simplify your answer.*

15. $f(x) = 3 - 7x.$

16. $f(x) = x^2 + 4.$

In Problems **17–30**, *graph the given equation or function.*

17. $y = -3x + 4.$

18. $x = y^2.$

19. $f(t) = 5t - 4.$

20. $y = f(x) = -1 - x.$

21. $f(x) = 5.$

22. $x = 4.$

23. $2x = y - 4.$

24. $3x + 2y - 6 = 0.$

25. $z = F(w) = \dfrac{1}{w^2}.$

26. $f(p) = \sqrt{p + 3}.$

27. $y = x^2 - 4.$

28. $y = -x^2 + 2x - 1.$

29. $f(x) = \begin{cases} 1 - x, & \text{if } x \le 0, \\ 1, & \text{if } x > 0. \end{cases}$

30. $f(x) = \frac{1}{2}|x|.$

31. A small aircraft weighs 17,800 N at takeoff. If the plane burns 1200 N of fuel each hour, express the weight W of the plane as a function of the time of flight t, in hours.

6

Trigonometric Functions of an Acute Angle

6.1 ANGLES AND ANGULAR MEASUREMENT

Trigonometry is concerned in part with the relationships between the angles and sides of triangles and in part with functions based on these relationships. It is an indispensable mathematical tool in many areas of science and engineering. For example, it is used in surveying, mechanics, electricity, optics, and the study of vibrating objects such as springs, strings, and membranes.

You may already be familiar with the advantages of trigonometric measurement. For instance, with trigonometry we can measure the height of a cliff without ever climbing it, and we can measure the width of a river without ever crossing it.

Our study of trigonometry starts with descriptions of angles and angular measurement. In the next section we shall define six special functions, called *trigonometric functions*, whose inputs are angles.

To begin, we consider the *half-line OA* shown in Fig. 6.1. The term **half-line**

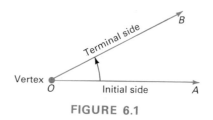

FIGURE 6.1

refs to that part of a straight line which extends indefinitely to one side of a point O on the line. If, in a plane, this half-line is rotated about its *endpoint* O to the position of OB, then an **angle** is said to be generated. We call OA the **initial side** of the angle, OB the **terminal side**, and point O the **vertex**.

We often denote angles by Greek letters such as θ (theta), α (alpha), β (beta), and γ (gamma). If an angle θ is generated by a counterclockwise rotation, then θ is called a **positive angle** [see Fig. 6.2(a)]. When the rotation is clockwise, θ is a **negative angle** [see Fig. 6.2(b)]. We use arrows to show the direction of rotation.

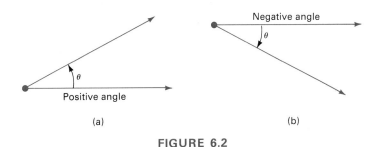

(a)   (b)

FIGURE 6.2

When an angle θ has its vertex at the origin of a rectangular coordinate system and its initial side lies on the positive x-axis, we say that the angle is in **standard position**. Some angles in standard position are shown in Fig. 6.3. If θ is in standard

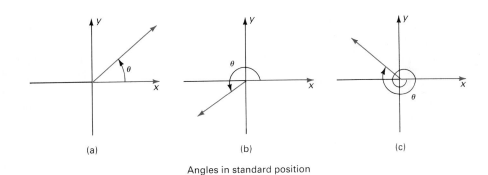

(a)   (b)   (c)

Angles in standard position

FIGURE 6.3

position and its terminal side lies in the first quadrant, then θ is called a **first-quadrant angle**. Similarly, there are **second-**, **third-**, and **fourth-quadrant angles**. Figure 6.3 shows a first-quadrant angle in (a), a third-quadrant angle in (b), and a second-quadrant angle in (c).

An angle in standard position whose terminal side lies on an axis is called a **quadrantal angle**. The angles in Fig. 6.4 are quadrantal. In (d), notice that the initial and terminal sides of θ coincide.

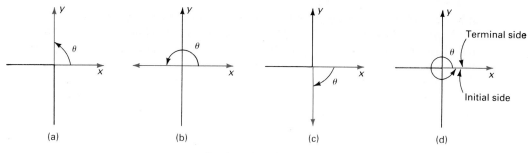

Quadrantal angles

FIGURE 6.4

When no confusion results, an angle may be referred to by the letter associated with its vertex. In Fig. 6.5(a) the angle can be called angle A, which is sometimes

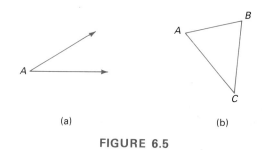

(a)   (b)

FIGURE 6.5

written with the *angle symbol* $\angle$ as $\angle A$. Similarly, in Fig. 6.5(b) the three angles of the triangle are angle A, angle B, and angle C. We can also name an angle by three letters, where the letter associated with the vertex is the middle letter. For example, in Fig. 6.5(b) angle A can also be called angle BAC or angle CAB. In Fig. 6.6 we can identify three angles: angle ABC, angle ABD, and angle CBD. The three-letter designation completely eliminates any confusion in identifying a particular angle. Angle ABC has also been labeled α, angle ABD has been labeled β, and angle CBD has been labeled γ.

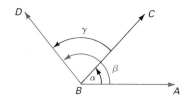

FIGURE 6.6

There are two commonly accepted units for measuring angles—*degrees* and *radians*. If a circle is divided into 360 equal arcs, then the central angle subtended by each arc is said to have a measure of *one degree*.* Equivalently, **one degree**, written 1°, is the measure of an angle generated by $\frac{1}{360}$ of a counterclockwise revolution. Thus an angle generated by one counterclockwise revolution has a measure of 360°. Figure 6.7 gives the degree measure of some angles in standard position. Notice in (c) and (f) that the measure of a negative angle includes a minus sign. In (a) the angle has a measure of 45°. Usually we avoid the phrase "has a measure of" and simply say that the angle is 45°. If θ represents this angle, then we write $\theta = 45°$.

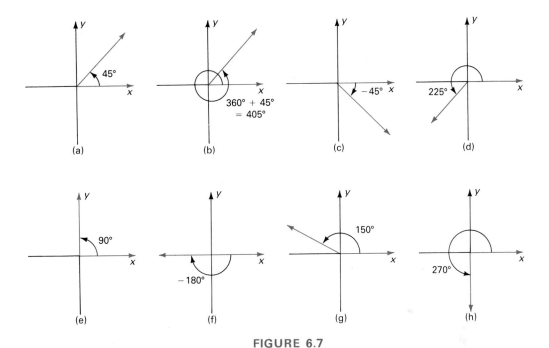

FIGURE 6.7

An angle α greater than 0° but less than 90° (that is, $0° < \alpha < 90°$) is called an **acute angle**. One between 90° and 180° (that is, $90° < \alpha < 180°$) is an **obtuse angle**. A 90° angle is called a **right angle**. See Fig. 6.8.

Because it is often necessary to measure an angle with a more precise unit than a degree, a degree is subdivided into 60 equal parts called **minutes**, and each minute is further subdivided into 60 equal parts called **seconds**. The symbol ′ is used to denote minutes, and the symbol ″ denotes seconds. Thus

$$60' = 1° \quad \text{and} \quad 60'' = 1'.$$

We can also use decimal form. For example, $30' = (\frac{30}{60})° = 0.5°$.

---

* Recall that a central angle of a circle is an angle formed by two radii; that is, its vertex is at the center of the circle.

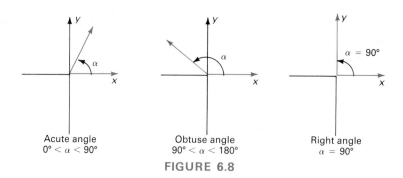

Acute angle
$0° < \alpha < 90°$

Obtuse angle
$90° < \alpha < 180°$

Right angle
$\alpha = 90°$

FIGURE 6.8

The second commonly used unit for measuring an angle is the *radian*.

One **radian** is the measure of a central angle of a circle which subtends an arc equal in length to the radius of the circle (see Fig. 6.9). The abbreviation for radian is rad.

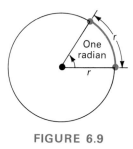

FIGURE 6.9

Thus for a circle of radius r, a central angle of 1 rad subtends an arc of length r, and a central angle of 2 rad subtends an arc of length $2r$. Similarly, a central angle of 2π radians subtends an arc of length $2\pi \cdot r$. But $2\pi r$ is the circumference of the circle and, hence, is the length of arc subtended by a central angle of 360°. We conclude that 2π is the radian measure of an angle of 360°:

$$2\pi \text{ rad} = 360°.$$

Dividing both sides by 2 gives

$$\boxed{\pi \text{ rad} = 180°,} \tag{1}$$

which is the basic relationship between radian measure and degree measure. Dividing both sides of Eq. 1 by π gives

$$\boxed{1 \text{ rad} = \frac{180°}{\pi}.} \tag{2}$$

Moreover, dividing both sides of the equation $180° = \pi$ rad by 180 gives

$$1° = \frac{\pi}{180} \text{ rad.} \qquad (3)$$

Computing $180/\pi$ and $\pi/180$, from Eqs. (2) and (3) we have

$$1 \text{ rad} \approx 57.296°$$

and

$$1° \approx 0.01745 \text{ rad.}$$

When an angle is measured in radians, the word *radian* is often omitted. That is, measuring an angle in radians can be thought of as measuring the angle in terms of a number. Thus an angle of π is understood to mean π rad; similarly, $\theta = 1$ means that $\theta = 1$ rad. When an angle is measured in degrees, however, the degree symbol must be included. We point out that the SI preferred unit for measuring angles is the radian, although degree measure is acceptable. When degrees are used, however, the use of calculators makes it convenient to give any parts of degrees in decimal form. We shall adopt that convention. For example, we shall write $23.15°$ rather than $23° 9'$.

By using Eq. 1 we can convert from degree measure to radian measure, and vice versa, as the following examples show.

Example 1

a. Convert $30°$ to radians.

Solution   Because π rad $= 180°$, dividing both sides by $180°$ gives $\dfrac{\pi \text{ rad}}{180°} = 1$. Thus

$$30° = 30°(1) = 30°\left(\frac{\pi \text{ rad}}{180°}\right).$$

On the right side the degree units cancel and we are left with radians:

$$30° = 30°\left(\frac{\pi \text{ rad}}{180°}\right) = \frac{30\pi}{180} \text{ rad} = \frac{\pi}{6} \text{ rad} \approx 0.5236 \text{ rad.}$$

Note that $\pi/6$ rad is an exact answer, whereas 0.5236 rad is an approximation. The π key on a calculator is useful when a decimal answer is required. In many instances an answer in terms of π is not only acceptable, it is preferred.

b. Convert $-240°$ to radians.

Solution

$$-240° = -240°\left(\frac{\pi}{180°}\right) = -\frac{4\pi}{3}.$$

Example 2

a. Convert $\dfrac{\pi}{4}$ rad to degrees.

Solution Because π rad $= 180°$, dividing both sides by π rad gives $1 = \dfrac{180°}{\pi \text{ rad}}$. Thus

$$\frac{\pi}{4} \text{ rad} = \frac{\pi}{4} \text{ rad} \cdot (1) = \frac{\pi}{4} \text{ rad} \cdot \frac{180°}{\pi \text{ rad}}.$$

On the right side the radian units cancel and we are left with degrees:

$$\frac{\pi}{4} \text{ rad} = \frac{\pi}{4} \text{ rad} \cdot \frac{180°}{\pi \text{ rad}} = 45°.$$

b. Convert $-\dfrac{5\pi}{6}$ rad to degrees.

Solution

$$-\frac{5\pi}{6} = -\frac{5\pi}{6}\left(\frac{180°}{\pi}\right) = -150°.$$

c. Convert an angle of 10 to degrees.

Solution

$$10 = 10\left(\frac{180°}{\pi}\right) = \left(\frac{1800}{\pi}\right)^{\!\circ}.$$

In decimal form we have 10 rad $\approx 572.96°$, which you should verify.

When two or more angles is standard position have the same terminal side, the angles are called **coterminal angles**. For example, $30°$ and $-330°$ are coterminal angles [see Fig. 6.10(a)]. In Fig. 6.10(b) you can see that $30°$ and $390°$ are also

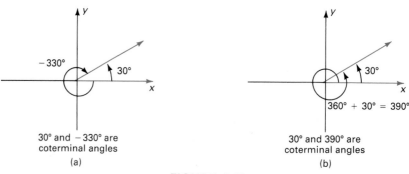

30° and −330° are
coterminal angles
(a)

30° and 390° are
coterminal angles
(b)

FIGURE 6.10

coterminal. Although 30° and 390° may look like the same angle, they are not; 390° is one revolution more than 30°. Clearly, if you add or subtract a multiple of 360° (or 2π) to or from an angle θ, the resulting angle is coterminal with θ. Thus there are infinitely many angles that are coterminal with θ.

Example 3

Find two positive and two negative angles that are coterminal with 140°.

Solution   Two such positive angles are [see Fig. 6.11(a)]

$$140° + 360° = 500°$$

and [see Fig. 6.11(b)]

$$140° + 2(360°) = 860°.$$

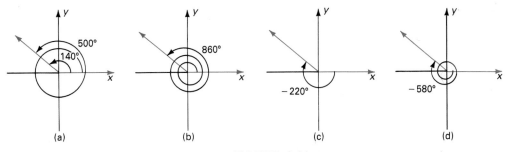

FIGURE 6.11

Two such negative angles are [see Fig. 6.11(c)]

$$140° - 360° = -220°$$

and [see Fig. 6.11(d)]

$$140° - 2(360°) = -580°.$$

Example 4

Find the angle θ that is coterminal with $16\pi/3$ and such that $0 \leq \theta < 2\pi$.

Solution   Because $16\pi/3 \geq 2\pi$, we shall keep subtracting 2π from $16\pi/3$ until we reach a coterminal angle between 0 and 2π.

$$\frac{16\pi}{3} - 2\pi = \frac{16\pi}{3} - \frac{6\pi}{3} = \frac{10\pi}{3}.$$

Because $\dfrac{10\pi}{3} \geq 2\pi$, we continue.

$$\frac{10\pi}{3} - 2\pi = \frac{10\pi}{3} - \frac{6\pi}{3} = \frac{4\pi}{3}.$$

Now, $0 \leq \dfrac{4\pi}{3} < 2\pi$ and $\dfrac{4\pi}{3}$ is coterminal with $\dfrac{16\pi}{3}$. Thus $\theta = \dfrac{4\pi}{3}$.

Using radian measure we can derive a formula for the length of a circular arc. Figure 6.12 shows a circle of radius r on which an arc of length s is subtended by

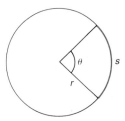

FIGURE 6.12

the central angle θ, where θ is in radians. Each angle of 1 rad subtends an arc of length r. Thus an angle of θ radians subtends an arc of length θr. But in Fig. 6.12 that length is s. Thus $s = \theta r$.

ARC LENGTH

The length s of the arc subtended on a circle of radius r by a central angle θ is given by

$$s = r\theta,$$

where θ is in radians.

Example 5

a. Find the length of the arc subtended by a central angle of $60°$ on a circle of radius 10 m. See Fig. 6.13.

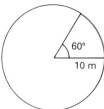

FIGURE 6.13

Solution To use the formula for arc length, we must first convert 60° to radians.

$$60° = 60° \cdot \frac{\pi}{180°} = \frac{\pi}{3}.$$

The arc length is given by

$$s = r\theta = (10 \text{ m})\left(\frac{\pi}{3}\right) = \frac{10\pi}{3} \text{ m} = 10.47 \text{ m}.$$

In the last line you can see that writing the value of θ as the dimensionless *number* $\pi/3$ rather than $\pi/3$ rad results in a unit that is meaningful.

b. On a circle of radius 6 m, the arc subtended by a central angle θ has length 24 m. Find θ.

Solution Because $s = r\theta$,

$$24 \text{ m} = (6 \text{ m})\theta$$

$$\theta = \frac{24 \text{ m}}{6 \text{ m}} = 4.$$

Problem Set 6.1

*In Problems **1–20**, convert each degree measure to radians and each radian measure to degrees. Give an exact answer. Draw each angle in standard position.*

1. 60°.

2. 135°.

3. $\dfrac{3\pi}{4}$.

4. $\dfrac{2\pi}{3}$.

5. $\dfrac{\pi}{2}$.

6. $-\dfrac{3\pi}{4}$.

7. 45°.

8. 210°.

9. $-330°$.

10. $-270°$.

11. $\dfrac{7\pi}{6}$.

12. 15°.

13. $-\dfrac{\pi}{8}$.

14. $\dfrac{5\pi}{6}$.

15. 720°.

16. $\dfrac{3\pi}{2}$.

17. 6π.

18. $\dfrac{7\pi}{8}$.

19. 4.

20. $\dfrac{1}{2}$.

*In Problems **21–36**, determine whether the angle is quadrantal, or a first-, second-, third-, or fourth-quadrant angle.*

21. 130°.

22. $\dfrac{\pi}{3}$.

23. $-45°$.

24. 90°.

25. $\dfrac{5\pi}{6}$.

26. 250°.

27. $-\pi$.

28. $-\dfrac{\pi}{4}$.

29. 370°.

30. 0.

31. 210°.

32. $\dfrac{2\pi}{3}$.

33. 4π.

34. $\pi°$.

35. $-\dfrac{\pi}{6}$.

36. 270°.

*In Problem **37–40**, determine θ in the given figure.*

37. See Fig. 6.14.

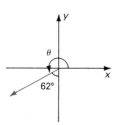

FIGURE 6.14

38. See Fig. 6.15.

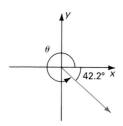

FIGURE 6.15

39. See Fig. 6.16.

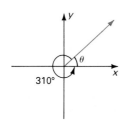

FIGURE 6.16

40. See Fig. 6.17.

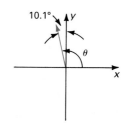

FIGURE 6.17

*In Problems **41–48** find two positive angles and two negative angles that are coterminal with the given angle.*

41. $\theta = 35°$.

42. $\theta = 480°$.

43. $\theta = 221.4°$.

44. $\theta = 90°$.

45. $\theta = -70°$.

46. $\theta = -115°$.

47. $\theta = \dfrac{\pi}{6}$.

48. $\theta = \dfrac{2\pi}{3}$.

*In Problems **49–54**, find the angle θ that is coterminal with the given angle and meets the given condition.*

49. $420°$; $\ 0° \le \theta < 360°$.

50. $1080°$; $\ 0° \le \theta < 360°$.

51. $\dfrac{13\pi}{4}$; $\ 0 \le \theta < 2\pi$.

52. $\dfrac{15\pi}{3}$; $\ 0 \le \theta < 2\pi$.

53. $\dfrac{17\pi}{2}$; $\ 0 \le \theta < 2\pi$.

54. $-\dfrac{5\pi}{6}$; $\ 0 \le \theta < 2\pi$.

*In Problems **55–62**, give all answers to three decimal places. Assume that π = 3.14159 or use the π key on your calculator.*

55. Convert $35°$ to radians.

56. Convert $212°$ to radians.

57. Convert $84.23°$ to radians.

58. Convert $19.601°$ to radians.

59. Convert 2.36 rad to degrees.

60. Convert -5.23 rad to degrees.

61. Convert -3.742 rad to degrees.

62. Convert 17.031 rad to degrees.

*In Problems **63–68**, determine the length of the arc subtended by the central angle θ of a circle of radius r. Give your answers to two decimal places.*

63. $\theta = \dfrac{\pi}{6}, \quad r = 13$ km.

64. $\theta = \dfrac{\pi}{3}, \quad r = 25$ m.

65. $\theta = 45°, \quad r = 18$ cm.

66. $\theta = 225°, \quad r = 5$ km.

67. $\theta = 300°, \quad r = 5$ cm.

68. $\theta = 150°, \quad r = 1$ ft.

69. An engineer is to design a cloverleaf for an exit on an interstate road. A part of the cloverleaf can be thought of as an arc of a circle of radius 300 m. The central angle of this arc is 120°. Find the length of this part of the cloverleaf.

70. If an automobile tire has a diameter of 28 in., through how many complete revolutions does the tire turn in 1 mi? (*Note:* 1 mi = 5280 ft.)

71. Convert 7.26 rad/s to revolutions per minute.

72. Through how many radians does the minute hand of a clock rotate in 40 min? In 40 h?

73. When a spectrometer is used to view the spectral lines of a light source, a doublet (two spectral lines representing light of nearly equal wavelengths) is observed at angular positions 121.99° and 122.04°. What is the angular separation of the two lines in radians?

74. A certain telescope is designed so that it can distinguish two points 1.0 cm apart at a distance of 1.6 km. The angular separation of the two points at that distance is 6.25×10^{-6} rad. What is the angular separation in degrees?

75. The angle θ (in radians) through which the rotor of a motor has turned is given by $\theta = \omega t$, where t is elapsed time in seconds and ω (omega) is the angular speed in radians per second. If a motor makes 600 revolutions per minute, find the angle through which the rotor turns in 2.0 s.

76. The sensitivity of a certain measuring instrument is given as 10 μA per degree deflection. What is the sensitivity in amperes per radian?

77. Suppose that a pendulum of length 20 cm swings through an arc of length 12 cm. Through how many degrees does the pendulum swing?

78. The minute hand of a clock is 6 cm long. In a time period of 50 min, through what distance does the tip of the minute hand move?

*In Problems **79** and **80**, determine θ in the given figure.*

79. See Fig. 6.18.

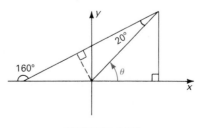

FIGURE 6.18

80. See Fig. 6.19.

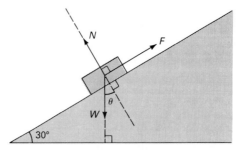

FIGURE 6.19

6.2 TRIGONOMETRIC FUNCTIONS OF ACUTE ANGLES

The entire realm of trigonometry is based on six special functions, called *trigonometric functions*, whose inputs are angles and whose outputs are numbers. We shall first consider these functions for an acute angle. In the next chapter we shall give a complete treatment for any angle.

Suppose we have an acute angle θ in standard position (see Fig. 6.20). On the terminal side we choose *any* point $P(x, y)$ except the origin. The distance from the

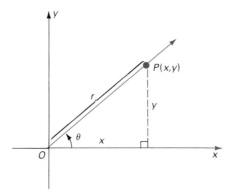

FIGURE 6.20

origin to P is denoted by r, where $r > 0$, and is called the **radius vector**. By constructing a perpendicular from P to the x-axis, we see that r is the length of the hypotenuse of a right triangle in which the other sides have lengths x and y. This triangle is called the **reference triangle** for θ. By the Pythagorean theorem,

$$r^2 = x^2 + y^2,$$

from which

$$r = \sqrt{x^2 + y^2}.$$

Thus with point P there are associated three numbers—its abscissa x, its ordinate y, and the radius vector r—and from these numbers six ratios can be formed. It is these ratios that define the trigonometric functions of θ. Their names (and abbreviations) are **sine** function (**sin**), **cosine** function (**cos**), **tangent** function (**tan**), **cotangent** function (**cot**), **secant** function (**sec**), and **cosecant** function (**csc**). Just as the output of a function f is written $f(x)$, the output of the sine function is written $\sin(\theta)$, or simply $\sin \theta$. The functions are defined as follows.

TRIGONOMETRIC FUNCTIONS

Suppose θ is in standard position. Let (x, y) be any point on the terminal side except $(0, 0)$, and let $r = \sqrt{x^2 + y^2}$. Then

$$\sin \theta = \frac{y}{r},$$

$$\cos \theta = \frac{x}{r},$$

$$\tan \theta = \frac{y}{x},$$

$$\cot \theta = \frac{x}{y},$$

$$\sec \theta = \frac{r}{x},$$

$$\csc \theta = \frac{r}{y}.$$

Notice that the inputs are angles and the outputs are numbers (provided the denominators are not zero).

The value of each trigonometric function depends only on the angle θ and not upon the choice of the point $P(x, y)$ on the terminal side of θ. To see why, let $P_1(x_1, y_1)$ be any other point (except the origin) on the terminal side of θ. In Fig. 6.21 the triangles OAP and OBP_1 are similar triangles, because their corresponding angles are equal. From geometry, the lengths of corresponding sides of similar triangles are proportional. It follows that for any ratio formed by the sides of triangle OAP, the corresponding ratio from triangle OBP_1 will be equal to it. Thus, for a given angle θ the ratios are independent of the choice of (x, y). As a result, the definition in the

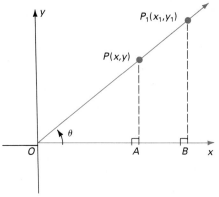

FIGURE 6.21

previous box associates, for example, the unique number $\sin \theta$ with a given angle θ and in this sense defines a function $f(\theta) = \sin \theta$.

Example 1

If the terminal side of an acute angle θ in standard position passes through the point $(6, 8)$, find the values of the six trigonometric functions of θ (see Fig. 6.22).

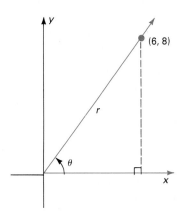

FIGURE 6.22

Solution   We use the point $(6, 8)$ as the point (x, y) on the terminal side of θ. Thus $x = 6$, $y = 8$, and

$$r = \sqrt{x^2 + y^2} = \sqrt{6^2 + 8^2} = \sqrt{100} = 10.$$

Applying the definitions of the trigonometric functions of θ, we have

$$\sin \theta = \frac{y}{r} = \frac{8}{10} = \frac{4}{5},$$

$$\cos \theta = \frac{x}{r} = \frac{6}{10} = \frac{3}{5},$$

$$\tan \theta = \frac{y}{x} = \frac{8}{6} = \frac{4}{3},$$

$$\cot \theta = \frac{x}{y} = \frac{6}{8} = \frac{3}{4},$$

$$\sec \theta = \frac{r}{x} = \frac{10}{6} = \frac{5}{3},$$

$$\csc \theta = \frac{r}{y} = \frac{10}{8} = \frac{5}{4}.$$

Many important relationships exist between the trigonometric functions. For any angle θ, $\sin \theta = y/r$ and $\csc \theta = r/y$. Provided these fractions are defined, we have

$$\csc \theta = \frac{r}{y} = \frac{1}{\dfrac{y}{r}} = \frac{1}{\sin \theta}.$$

Similarly, we can obtain the following reciprocal relations.

RECIPROCAL RELATIONS

$$\csc \theta = \frac{1}{\sin \theta} \quad \text{and} \quad \sin \theta = \frac{1}{\csc \theta}.$$

$$\sec \theta = \frac{1}{\cos \theta} \quad \text{and} \quad \cos \theta = \frac{1}{\sec \theta}.$$

$$\cot \theta = \frac{1}{\tan \theta} \quad \text{and} \quad \tan \theta = \frac{1}{\cot \theta}.$$

To describe these results we say that the sine and cosecant, the cosine and secant, and the tangent and cotangent functions are pairs of **reciprocal functions**. This means, for example, that if $\sin \theta = 0.5$, then $\csc \theta = 1/0.5 = 2$. Similarly, if $\cos \theta = \frac{2}{3}$, then $\sec \theta = \frac{3}{2}$. We encourage you to verify the reciprocal relations for the results of Example 1.

We can obtain two more important relationships. Because $\sin \theta = y/r$ and $\cos \theta = x/r$, we have

$$\frac{\sin \theta}{\cos \theta} = \frac{\dfrac{y}{r}}{\dfrac{x}{r}} = \frac{y}{r} \cdot \frac{r}{x} = \frac{y}{x} = \tan \theta,$$

provided $\cos \theta \neq 0$. Similarly, $(\cos \theta)/(\sin \theta) = \cot \theta$.

$$\frac{\sin \theta}{\cos \theta} = \tan \theta \quad \text{and} \quad \frac{\cos \theta}{\sin \theta} = \cot \theta.$$

Thus if $\sin \theta = 0.3420$ and $\cos \theta = 0.9397$, then $\tan \theta = 0.3420/0.9397 = 0.3639$.

Example 2

If $\sin \theta = \frac{1}{3}$ and $\cos \theta = 2\sqrt{2}/3$, find the other trigonometric values of θ. Give answers with rationalized denominators.

Solution

$$\tan \theta = \frac{\sin \theta}{\cos \theta} = \frac{\frac{1}{3}}{\frac{2\sqrt{2}}{3}} = \frac{1}{2\sqrt{2}} = \frac{\sqrt{2}}{4},$$

$$\cot \theta = \frac{1}{\tan \theta} = \frac{1}{\frac{1}{2\sqrt{2}}} = 2\sqrt{2},$$

$$\sec \theta = \frac{1}{\cos \theta} = \frac{1}{\frac{2\sqrt{2}}{3}} = \frac{3}{2\sqrt{2}} = \frac{3\sqrt{2}}{4},$$

$$\csc \theta = \frac{1}{\sin \theta} = \frac{1}{\frac{1}{3}} = 3.$$

The values of the trigonometric functions of an acute angle θ can be conveniently interpreted as ratios of the lengths of the sides of a right triangle. Suppose θ is an angle in the right triangle shown in Fig. 6.23. We call side AB the **opposite side** for θ, side OA the **adjacent side**, and side OB the **hypotenuse**. For convenience we shall abbreviate the lengths of these sides by **opp**, **adj**, and **hyp**, respectively. We can superimpose a rectangular coordinate system on this triangle so that θ is a first-quadrant angle and we can talk about lengths (see Fig. 6.24). Observe that the

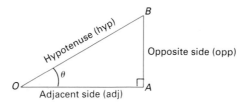

FIGURE 6.23

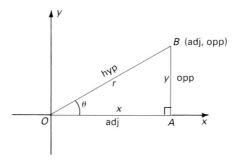

FIGURE 6.24

abscissa of point B on the terminal side of θ is the length (x) of the adjacent side for θ, the ordinate is the length (y) of the opposite side, and the length of the hypotenuse is r. Thus the value of each trigonometric function of θ can be given as a ratio of the lengths of these sides. For example,

$$\sin \theta = \frac{y}{r} = \frac{\text{opp}}{\text{hyp}}.$$

In summary, we have the following.

FUNCTIONS OF ACUTE ANGLE

If θ is an acute angle of a right triangle, then

$$\sin \theta = \frac{\text{opp}}{\text{hyp}}, \qquad \csc \theta = \frac{\text{hyp}}{\text{opp}},$$

$$\cos \theta = \frac{\text{adj}}{\text{hyp}}, \qquad \sec \theta = \frac{\text{hyp}}{\text{adj}},$$

$$\tan \theta = \frac{\text{opp}}{\text{adj}}, \qquad \cot \theta = \frac{\text{adj}}{\text{opp}}.$$

With these relationships we can determine the values of the trigonometric functions of an acute angle θ in a right triangle without directly considering θ is standard position.

Example 3

Find the values of the six trigonometric functions of angles A and B in Fig. 6.25.

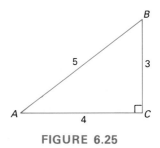

FIGURE 6.25

Solution For angle A, the opposite side has length 3 (units) and the adjacent side has length 4.

$$\sin A = \frac{\text{opp}}{\text{hyp}} = \frac{3}{5}, \qquad \csc A = \frac{\text{hyp}}{\text{opp}} = \frac{5}{3},$$

$$\cos A = \frac{\text{adj}}{\text{hyp}} = \frac{4}{5}, \qquad \sec A = \frac{\text{hyp}}{\text{adj}} = \frac{5}{4},$$

$$\tan A = \frac{\text{opp}}{\text{adj}} = \frac{3}{4}, \qquad \cot A = \frac{\text{adj}}{\text{opp}} = \frac{4}{3}.$$

For angle B, the opposite side is 4 and the adjacent side is 3.

$$\sin B = \frac{\text{opp}}{\text{hyp}} = \frac{4}{5}, \qquad \csc B = \frac{\text{hyp}}{\text{opp}} = \frac{5}{4},$$

$$\cos B = \frac{\text{adj}}{\text{hyp}} = \frac{3}{5}, \qquad \sec B = \frac{\text{hyp}}{\text{adj}} = \frac{5}{3},$$

$$\tan B = \frac{\text{opp}}{\text{adj}} = \frac{4}{3}, \qquad \cot B = \frac{\text{adj}}{\text{opp}} = \frac{3}{4}.$$

For any right triangle, the sum of the two acute angles is $90°$; that is, the acute angles are **complementary**. In Example 3, angles A and B are complementary. Notice in Fig. 6.25 that the side opposite A is adjacent to B; the side adjacent to A is opposite B. Thus $\sin A = \cos B$, $\tan A = \cot B$, $\sec A = \csc B$, and so on. The pairs of functions sine and cosine, tangent and cotangent, and secant and cosecant are called **cofunctions**. We draw the following conclusion.

> The value of a trigonometric function of an acute angle θ is equal to value of the corresponding cofunction of the complementary angle, $90° - \theta$.

For example, $\sin 10° = \cos 80°$ and $\sec 30° = \csc 60°$. Do not confuse cofunction relationships with reciprocal relationships. For example, $\sin 30° = \cos 60°$ and $\sin 30° = 1/\csc 30°$.

Example 4

Suppose θ is an acute angle such that $\cos \theta = \frac{5}{7}$. Find the values of $\tan \theta$ and $\cot \theta$.

Solution In Fig. 6.26 we drew a right triangle for θ. Because $\cos \theta = \text{adj/hyp}$ and here we

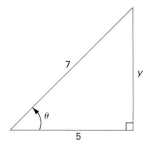

FIGURE 6.26

want $\cos \theta = \frac{5}{7}$, we used adj $= 5$ and hyp $= 7$.* To find the length y of the opposite side, we have

$$5^2 + y^2 = 7^2$$
$$y^2 = 7^2 - 5^2 = 49 - 25 = 24$$
$$y = \sqrt{24} = \sqrt{4}\sqrt{6} = 2\sqrt{6}.$$

Thus

$$\tan \theta = \frac{\text{opp}}{\text{adj}} = \frac{2\sqrt{6}}{5}.$$

Because the tangent and cotangent functions are reciprocal functions,

$$\cot \theta = \frac{5}{2\sqrt{6}} = \frac{5\sqrt{6}}{12}.$$

Notice that we wrote this answer with a rationalized denominator.

* We could also choose for example, adj $= 10$ and hyp $= 14$.

Problem Set 6.2

In Problems 1–14, determine the values of the six trigonometric functions of an angle θ in standard position if the terminal side of θ passes through the given point. Give all answers in rationalized form.

1. (8, 6).   2. (12, 5).   3. (3, 4).   4. (4, 3).

5. $(1, \sqrt{3})$.   6. $(\sqrt{3}, 1)$.   7. $(\sqrt{5}, 2)$.   8. (2, 3).

9. $(3, \sqrt{3})$.   10. (1, 7).   11. (1,1).   12. (5, 12).

13. $(\sqrt{2}, 1)$.   14. $(\sqrt{15}, 1)$.

In Problems 15 and 16, find the values of the six trigonometric functions of angles A and B in the given figure.

15. See Fig. 6.27.   16. See Fig. 6.28.

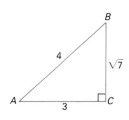

FIGURE 6.27

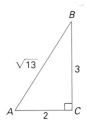

FIGURE 6.28

In Problems 17 and 18, find the values of the six trigonometric functions of θ in the given figure.

17. See Fig. 6.29.   18. See Fig. 6.30.

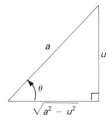

FIGURE 6.29

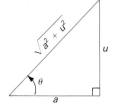

FIGURE 6.30

19. If $\sin \theta = 0.6$ and $\cos \theta = 0.8$, find $\tan \theta$ and $\sec \theta$ in an easy way.

20. If $\sin \theta = \sqrt{2}/2$ and $\cos \theta = \sqrt{2}/2$, find $\cot \theta$ in an easy way.

In Problems 21–26, give a reason why the given statement is true.

21. $\sin 35° = \cos 55°$.   22. $\sin \dfrac{\pi}{4} = \cos \dfrac{\pi}{4}$.   23. $\tan 2° = \dfrac{1}{\cot 2°}$.

24. $\sin \dfrac{\pi}{4} = \dfrac{1}{\csc (\pi/4)}$.   25. $\tan 0.32 = \dfrac{\sin 0.32}{\cos 0.32}$.   26. $\sec 1 = \dfrac{1}{\cos 1}$.

*In Problems **27–30**, the value of a trigonometric function of an acute angle θ is given. Find the values of the remaining trigonometric functions of θ.*

27. $\sin \theta = \dfrac{2}{5}$.

28. $\cos \theta = \dfrac{2}{3}$.

29. $\tan \theta = 4 = \dfrac{4}{1}$.

30. $\sec \theta = \dfrac{5}{4}$.

6.3 VALUES OF TRIGONOMETRIC FUNCTIONS

Using geometry we can find the *exact* trigonometric function values of 30° (or π/6), 45° (or π/4), and 60° (or π/3), which are called **special angles**. Figure 6.31(a) shows an equilateral triangle in which all sides have length 2. Each angle must be 60°. The bisector of any of these angles is also the perpendicular bisector of the side opposite

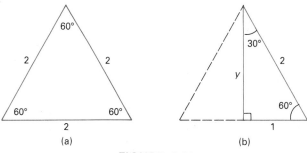

FIGURE 6.31

the angle. Figure 6.31(b) shows that the result of constructing such a bisector consists of two right triangles, where each triangle has angles of 30° and 60°. The length y of the side opposite 60° is found by the Pythagorean theorem:

$$1^2 + y^2 = 2^2$$
$$1 + y^2 = 4$$
$$y^2 = 3$$
$$y = \sqrt{3}.$$

From the 30°-60°-90° triangle in Fig. 6.32, which was obtained from Fig. 6.31(b), we have

$$\sin 30° = \frac{\text{opp}}{\text{hyp}} = \frac{1}{2}.$$

$$\cos 30° = \frac{\text{adj}}{\text{hyp}} = \frac{\sqrt{3}}{2}.$$

$$\tan 30° = \frac{\text{opp}}{\text{adj}} = \frac{1}{\sqrt{3}} = \frac{\sqrt{3}}{3}.$$

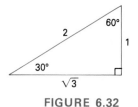

FIGURE 6.32

and

$$\sin 60° = \frac{\text{opp}}{\text{hyp}} = \frac{\sqrt{3}}{2}$$

$$\cos 60° = \frac{\text{adj}}{\text{hyp}} = \frac{1}{2}$$

$$\tan 60° = \frac{\text{opp}}{\text{adj}} = \frac{\sqrt{3}}{1} = \sqrt{3}.$$

The values of the cotangent, secant, and cosecant of these angles can be found by using the reciprocal relationships.

To obtain the trigonometric function values of 45°, we use the isosceles right triangle in Fig. 6.33. Two sides have length 1 and, from geometry, the acute angles

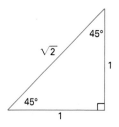

FIGURE 6.33

must each be 45°. The length r of the hypotenuse is found with the Pythagorean theorem:

$$r^2 = 1^2 + 1^2 = 2$$

$$r = \sqrt{2}.$$

Thus

$$\sin 45° = \frac{\text{opp}}{\text{hyp}} = \frac{1}{\sqrt{2}} = \frac{\sqrt{2}}{2}$$

$$\cos 45° = \frac{\text{adj}}{\text{hyp}} = \frac{1}{\sqrt{2}} = \frac{\sqrt{2}}{2}$$

$$\tan 45° = \frac{\text{opp}}{\text{adj}} = \frac{1}{1} = 1, \qquad \text{and so on.}$$

The results of the preceding discussion of special angles are summarized in Table 6.1. These angles occur so frequently that you should become totally familiar with these results. Rather than memorize the table, however, we suggest that you memorize the triangles from which the results were obtained and use them in conjunction with the definitions of the trigonometric functions. The important triangles are repeated for your convenience in Fig. 6.34. Because 30° and 60° are complementary and the values of cofunctions of complementary angles are equal, we have sin 30° = cos 60°, and so on.

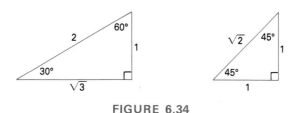

FIGURE 6.34

TABLE 6.1

		$30°\left(\text{or } \dfrac{\pi}{6}\right)$	$45°\left(\text{or } \dfrac{\pi}{4}\right)$	$60°\left(\text{or } \dfrac{\pi}{3}\right)$
sin	$\dfrac{\text{opp}}{\text{hyp}}$	$\dfrac{1}{2}$	$\dfrac{1}{\sqrt{2}} = \dfrac{\sqrt{2}}{2}$	$\dfrac{\sqrt{3}}{2}$
cos	$\dfrac{\text{adj}}{\text{hyp}}$	$\dfrac{\sqrt{3}}{2}$	$\dfrac{1}{\sqrt{2}} = \dfrac{\sqrt{2}}{2}$	$\dfrac{1}{2}$
tan	$\dfrac{\text{opp}}{\text{adj}}$	$\dfrac{1}{\sqrt{3}} = \dfrac{\sqrt{3}}{3}$	$\dfrac{1}{1} = 1$	$\dfrac{\sqrt{3}}{1} = \sqrt{3}$
cot	$\dfrac{\text{adj}}{\text{opp}}$	$\dfrac{\sqrt{3}}{1} = \sqrt{3}$	$\dfrac{1}{1} = 1$	$\dfrac{1}{\sqrt{3}} = \dfrac{\sqrt{3}}{3}$
sec	$\dfrac{\text{hyp}}{\text{adj}}$	$\dfrac{2}{\sqrt{3}} = \dfrac{2\sqrt{3}}{3}$	$\dfrac{\sqrt{2}}{1} = \sqrt{2}$	$\dfrac{2}{1} = 2$
csc	$\dfrac{\text{hyp}}{\text{opp}}$	$\dfrac{2}{1} = 2$	$\dfrac{\sqrt{2}}{1} = \sqrt{2}$	$\dfrac{2}{\sqrt{3}} = \dfrac{2\sqrt{3}}{3}$

Trigonometric Functions on a Calculator

We will obtain the values of the trigonometric functions of most angles by using a calculator. Some calculators have only a decimal-degree mode; others can be set either for decimal-degree measure or for radian measure. If yours accepts degree measure only and an angle is given in radians, it will be necessary for you first to convert from radians to degrees. Now is a good time to determine exactly what modes your calculator can handle and to learn how to use them. The information in Appendix A (Sec. A.6) may be a helpful supplement to the manual that came with your calculator. In what follows we shall assume that your calculator has both a decimal-degree mode and a radian mode.

Assuming that the calculator is in decimal-degree mode, we find the values of the trigonometric functions of $28.4°$ by keying in 28.4 and then pressing the key for the desired trigonometric function. For example, to find $\sin 28.4°$ we have (to four decimal places)

$$\boxed{28.4}\ \boxed{\text{SIN}} \qquad \text{Display:} \quad 0.4756$$

Thus

$$\sin 28.4° = 0.4756.$$

(For mathematical exactness, the statement should be $\sin 28.4° \approx 0.4756$.) Similarly,

$$\cos 28.4° = 0.8796 \quad \text{and} \quad \tan 28.4° = 0.5407,$$

which you should verify.

Most calculators do not have keys for the cotangent, secant, and cosecant functions. Instead, you must use the reciprocal relationships, such as $\cot \theta = 1/\tan \theta$. The reciprocal key $\boxed{1/x}$ is useful here. For example, to find $\cot 28.4°$ we have

$$\boxed{28.4}\ \boxed{\text{TAN}}\ \boxed{1/x} \qquad \text{Display:} \quad 1.8495$$

Thus

$$\cot 28.4° = 1.8495.$$

Similarly,

$$\sec 28.4° = \frac{1}{\cos 28.4°} = 1.1368$$

$$\csc 28.4° = \frac{1}{\sin 28.4°} = 2.1025.$$

To find a trigonometric function value for an angle in radians, *place the calculator in radian mode* and proceed as before. For example, to find $\sin \pi/3$ we have

$$\boxed{\pi}\ \boxed{\div}\ \boxed{3}\ \boxed{=}\ \boxed{\text{SIN}} \qquad \text{Display:} \quad 0.8660$$

Thus

$$\sin \frac{\pi}{3} = 0.8660.$$

[Notice that $\pi/3$ is a special angle ($60°$) and an exact answer is $\sqrt{3}/2$.] You should verify that

$$\sin 0.3053 = 0.3006 \quad \text{and} \quad \tan 1.1694 = 2.3560.$$

To find an angle when you are given one of its trigonometric values, use the keys marked $\boxed{\text{SIN}^{-1}}$, $\boxed{\text{COS}^{-1}}$, $\boxed{\text{TAN}^{-1}}$, which are read *inverse sine* (or *arcsine*), *inverse cosine* (or *arccosine*), and *inverse tangent* (or *arctangent*), respectively. (These inverse functions are discussed more fully in Chapter 16.) With some calculators having, for example, no $\boxed{\text{SIN}^{-1}}$ key, an $\boxed{\text{INV}}$ key is used with the $\boxed{\text{SIN}}$ key.

For example, suppose we want to find the acute angle θ such that

$$\sin \theta = 0.5873.$$

In degree mode we have

$$\boxed{0.5873}\ \boxed{\text{SIN}^{-1}} \qquad \text{Display:} \quad 35.97$$

or

$$\boxed{0.5873}\ \boxed{\text{INV}}\ \boxed{\text{SIN}} \qquad \text{Display:} \quad 35.97$$

Thus (to four significant digits)

$$\theta = 35.97°.$$

Radian mode gives $\theta = 0.6277$. Similarly, if

$$\cot \theta = 1.419,$$

then $\tan \theta = 1/1.419$, from which $\theta = 35.17°$. The keystroke sequence is (degree mode)

$$\boxed{1.419}\ \boxed{1/x}\ \boxed{\text{TAN}^{-1}} \qquad \text{Display:} \quad 35.17$$

Example 1

Figure 6.35 shows the reflected and refracted rays when light strikes (or is *incident on*) a glass plate of refractive index n. Snell's law states that $\sin \theta_1 = n \sin \theta_2$, where θ_1 and θ_2 are measured from the line perpendicular to the glass surface as indicated. If $\theta_1 = 53.4°$ and $\theta_2 = 32.3°$, find the refractive index of the glass.

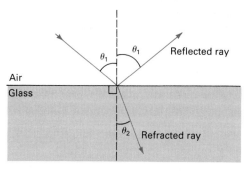

FIGURE 6.35

Solution

$$\sin \theta_1 = n \sin \theta_2$$

$$\sin 53.4° = n \sin 32.3°$$

$$n = \frac{\sin 53.4°}{\sin 32.3°} = \frac{0.80282}{0.53435}$$

$$= 1.50.$$

With efficient use of a calculator, in Example 1 the rounded trigonometric values 0.80282 and 0.53435 would not have been written down and then keyed into the calculator for the next calculation. Instead, the displayed values of sin 53.4° and sin 32.3° would be retained in the calculator for later use. This minimizes rounding errors. To illustrate, a calculator solution for Example 1 would appear as follows.

$$n = \frac{\sin 53.4°}{\sin 32.3°} = 1.50.$$

The answer was obtained from the keystroke sequence

$\boxed{53.4}$ $\boxed{\text{SIN}}$ $\boxed{\div}$ $\boxed{32.3}$ $\boxed{\text{SIN}}$ $\boxed{=}$ Display: 1.50

In future examples we may at times show intermediate steps by writing down displayed values that have been rounded, because this could help you gain a feel for numbers. However, these rounded values will not be used for later steps. Instead, all displayed digits will be retained in the calculator to be used for the final result.

Example 2

When a beam of electrons moving in a transparent medium travels at a speed that is greater than the speed of light in that medium, visible radiation, called Cerenkov radiation, is emitted at an angle θ to the direction of motion of the electrons, where

$$\cos \theta = \frac{c}{nv}.$$

Here c is the speed of light in vacuum, n is the index of refraction of the medium, and v is the speed of the electrons. For a particular crystal, suppose that $n = 1.785$ and $v = 0.5711c$. Find θ.

Solution   By substitution we have

$$\cos \theta = \frac{c}{nv} = \frac{c}{(1.785)(0.5711c)}$$

$$= \frac{1}{(1.785)(0.5711)} \quad \text{[cancellation]}.$$

Thus

$$\theta = 11.20°,$$

which was obtained by the keystroke sequence

$$\boxed{1.785} \quad \boxed{\times} \quad \boxed{0.5711} \quad \boxed{=} \quad \boxed{1/x} \quad \boxed{\text{COS}^{-1}} \qquad \text{Display:} \quad 11.20$$

Finally, we point out that we abbreviate $(\sin \theta)^2$ by writing $\sin^2 \theta$ (read *sine squared theta*). Similar abbreviations are used with the other trigonometric functions. Thus

$$\sin^2 \frac{\pi}{6} = \left(\frac{1}{2}\right)^2 = \frac{1}{4}.$$

Problem Set 6.3

*In Problems **1–18**, find the values of the given trigonometric functions. Give all answers to four decimal places.*

1. $\sin 32°$.	**2.** $\cos 75°$.	**3.** $\cos 14.3°$.
4. $\sin 32.4°$.	**5.** $\tan 53.04°$.	**6.** $\tan 0.66°$.
7. $\cot 12.4°$.	**8.** $\sec 79.16°$.	**9.** $\sec 45.11°$.
10. $\csc 80.25°$.	**11.** $\csc 33.01°$.	**12.** $\cot 74.129°$.
13. $\sin 0.6109$.	**14.** $\csc 0.6109$.	**15.** $\cos 1.1606$.
16. $\sec 1.1606$.	**17.** $\tan \dfrac{\pi}{8}$.	**18.** $\cos \dfrac{\pi}{16}$.

*In each of Problems **19–24**, find the acute angle θ, in degrees, subject to the given condition. Give your answer to one decimal place.*

19. $\sin \theta = 0.7269$.	**20.** $\csc \theta = 8.206$.	**21.** $\cos \theta = 0.1249$.
22. $\sec \theta = 1.054$.	**23.** $\cot \theta = 1.6247$.	**24.** $\tan \theta = 0.8847$.

In each of Problems **25–30** , find the acute angle θ, in radians, subject to the given condition. In Problems **25–28**, round your answer to two decimal places. Do not use a calculator in Problems **29** and **30**.

25. $\sin \theta = 0.2979$.

26. $\cos \theta = 0.6648$.

27. $\tan \theta = 2.723$.

28. $\cot \theta = 0.3346$.

29. $\cos \theta = \dfrac{\sqrt{3}}{2}$.

30. $\sin \theta = \dfrac{\sqrt{2}}{2}$.

31. Find the exact value of $\cos^2 \dfrac{\pi}{4}$.

32. Find the exact value of $\left(\sin \dfrac{\pi}{3} \right)\left(\cos \dfrac{\pi}{3} \right)$.

33. Find the exact value of $(\sin 30°)(\cos 60°) + (\cos 30°)(\sin 60°)$.

34. Find the exact value of $\sin^2 \dfrac{\pi}{4} + \cos^2 \dfrac{\pi}{4}$.

35. If light is incident on glass of refractive index $n = 1.60$ at an angle $\theta_1 = 30.3°$, find the angle θ_2 that the refracted ray makes with the perpendicular to the glass surface. (See Example 1.)

36. When a beam of electrons moves at a speed of $0.6c$ in a crystal whose index of refraction is 1.75, at what angle to the electron beam is the Cerenkov radiation emitted? (See Example 2.)

37. When vertically polarized light of intensity I_0 is incident on a polarizing filter that has its polarizing axis at an acute angle θ with the vertical, the intensity I_t of the transmitted light and θ are related by

$$\cos \theta = \sqrt{\frac{I_t}{I_0}}.$$

If $I_t = \frac{1}{4}I_0$, find θ by using your knowledge of special angles instead of a calculator.

38. For what value of θ will the intensity of the incident light in Problem 37 be reduced by one-half in passing through a polarizing filter? Use your knowledge of special angles instead of a calculator.

39. In an ac circuit, the potential difference V_R across a resistor as a function of time t, in seconds, is given by

$$V_R = 12 \sin (40\pi t),$$

where V_R is in volts. Find the exact value of V_R when $t = \frac{1}{120}$.

40. When a plane flies with a speed v_p that is greater than the speed v of sound in air, the envelope of the waves created is a cone, the surface of which makes an angle θ with the direction of motion of the plane. It can be shown that

$$\sin \theta = \frac{v}{v_p},$$

where the ratio v_p/v is called the *Mach number*. Find θ if the speed of a plane is Mach 1.3.

41. The range R of a projectile with an initial speed v_0 and an angle of projection θ is given by

$$R = \frac{v_0^2}{9.8} \sin(2\theta),$$

where v_0 is in meters per second and R is in meters.
(a) Find the range of a projectile fired with an initial speed of 40 m/s at an angle of 45°.
(b) Find the range of the same projectile if the angle of projection is lowered by 10°.

42. When a ball is thrown with an initial velocity of v_0 at an acute angle θ with the horizontal, the upward component of the velocity is given by the expression $v_0 \sin \theta$. If the initial velocity of the ball is 20 m/s, use your knowledge of special angles to find the value of θ that results in an upward component of 10 m/s.

43. The angle of deviation θ for which maxima occur in the diffraction pattern produced by a plane grating can be found from the equation

$$\sin \theta = \frac{m\lambda}{d},$$

where m is the order of the observed spectrum, λ is the wavelength of light, and d is the grating spacing. If a particular grating has a spacing of $d = 1.69 \times 10^{-6}$ m, what is the angular deviation of violet light of wavelength $\lambda = 4 \times 10^{-7}$ m in the first-order $(m = 1)$ visible spectrum?

44. In 1812, Brewster showed that when the refracted and reflected rays are perpendicular (see Example 1), the reflected light is completely polarized. Show that for this condition, Snell's law can be written $\tan \theta_1 = n$. *Hint:* First show that θ_1 and θ_2 are complementary.

45. When light is incident on glass with a refractive index $n = 1.57$, the reflected ray is found to be completely polarized; that is, the reflected and refracted rays are perpendicular.
(a) Find the angle (of incidence) θ_1.
(b) Find the angle that the refracted ray makes with the perpendicular to the glass surface. (See Example 1 and Problem 44.)

46. If θ is an angle near zero, then $\sin \theta \approx \tan \theta \approx \theta$, where θ is in radians. Confirm this by computing $\sin \theta$ and $\tan \theta$ if θ is (a) 0.4, (b) 0.1, (c) 0.01, (d) 0.001, and (e) 0.00034. Give your answers to five decimal places. The statement $\sin \theta \approx \tan \theta \approx \theta$ for θ near zero is often called the **small-angle approximation**.

47. The index of refraction n of a prism with apex angle A is given by

$$n = \frac{\sin\left(\dfrac{A + \delta}{2}\right)}{\sin\left(\dfrac{A}{2}\right)},$$

where δ (delta) is called the angle of minimum deviation. If a prism has an angle of minimum deviation of 27° and an apex angle of 55°, determine its index of refraction.

6.4 SOLUTION OF A RIGHT TRIANGLE

Six parts are associated with a triangle: three sides and three angles. The angles of a right triangle are commonly labeled A, B, and C, where C is the right angle (see Figure 6.36), and the sides opposite these angles are labeled a, b, and c, respectively. To *solve*

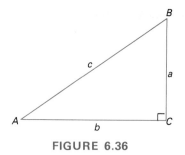

FIGURE 6.36

a right triangle means to find all unknown sides and angles when we know either

The length of one side and the measure of any acute angle

or

The lengths of any two sides.

Certain facts are useful when we are solving a right triangle. Recall that the sum of the two acute angles is 90°; that is, the acute angles of a right triangle are *complementary angles*. Also, when two sides are known, the third side can be found by applying the Pythagorean theorem. The examples that follow illustrate basic techniques in solving a right triangle. We shall omit units of length for convenience.

Example 1

Given one side and an acute angle.
Solve right triangle ABC given that $A = 32.2°$ and $b = 16.4$.

Solution   In Fig. 6.37 we have labeled a right triangle based on the given data. To solve for

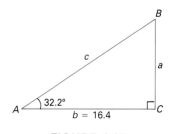

FIGURE 6.37

angle B, we use the fact that $A + B = 90°$. Thus

$$B = 90° - 32.2° = 57.8°.$$

To solve for side a we use the equation $\tan A = \text{opp}/\text{adj} = a/b$ because it involves the given data (A and b) and the unknown side (a):

$$\tan 32.2° = \frac{a}{16.4}.$$

Multiplying both sides by 16.4 gives

$$a = 16.4(\tan 32.2°) = 16.4(0.62973) = 10.3.$$

Here the keystroke sequence is (degree mode)

| 16.4 | × | 32.2 | TAN | = | Display: 10.3

To solve for side c we have

$$\cos A = \frac{\text{adj}}{\text{hyp}} = \frac{b}{c}$$

$$\cos 32.2° = \frac{16.4}{c}$$

$$c \cos 32.2° = 16.4.$$

Dividing both sides by $\cos 32.2°$ gives

$$c = \frac{16.4}{\cos 32.2°} = \frac{16.4}{0.84619} = 19.4.$$

Here the keystroke sequence is

| 16.4 | ÷ | 32.2 | COS | = | Display: 19.4

Thus the missing parts are

$$a = 10.3, \quad c = 19.4, \quad \text{and} \quad B = 57.8°.$$

We chose not to use the Pythagorean theorem to find c because that would have required us to use a calculated value of a. Moreover, if we had made an error in finding a, we would not have gotten a correct value for c. The moral is: Wherever practical, use the *given* parts to find the missing parts of a right triangle.

Here is a reminder. In Example 1, it is unnecessary to show (or write down) the rounded numbers 0.62973 and 0.84619. These intermediate values are shown here because of the mental pictures they convey.

Example 2

Given two sides.
Solve right triangle ABC given that $c = 13$ and $b = 5$.

Solution Figure 6.38 shows a right triangle labeled with the given data. To solve for angle A, we have

$$\cos A = \frac{\text{adj}}{\text{hyp}} = \frac{5}{13} = 0.38462$$

$$A = 67.4°,$$

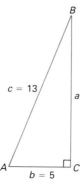

FIGURE 6.38

which is obtained from the keystroke sequence (degree mode)

$$\boxed{5} \quad \boxed{\div} \quad \boxed{13} \quad \boxed{=} \quad \boxed{\text{COS}^{-1}} \qquad \text{Display: } 67.4$$

The entire displayed value of A should be stored in memory for later use. Because $A + B = 90°$,

$$B = 90° - A = 90° - 67.4° = 22.6°.$$

Alternatively, we can find B without making use of the computed value of A.

$$\sin B = \frac{\text{opp}}{\text{hyp}} = \frac{5}{13}.$$

$$\boxed{5} \quad \boxed{\div} \quad \boxed{13} \quad \boxed{=} \quad \boxed{\text{SIN}^{-1}} \qquad \text{Display: } 22.6$$

To find side a we can use the Pythagorean theorem.

$$a^2 + 5^2 = 13^2$$
$$a^2 = 13^2 - 5^2$$
$$a = \sqrt{13^2 - 5^2} = \sqrt{144} = 12.$$

Here the keystroke sequence is

$$\boxed{13} \quad \boxed{x^2} \quad \boxed{-} \quad \boxed{5} \quad \boxed{x^2} \quad \boxed{=} \quad \boxed{\sqrt{}} \qquad \text{Display: } 12$$

Alternatively, we can find side a as follows.

$$\sin A = \frac{a}{c}$$

$$a = c \sin A = 13 \sin A.$$

Because A is in memory, the keystroke sequence is

$$\boxed{13} \quad \boxed{\times} \quad \boxed{\text{RCL}} \quad \boxed{\text{SIN}} \quad \boxed{=} \qquad \text{Display: } 12$$

Thus the missing parts are

$$a = 12, \quad A = 67.4°, \quad \text{and} \quad B = 22.6°.$$

Example 3

The right triangle in Fig. 6.39(a) is called a **power triangle** and is used in solving ac circuit problems. The hypotenuse represents the **apparent power** P_A, measured in volt-amperes (VA). The other sides represent the **real power** P, measured in watts (W), and the **reactive power** P_R, measured in vars. The angle θ is called the **phase angle**. In a particular circuit, the real power is $P = 23$ W and the reactive power is $P_R = 5.1$ vars. Find the apparent power P_A and the phase angle.

(a) (b)

FIGURE 6.39

Solution Figure 6.39(b) shows the values for our particular problem. We have

$$\tan \theta = \frac{5.1}{23} = 0.22174$$

$$\theta = 12.5°.$$

Also

$$P_A = \sqrt{23^2 + 5.1^2} = 23.6 \text{ VA}.$$

Example 4

An ac circuit has an inductive reactance $X_L = 10.1$ Ω and a resistance $R = 22.3$ Ω. In such a circuit the **impedance** Z, in ohms, can be represented by the hypotenuse of a right triangle with X_L and R as its other sides, as shown in Fig. 6.40. The angle θ is called the **phase angle**. Find the phase angle and the impedance.

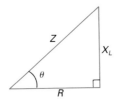

FIGURE 6.40

Solution   From Fig. 6.40, we have

$$\tan \theta = \frac{X_L}{R} = \frac{10.1}{22.3} = 0.45291.$$

Thus

$$\theta = 24.4°.$$

Also

$$Z = \sqrt{22.3^2 + 10.1^2} = 24.5 \ \Omega.$$

Example 5

A plane flying at an altitude of 1 mi begins to climb at a constant angle of 28°. To the nearest second, how long will it take the plane to reach an altitude of 1.75 mi if its constant speed throughout the climb is 200 mi/h?

Solution   The plane must increase its altitude by 0.75 mi. See the right triangle in Fig. 6.41. We first find the distance d (in miles) of the climb.

$$\sin 28° = \frac{0.75}{d}$$

$$d \sin 28° = 0.75$$

$$d = \frac{0.75}{\sin 28°} = \frac{0.75}{0.46947} = 1.5975 \ \text{mi}.$$

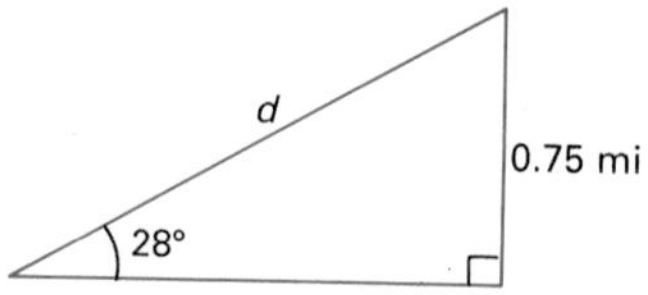

FIGURE 6.41

The time to travel this 1.5975 mi is given by

$$\text{Time} = \frac{\text{distance}}{\text{rate}} = \frac{1.5975 \ \text{mi}}{200 \ \text{mi/h}} = 0.008 \ \text{h}.$$

Finally, we need to convert 0.008 h to seconds. Because 1 h = 3600 s,

$$0.008 \ \text{h} = (0.008 \ \text{h})\left(\frac{3600 \ \text{s}}{1 \ \text{h}}\right) = 29 \ \text{s}.$$

Two expressions that occur at times in word problems are *angle of elevation* and *angle of depression*. In Fig. 6.42, if an observer at point P looks up at a point Q, then

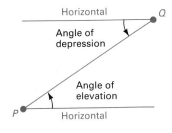

FIGURE 6.42

the angle that the line PQ makes with the horizontal line containing P is called the **angle of elevation** of Q from P. If an observer at Q looks down at P, then the angle that PQ makes with the horizontal line containing Q is called the **angle of depression** of P from Q (see Fig. 6.42). The angle of elevation of Q from P is equal to the angle of depression of P from Q because these are alternate interior angles formed by two parallel lines cut by a transversal.

Example 6

The length of a kite string is 228 ft and the angle of elevation of the kite is 60°. Find the height of the kite (see Fig. 6.43).

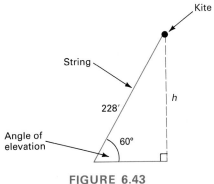

FIGURE 6.43

Solution   If h is the height in feet, then from the right triangle in Fig. 6.43,

$$\sin 60° = \frac{\text{opp}}{\text{hyp}} = \frac{h}{228}$$

$$h = 228 \sin 60°$$

$$= 228\left(\frac{\sqrt{3}}{2}\right) = 114\sqrt{3} \text{ ft.}$$

Alternatively, a calculator gives

$$h = 228 \sin 60° = 197.5 \text{ ft.}$$

Example 7

A boat leaves a dock and travels 200 mi at an angle of 23° east of north (we say that the *bearing* is N 23° E). It then travels 100 mi due east. What is the distance of the boat from the dock?

Solution Figure 6.44 describes the situation, where the dock is at A.

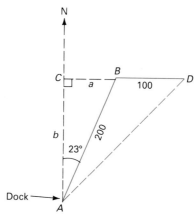

FIGURE 6.44

In right triangle ADC we shall find $\overline{CD}$ and $\overline{AC}$ and then use these sides to find $\overline{AD}$, the desired distance. From right triangle ABC we have

$$\sin 23° = \frac{a}{200}$$

$$200 \sin 23° = a.$$

Also,

$$\cos 23° = \frac{b}{200}$$

$$200 \cos 23° = b.$$

In triangle ADC,

$$\overline{CD} = a + 100 = 200 \sin 23° + 100$$

and

$$\overline{AC} = b = 200 \cos 23°.$$

Thus

$$\overline{AD} = \sqrt{\overline{CD}^2 + \overline{AC}^2}$$

$$= \sqrt{(200 \sin 23° + 100)^2 + (200 \cos 23°)^2}$$

$$= 256.2 \text{ mi.}$$

Problem Set 6.4

In Problems 1–6, ABC is a right triangle with right angle C. Solve for the remaining parts of the triangle. Do not use a calculator.

1. $B = 60°$, $a = 3$.
4. $B = 60°$, $c = 1$.

2. $A = 30°$, $b = 3$.
5. $b = 4$, $c = 8$.

3. $A = 45°$, $c = 6$.
6. $a = 2$, $b = 2$.

In Problems 7–16, solve the right triangle ABC for the remaining parts.

7. $c = 12.4$, $A = 33.5°$.
10. $a = 4$, $c = 15$.
13. $b = 24.7$, $c = 60.3$.
16. $a = 82.5$, $B = 65.7°$.

8. $a = 12$, $A = 12°$.
11. $b = 24.6$, $B = 27.4°$.
14. $c = 80.4$, $B = 86.5°$.

9. $a = 23.4$, $b = 18.1$.
12. $a = 3.5$, $b = 2.8$.
15. $b = 90.8$, $A = 12.6°$.

17. An ac circuit has an impedance $Z = 36\ \Omega$. If the phase angle θ is $32.6°$, find the resistance R and the inductive reactance X_L. (See Example 4.)

18. An ac circuit has an impedance $Z = 22\ \Omega$. If the circuit has a resistance $R = 18.9\ \Omega$, find the inductive reactance X_L and the phase angle θ. (See Example 4.)

19. The apparent power in a circuit is 1.65 VA and the phase angle is $59°$. Find the real power and the reactive power. (See Example 3.)

20. If the real power in a circuit is 1.73 W and the phase angle θ is $66.5°$, find the apparent power and the reactive power. (See Example 3.)

21. From the top of a 50-m cliff that overlooks a bay, the angle of depression of a buoy is $14°$. To the nearest meter, what is the distance along the water of the buoy from the cliff?

22. The length of a kite string is 600 ft, and the angle of elevation of the kite is $34°$. How high is the kite? Give your answer to the nearest foot.

23. A tree casts a shadow 17 m long when the angle of elevation of the sun from the tip of the shadow is $50°$. Find the height of the tree.

24. From the top of a 100-m tower, the angle of depression of a person on the ground is $39°$. How far from the base of the tower is the person?

25. An airplane pilot wants to increase his altitude by 3 mi. He plans to climb at a constant angle of $10°$ and a constant rate of 200 mi/h. How many minutes will it take to reach that altitude?

26. A wire bracing an antenna is 15 m long. One end is attached to the top of the antenna, and the other end is attached to level gound at a distance of 12 m from the base of the antenna. Find the angle that the wire makes with the ground.

27. A bridge will be constructed at a certain point along a river. To gather engineering data, a surveyor is sent out to find the width w of the river (see Fig. 6.45). With her transit at C, the surveyor determines the right angle C. She measures the distance from point C to point A and finds it to be 33 m. Then, with her transit at point A, she determines that angle A is $69.6°$. What is the width of the river?

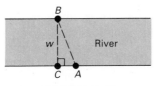

FIGURE 6.45

28. An escalator inclined at an angle of 42.6° moves at a constant speed of 0.6 m/s. A person steps on the escalator and reaches the top in 42 s. Through what vertical height has the person been lifted?

29. To determine the height of a building, a surveyor determines that the angle of elevation to the top is 53.5°. If the sighting instrument was 30.3 m from the building and 1.5 m above the ground, find the height of the building.

30. The leaning tower of Pisa was designed to be 56 m high, and its top is now 5.2 m out of plumb. Find the angle that the axis of the tower makes with the vertical.

31. A handrail is to be installed along a stairway that has a horizontal distance of 9.1 m and that rises 6.1 m vertically. (a) How long must the handrail be? (b) At what angle with the horizontal should it be mounted?

32. After passing a toll gate, a car travels on a road that rises until it reaches a bridge. If the angle of elevation of the road is 3.5° and the length of the road is 170 meters, how high above ground level is the bridge?

33. The Arctic Circle is approximately at latitude 66.5 N (see Fig. 6.46). Given that the radius of the earth is 6370 km, find the radius of the Arctic Circle.

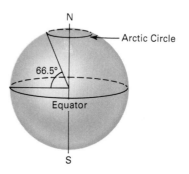

FIGURE 6.46

34. An airplane, flying at 90 km/h, travels for 1.5 h on a straight-line course heading 15.8° west of north. How far west and how far north of its starting point is it?

35. An antenna is situated on the edge of a cliff. From a point on level ground 480 m from the base of the cliff, the angle of elevation of the top of the antenna is 15°; the angle of elevation of the bottom of the antenna is 13°. How tall is the antenna?

36. The floor of a wedge-shaped storage shed is a square that is 5.0 m on a side. If the roof contacts the back edge of the floor at an angle of 16°, find the dimensions of the roof.

37. A man 2 m tall stands 4 m from the base of a street light. He can observe the light when his line of sight makes an angle of 42° with the horizontal. Find the length of the shadow cast by the man.

38. A degree may seem like a very small unit, but an error of 1° in measuring an angle may be very significant. For example, suppose that a laser beam is directed toward the visible center of the moon and that it misses its assigned target by 0.008°. How far is it (in miles) from its assigned target? (Treat the part of the moon's surface involved in this problem as a piece of a plane perpendicular to the laser beam, and take the distance from the surface of the earth to the assigned target on the moon to be 234,000 mi.)

39. Two office buildings on level ground are 500 ft apart. The angles of depression from the top of the taller building to the top and the base of the other building are 10° and 25°, respectively. Find the heights of the two buildings. Give your answers to the nearest foot.

40. A pilot flying over an ocean at an altitude of 4000 ft sights two ships at points A and B directly ahead of him. He determines that the angle of depression of point A is 35°, and for point B it is 15°. Find the distance (in feet) between A and B.

41. A nut in the shape of a regular hexagon measures 2 cm, as shown in Fig. 6.47. What is the smallest diameter of a circular pipe that will fit over the nut?

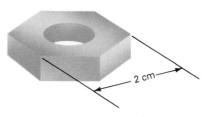

FIGURE 6.47

42. It is a common experience to hear an approaching jet aircraft, especially a low-flying one, and—upon looking in the direction of the sound—to find the aircraft in a very different part of the sky. Suppose that the sound of a plane traveling at 200 mi/h and at an altitude of 3000 ft is heard by an observer, and the sound emanates from a position in the sky that is 20° above the horizon. If the plane passes directly over the observer, at what angle of elevation should the observer look to see the aircraft when it is heard? Give your answer to the nearest degree. (*Note:* 1 mi = 5280 ft. Take the speed of sound to be 1100 ft/s.)

43. A boat leaves a port and travels 200 km due west. It then travels 100 km at an angle of 40° west of south. Find the distance of the boat from the port.

6.5 REVIEW

Important Terms and Symbols

Section 6.1 Angle, initial side, terminal side, vertex, positive angle, negative angle, standard position, quadrant of an angle, quadrantal angle, degree, minute, second, acute angle, obtuse angle, right angle, radian, coterminal angles, arc length.

Section 6.2 Radius vector (r), reference triangle, trigonometric functions, sine (sin), sin θ, cosine (cos), tangent (tan), cotangent (cot), secant (sec), cosecant (csc), reciprocal relations, reciprocal functions, opposite side, adjacent side, hypotenuse, complementary angles, cofunctions.

Section 6.3 Special angles.

Section 6.4 Angle of elevation, angle of depression.

Formula Summary

Degrees and Radians

π rad $= 180°$

1 rad $= \left(\dfrac{180}{\pi}\right)°$

$1° = \dfrac{\pi}{180}$ rad

Arc Length

$s = r\theta$, θ in radians

Trigonometric Functions For Acute Angle

$\sin \theta = \dfrac{y}{r}$ $\csc \theta = \dfrac{r}{y}$

$\cos \theta = \dfrac{x}{r}$ $\sec \theta = \dfrac{r}{x}$

$\tan \theta = \dfrac{y}{x}$ $\cot \theta = \dfrac{x}{y}$

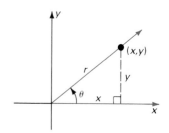

Reciprocal Relations

$\csc \theta = \dfrac{1}{\sin \theta}$

$\sec \theta = \dfrac{1}{\cos \theta}$

$\cot \theta = \dfrac{1}{\tan \theta}$

Additional Relations

$\dfrac{\sin \theta}{\cos \theta} = \tan \theta$

$\dfrac{\cos \theta}{\tan \theta} = \cot \theta$

Right Triangle Trigonometry

$\sin \theta = \dfrac{\text{opp}}{\text{hyp}}$ $\csc \theta = \dfrac{\text{hyp}}{\text{opp}}$

$\cos \theta = \dfrac{\text{adj}}{\text{hyp}}$ $\sec \theta = \dfrac{\text{hyp}}{\text{adj}}$

$\tan \theta = \dfrac{\text{opp}}{\text{adj}}$ $\cot \theta = \dfrac{\text{adj}}{\text{opp}}$

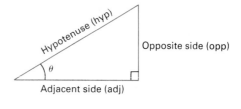

Special Angles

$30° = \dfrac{\pi}{6}$

$45° = \dfrac{\pi}{4}$

$60° = \dfrac{\pi}{3}$

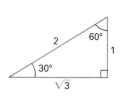

Review Questions

1. Which of the following statements is (are) true? _____.
   a. $\sin 21° = 1/\sec 69°$.   b. $-90°$ is a quadrantal angle.
   c. $90° = \pi$ rad   d. $\tan 60° = \cot 30°$.
   e. $\sin 45° = \cos 45°$.   f. $130°$ is an obtuse angle.
2. In 3 min the second hand of a clock rotates through _____ radians.
3. The angles $10°$, $370°$, and $-350°$ are _____ with one another.
4. In Fig. 6.48, $a =$ _____.

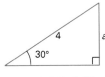

FIGURE 6.48

5. In Fig. 6.49, $\theta =$ _____.

FIGURE 6.49

6. To convert x radians to degrees, we multiply x by _____.
7. If $\sin \theta = \frac{3}{4}$, then $\csc \theta =$ _____.
8. If $\sin \theta = a$ and $\cos \theta = b$, then $\cot \theta =$ _____.
9. $720°$ equals how many radians? _____.

Answers to Review Questions

1. a, b, d, e, f.   2. 6π.   3. Coterminal.   4. 2.   5. $45°$.   6. $180°/\pi$.   7. $\frac{4}{3}$.
8. b/a.   9. 4π.

Review Problems

In Problems 1–8, convert each degree measurement to radians and each radian measurement to degrees.

1. $300°$.   2. $750°$.   3. $\dfrac{5\pi}{6}$.   4. $-\dfrac{9\pi}{4}$.

5. $-50°$.   6. $36°$.   7. $\dfrac{\pi}{18}$.   8. $\dfrac{3\pi}{10}$.

*In Problems **9–12**, find the quadrants of the angles.*

9. 224°.   **10.** −370°.   **11.** $\dfrac{7\pi}{3}$.   **12.** $\dfrac{9\pi}{10}$.

*In Problems **13–16**, find the exact arc length cut off by the given central angle θ on a circle with the given radius r.*

13. $\theta = \dfrac{7\pi}{6}$, $r = 12$.

14. $\theta = \dfrac{5\pi}{4}$, $r = 10$.

15. $\theta = 135°$, $r = 2$.

16. $\theta = 60°$, $r = 3$.

17. Find an angle θ that is coterminal with 500°, such that $0° \le \theta < 360°$.

18. Find an angle θ that is coterminal with $-5\pi/4$, such that $0 \le \theta < 2\pi$.

*In Problems **19** and **20**, the given point lies on the terminal side of an angle θ in standard position. Find the values of the six trigonometric functions of θ.*

19. (1, 4).   **20.** (2, 2).

21. If $\sin \theta = \dfrac{1}{5}$ and θ is acute, find the other trigonometric values of θ.

22. If $\tan \theta = \dfrac{2}{3}$ and θ is acute, find the other trigonometric values of θ.

*In Problems **23–26**, find the given value. Do not use a calculator.*

23. $\sin 60°$.   **24.** $\cos 30°$.

25. $\sec \dfrac{\pi}{4}$.   **26.** $\cot \dfrac{\pi}{6}$.

*In Problems **27–34**, solve the right triangle ABC for the remaining parts.*

27. $a = 4$, $b = 10$.   **28.** $a = 5$, $c = 21$.   **29.** $a = 6$, $B = 15°$.

30. $b = 9$, $A = 24°$.   **31.** $b = 7$, $A = 46°$.   **32.** $c = 12$, $B = 35.2°$.

33. $c = 20$, $A = 32.6°$.   **34.** $a = 45$, $A = 60°$.

35. An antenna mast, which had its top section blown over by the wind, forms a right triangle with the ground. The broken part makes an angle of 57° with the ground. The top of the antenna touches the ground at a point 30 m from the base of the antenna. Find the original height of the antenna.

36. A wrecking company plans to make a bid for demolishing a smokestack and needs to know its height. At a point 34 m from the base of the stack, the wreckers find that the angle of elevation of its top is 32°. What is the height of the smokestack?

37. One university boasts that a tower clock has never stopped in the past 20 years. The minute hand is 1 ft long. How many miles (to the nearest mile) has the tip of the minute hand traveled in that time? Assume that there are 365 d in a year. (*Note:* 1 mi = 5280 ft.)

38. On level ground, a vertical pole casts a shadow 20 m long. If the rays of the sun make an angle of 55° with the horizontal, how tall is the pole?

39. The inclination of an agricultural grain conveyor can be set between 15° and 50°. If the length of the conveyor is 7 m, what are the minimum and maximum heights the conveyor can reach?

40. A plane is flying at an altitude of 1 mi and its pilot observes an airport. If the angle of depression of the airport is 10°, determine the number of miles between the plane and the airport.

41. A V-gauge is a device for measuring the diameter of a wire. Suppose that in the V-gauge shown in Fig. 6.50, angle AVB measures 54°. Find the length $\overline{VP}$ for a wire of diameter 2.0 mm.

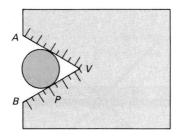

FIGURE 6.50

7

Reference Angles
and Vectors

7.1   TRIGONOMETRIC FUNCTIONS OF ANY ANGLE

In the preceding chapter we discussed trigonometric function values for acute angles only. More generally, for *any* angle θ we use the same definition as before:

1. Place θ in standard position.
2. Let (x, y) be any point on the terminal side of θ (except the origin).
3. Let $r = \sqrt{x^2 + y^2}$. Then

$$\sin \theta = \frac{y}{r}, \qquad \csc \theta = \frac{r}{y},$$

$$\cos \theta = \frac{x}{r}, \qquad \sec \theta = \frac{r}{x},$$

$$\tan \theta = \frac{y}{x}, \qquad \cot \theta = \frac{x}{y}.$$

Example 1

Find the values of the trigonometric functions of θ if θ is in standard position and the point $(-3, 2)$ lies on its terminal side (see Fig. 7.1).

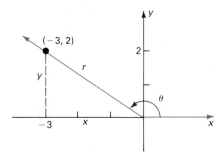

FIGURE 7.1

Solution   Letting $x = -3$ and $y = 2$, we have

$$r = \sqrt{x^2 + y^2} = \sqrt{(-3)^2 + 2^2} = \sqrt{13}.$$

Thus

$$\sin \theta = \frac{y}{r} = \frac{2}{\sqrt{13}} = \frac{2\sqrt{13}}{13}$$

$$\cos \theta = \frac{x}{r} = \frac{-3}{\sqrt{13}} = -\frac{3\sqrt{13}}{13}$$

$$\tan \theta = \frac{y}{x} = \frac{2}{-3} = -\frac{2}{3}$$

$$\cot \theta = \frac{x}{y} = \frac{-3}{2} = -\frac{3}{2}$$

$$\sec \theta = \frac{r}{x} = \frac{\sqrt{13}}{-3} = -\frac{\sqrt{13}}{3}$$

$$\csc \theta = \frac{r}{y} = \frac{\sqrt{13}}{2}.$$

You should realize that *the reciprocal relations of Sec. 6.2 are true for any angle θ for which they are defined.* As a result, we could have used them here to find $\cot \theta$, $\sec \theta$, and $\csc \theta$ from $\tan \theta$, $\cos \theta$, and $\sin \theta$, respectively.

In Example 1 some function values are positive and others are negative, because their signs depend on the signs of x, y, and r. Suppose we consider the signs of the trigonometric functions of angles in the various quadrants. Let (x, y) be a point on the terminal side of a first-quadrant angle. Both x and y are positive and r is *always* positive. We conclude that all trigonometric functions of a first-quadrant angle are positive. For a second-quadrant angle, x is negative, whereas y and r are positive. Thus only the sine and cosecant functions have positive values. Using similar reasoning you should determine the signs of the functions for third- and fourth-quadrant angles. The chart in Fig. 7.2 gives a summary of the functions that have positive values for angles in the various quadrants.

Positive trigonometric functions

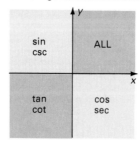

FIGURE 7.2

Example 2

Use the chart in Fig. 7.2 to find the signs of the following function values.

a. tan 200°.

Solution   A 200° angle is a third-quadrant angle (see Fig. 7.3). From the chart, tan 200° is positive.

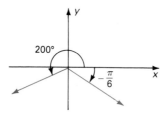

FIGURE 7.3

b. $\sin\left(-\dfrac{\pi}{6}\right)$.

Solution   An angle of $-\dfrac{\pi}{6}$ is a fourth-quadrant angle (see Fig. 7.3). From the chart, the only positive functions here are the cosine and secant. Thus $\sin\left(-\dfrac{\pi}{6}\right)$ is negative.

We now consider the values of the trigonometric functions of some quadrantal angles. You will see that such values are not always defined.

Example 3

Find the six trigonometric function values of $0°$.

Solution Figure 7.4(a) shows the angle $0°$. For a point (x, y) on the terminal side, we shall choose $(1, 0)$. Then $r = 1$, and we have

$$\sin 0° = \frac{y}{r} = \frac{0}{1} = 0,$$

$$\cos 0° = \frac{x}{r} = \frac{1}{1} = 1,$$

$$\tan 0° = \frac{y}{x} = \frac{0}{1} = 0,$$

$$\cot 0° = \frac{x}{y} = \frac{1}{0}, \qquad \text{which is \textbf{not defined}},$$

$$\sec 0° = \frac{r}{x} = \frac{1}{1} = 1,$$

$$\csc 0° = \frac{r}{y} = \frac{1}{0}, \qquad \text{which is \textbf{not defined}}.$$

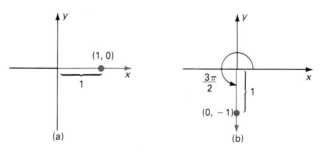

FIGURE 7.4

Example 4

Find the six trigonometric function values of $\dfrac{3\pi}{2}$.

Solution   See Fig. 7.4(b). We shall choose $(0, -1)$ as a point (x, y) on the terminal side. Then $r = 1$ and we have

$$\sin \frac{3\pi}{2} = \frac{y}{r} = \frac{-1}{1} = -1,$$

$$\cos \frac{3\pi}{2} = \frac{x}{r} = \frac{0}{1} = 0,$$

$$\tan \frac{3\pi}{2} = \frac{y}{x} = \frac{-1}{0}, \qquad \text{which is \textbf{not defined},}$$

$$\cot \frac{3\pi}{2} = \frac{x}{y} = \frac{0}{-1} = 0,$$

$$\sec \frac{3\pi}{2} = \frac{r}{x} = \frac{1}{0}, \qquad \text{which is \textbf{not defined},}$$

$$\csc \frac{3\pi}{2} = \frac{r}{y} = \frac{1}{-1} = -1.$$

Function values for $90°$ and $180°$ can be found in a similar way by using the points $(0, 1)$ and $(-1, 0)$, respectively. Table 7.1 summarizes trigonometric function values for quadrantal angles. A dash means that the value is not defined.

TABLE 7.1

	$0°$ (or 0)	$90°$ $\left(\text{or } \dfrac{\pi}{2}\right)$	$180°$ (or π)	$270°$ $\left(\text{or } \dfrac{3\pi}{2}\right)$
sin	0	1	0	-1
cos	1	0	-1	0
tan	0	—	0	—
cot	—	0	—	0
sec	1	—	-1	—
csc	—	1	—	-1

If α and β are coterminal angles, then we can choose a point that lies both on the terminal side of α and the terminal side of β (see Fig. 7.5). It follows from the definitions of the trigonometic functions that $\sin \alpha = \sin \beta$, $\cos \alpha = \cos \beta$, and so on. In general, **corresponding trigonometric function values of coterminal angles are equal**. For example, $-\pi/2$ and $3\pi/2$ are coterminal (see Fig. 7.6), so $\sin(-\pi/2) = \sin(3\pi/2) = -1$ (from Example 4).

With each angle θ that is not quadrantal, we can associate another angle, called the **reference angle** of θ. It is the positive *acute* angle between the terminal side of θ and the x-axis (not the y-axis). Figure 7.7 shows different situations.

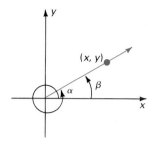

FIGURE 7.5

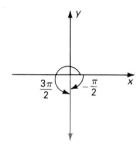

FIGURE 7.6

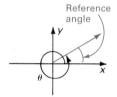 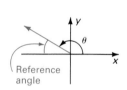 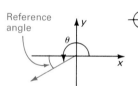 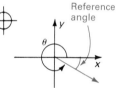

FIGURE 7.7

Example 5

Find the reference angles of $150°$, $\dfrac{5\pi}{4}$, $300°$, and $400°$.

Solution   In Fig. 7.8 each angle is shown with its reference angle, denoted θ_R. For $150°$ (a second-quadrant angle), we find the reference angle by subtracting $150°$ from $180°$.

$$\theta_R = 180° - 150° = 30°.$$

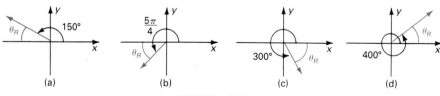

FIGURE 7.8

For $5\pi/4$ (a third-quadrant angle), we subtract π from $5\pi/4$.

$$\theta_R = \frac{5\pi}{4} - \pi = \frac{\pi}{4}.$$

For $300°$ (a fourth-quadrant angle), we subtract $300°$ from $360°$.

$$\theta_R = 360° - 300° = 60°.$$

For $400°$,

$$\theta_R = 400° - 360° = 40°.$$

If an angle is not quadrantal, we can find its trigonometric function values from those of its reference angle. For example, consider $150°$ [see Fig. 7.9(a)]. Its reference

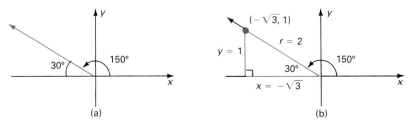

FIGURE 7.9

angle is $30°$. To locate a point on the terminal side of $150°$, we shall make use of a $30°$-$60°$-$90°$ triangle. From Fig. 7.9(b), the point $(-\sqrt{3}, 1)$ lies on the terminal side and $r = 2$. Thus

$$\sin 150° = \frac{y}{r} = \frac{1}{2}$$

$$\cos 150° = \frac{x}{r} = \frac{-\sqrt{3}}{2} = -\frac{\sqrt{3}}{2},$$

and so on. We can obtain these values in a simpler way. For the reference angle, we have

$$\sin 30° = \frac{1}{2}, \qquad \cos 30° = \frac{\sqrt{3}}{2},$$

and so forth. Notice that the function values of $150°$ are numerically equal to those of its reference angle, $30°$. These values may differ only in their signs. Thus to find $\cos 150°$ we can find $\cos 30°$ and put a minus sign in front of it. The minus sign is needed because the cosine of a second-quadrant angle ($150°$) is negative. Similarly, tan 150° must be negative, so tan 150° = −(tan 30°) = −1/√3 = −√3/3. In general, we have the following.

> To find a trigonometric function value of an angle θ, find the same trigonometric function value of the reference angle of θ and attach the proper sign. This sign depends on the function involved and the quadrant of θ.

Example 6

a. Find sin 135°.

Solution   Because 135° is a second-quadrant angle [Fig. 7.10(a)], sin 135° is positive. The reference angle is 180° − 135° = 45°. Thus

$$\sin 135° = +\sin 45° = \frac{1}{\sqrt{2}} = \frac{\sqrt{2}}{2}.$$

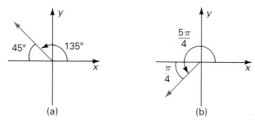

(a)   (b)

FIGURE 7.10

b. Find $\cos \dfrac{5\pi}{4}$.

Solution   Because $\dfrac{5\pi}{4}$ is a third-quadrant angle [Fig. 7.10(b)], $\cos \dfrac{5\pi}{4}$ is negative. The reference angle is $\dfrac{5\pi}{4} - \pi = \dfrac{\pi}{4}$ (or 45°). Thus

$$\cos \frac{5\pi}{4} = -\cos \frac{\pi}{4} = -\frac{1}{\sqrt{2}} = -\frac{\sqrt{2}}{2}.$$

Example 7

a. Find $\cot\left(-\dfrac{\pi}{6}\right)$.

Solution Because $-\pi/6$ is a fourth-quadrant angle [Fig. 7.11(a)], $\cot(-\pi/6)$ is negative. The reference angle is $\pi/6$ (or $30°$). Thus

$$\cot\left(-\frac{\pi}{6}\right) = -\cot\frac{\pi}{6} = -\frac{\sqrt{3}}{1} = -\sqrt{3}.$$

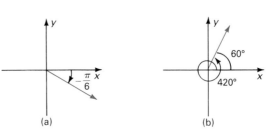

(a)   (b)

FIGURE 7.11

b. Find $\sec 420°$.

Solution Because $420°$ is a first-quadrant angle [Fig. 7.11(b)], $\sec 420°$ is positive. The reference angle is $420° - 360° = 60°$. Thus

$$\sec 420° = +\sec 60° = \frac{2}{1} = 2.$$

Example 8

Find $\tan 164.6°$.

Solution Because $164.6°$ is a second-quadrant angle (see Fig. 7.12), $\tan 164.6°$ is negative.

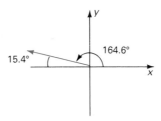

FIGURE 7.12

The reference angle is $180° - 164.6° = 15.4°$. Thus

$$\tan 164.6° = -\tan 15.4° = -0.2754.$$

You should verify our result by finding $\tan 164.6°$ directly with a calculator without using a reference angle.

Examples 9 and 10 show how to find an angle when you are given one of its trigonometric function values.

Example 9

Find all angles θ, where $0° \leq \theta < 360°$, such that $\tan \theta = -0.3640$.

Solution   Forget about the minus sign for a moment. Using the TAN^{-1} key on a calculator, we find that the acute angle with tangent of 0.3640 is 20°. Thus the reference angle of θ is 20°. Now, back to the minus sign. The tangent function is negative for second- and fourth-quadrant angles. Thus there must be *two* values of θ that meet the given conditions (see Fig. 7.13). We have

$$\theta = 180° - 20° = 160° \quad \text{and} \quad \theta = 360° - 20° = 340°,$$

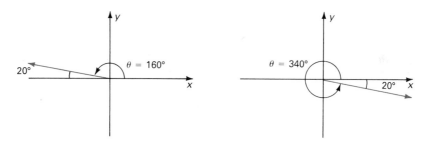

FIGURE 7.13

In Example 9, the condition that $\tan \theta = -0.3640$ was given and the *acute* reference angle with tangent of 0.3640 was initially found with the use of the TAN^{-1} key. If, instead, this key is used after entering the given value, -0.3640, a typical calculator result is $-20°$. You should try it on your calculator before you read on. The calculator result is not incorrect, because $\tan(-20°) = -0.3640$, but it does have to be interpreted in the context of the problem. Calculators usually give values of an angle that are within a defined range of values. For TAN^{-1}, these **principal values** are between $-90°$ and $90°$.* You can see that the calculator result is within that range. However, the problem requires that $0° \leq \theta < 360°$, so $\theta = -20°$ is unacceptable. The coterminal angle $360° - 20° = 340°$ is appropriate, as is the second-quadrant angle 160°.

Example 10

Find θ if $\cos \theta = -0.8360$ and $0° \leq \theta < 360°$.

---

* We shall have more to say about principal values in Chapter 16.

Solution   The cosine function is negative for second- and third-quadrant angles. The acute angle with cosine of 0.8360 is 33.3°. Thus (see Fig. 7.14)

$$\theta = 180° - 33.3° = 146.7°,$$

and

$$\theta = 180° + 33.3° = 213.3°.$$

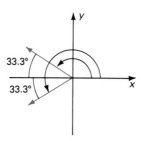

FIGURE 7.14

Example 11

Suppose that $\sin \theta = -\frac{1}{4}$. Find the other trigonometric function values of θ if $\tan \theta$ is positive.

Solution   First we find the quadrant of θ. Because $\sin \theta$ is negative, θ is a third- or fourth-quadrant angle. But since $\tan \theta$ is positive, θ is a first- or third-quadrant angle. Both conditions are met if θ is a third-quadrant angle. Because $\sin \theta = y/r = -\frac{1}{4}$, we may choose $y = -1$ and $r = 4$ (remember, r must be positive). See Fig. 7.15.

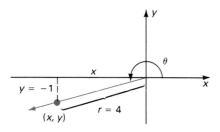

FIGURE 7.15

To find x we apply the Pythagorean theorem to the right triangle formed by the terminal side of θ, the x-axis, and the dashed line.

$$x^2 + y^2 = r^2$$
$$x^2 + (-1)^2 = 4^2$$
$$x^2 + 1 = 16$$
$$x^2 = 15.$$

Because θ is a third-quadrant angle, x must be negative. Thus

$$x = -\sqrt{15}, \qquad y = -1, \qquad r = 4.$$

Therefore,

$$\cos \theta = \frac{x}{r} = \frac{-\sqrt{15}}{4} = -\frac{\sqrt{15}}{4}$$

$$\tan \theta = \frac{y}{x} = \frac{-1}{-\sqrt{15}} = \frac{\sqrt{15}}{4}$$

$$\cot \theta = \frac{x}{y} = \frac{-\sqrt{15}}{-1} = \sqrt{15}$$

$$\sec \theta = \frac{r}{x} = \frac{4}{-\sqrt{15}} = -\frac{4\sqrt{15}}{15}$$

$$\csc \theta = \frac{r}{y} = \frac{4}{-1} = -4.$$

Problem Set 7.1

In Problems **1–6**, *the given point lies on the terminal side of an angle θ in standard position. Find the six trigonometric function values of θ.*

1. $(-1, -1)$.   2. $(-3, -3)$.   3. $(-1, 7)$.

4. $(-2, -10)$.   5. $(1, -\sqrt{3})$.   6. $(-3\sqrt{3}, 3)$.

In Problems **19–24**, *give the sign of the indicated function value.*

7. $\sin 100°$.   8. $\tan 200°$.   9. $\csc 280°$.

10. $\cos 72°$.   11. $\cot 175°$.   12. $\sec 240°$.

13. $\tan \dfrac{\pi}{4}$.   14. $\sin \dfrac{7\pi}{6}$.   15. $\cos \dfrac{4\pi}{3}$.

16. $\csc \dfrac{5\pi}{6}$.   17. $\sec \dfrac{11\pi}{6}$.   18. $\cot \dfrac{7\pi}{4}$.

19. $\cos 460°$.   20. $\sin (-20°)$.   21. $\tan \left(-\dfrac{3\pi}{4}\right)$.

22. $\csc \dfrac{8\pi}{3}$.   23. $\sec \left(-\dfrac{7\pi}{6}\right)$.   24. $\cot \dfrac{25\pi}{6}$.

In Problems **25–32**, *find the reference angle of the given angle.*

25. $115°$.   26. $310°$.   27. $227°$.   28. $405°$.

29. $-22.6°$.   30. $\dfrac{5\pi}{6}$.   31. $\dfrac{13\pi}{6}$.   32. $-\dfrac{3\pi}{4}$.

In Problems **33-52**, *find the given function values without using a calculator.*

33. sin 150°.   **34.** sin 120°.   **35.** cos 240°.   **36.** cos 225°.

37. tan (−225°).   **38.** tan 300°.   **39.** cot 315°.   **40.** cot (−210°).

41. sec 210°.   **42.** sec 330°.   **43.** csc (−90°).   **44.** csc 150°.

45. sin 450°.   **46.** cos 720°.   **47.** sin $\dfrac{5\pi}{4}$.   **48.** cos $\dfrac{2\pi}{3}$.

49. tan $\dfrac{4\pi}{3}$.   **50.** sin $\dfrac{7\pi}{4}$.   **51.** cos $\dfrac{11\pi}{6}$.   **52.** cos $\left(-\dfrac{11\pi}{6}\right)$.

In Problems **53-72**, *find the given function values.*

53. tan 337°.   **54.** sin 190.5°.   **55.** csc 534°.   **56.** cot 295°.

57. sin 100.3°.   **58.** cos 349.2°.   **59.** cot (−158.4°).   **60.** sec (−255°).

61. sec (−84°).   **62.** tan 500°.   **63.** cos (−170°).   **64.** cos 140.2°.

65. tan 570°.   **66.** csc 450°.   **67.** tan 161.6°.   **68.** cos 187.4°.

69. sin (−740°).   **70.** tan (−98°).   **71.** sin (−3.3).   **72.** cos 3.2.

In Problems **73-92**, *determine all values of θ, where* 0° ≤ θ ≤ 360°, *such that the given conditions are met.*

73. tan θ = −1.   **74.** cos θ = −0.8616.   **75.** sin θ = −0.3907.

76. csc θ = −1.390.   **77.** cos θ = $\dfrac{1}{2}$.   **78.** sin θ = 0.4512.

79. sin θ = $\dfrac{\sqrt{2}}{2}$.   **80.** tan θ = 0.4245.   **81.** sec θ = −1.122.

82. sec θ = 1.122.   **83.** sin θ = 0.6428.   **84.** cos θ = 0.8192.

85. tan θ = 7.953.   **86.** cot θ = −1.799.

87. sin θ = −0.6461 and tan θ is negative.   **88.** cos θ = −0.8616 and tan θ is negative.

89. cos θ = 0.2345 and sin θ is positive.   **90.** cos θ = −0.9877 and sin θ is negative.

91. tan θ = −0.7813 and sin θ is negative.   **92.** sin θ = 0.4566 and cos θ is positive.

In Problems **93-98**, *a function value of θ and a condition on θ are given. Find the five other trigonometric function values. Do not use a calculator.*

93. sin θ = $\dfrac{1}{6}$ and cos θ is negative.   **94.** cos θ = $-\dfrac{1}{3}$ and tan θ is positive.

95. tan θ = $-\dfrac{\sqrt{33}}{4}$ and sin θ is negative.   **96.** cot θ = $\sqrt{15}$ and sec θ is positive.

97. csc θ = −2 and cos θ is negative.   **98.** sec θ = $-\dfrac{\sqrt{21}}{3}$ and cot θ is negative.

7.2 VECTORS

Suppose you were given a rectangular box and a measuring stick and were asked to determine the length and area of a certain side of the box. You would probably be able to supply the answers in short order. Your might find the length of the side to be 1.2 m and the area to be 2.4 m². However, you might overlook the fact that length and width have a common property. Each is completely specified by a *number* and a *unit*. Such physical quantities—others are mass, speed, distance, and temperature—are called **scalar quantities**.

In science and engineering, however, there are other physical quantities that cannot be completely specified that simply. Each of these quantities has associated with it not only a number and a unit, which indicate magnitude, but also a *direction*. Such quantities are called **vector quantities**. Examples are displacement, force, velocity, acceleration, and torque. For example, we speak of a displacement (or change in position) of 10 m *south*.

To represent a vector quantity, we can use a directed line segment or arrow, which is called a **vector**. The length of the arrow represents, to some suitable scale, the magnitude of the quantity. The direction of the arrow gives the direction of the quantity. For example, with a scale of 1 cm = 4 m/s, a velocity of 20 m/s south can be represented by a 5-cm arrow pointing south. Similarly, the arrow labeled **V** in Fig. 7.16 is a vector. Its **initial point**, or **tail**, is at *A* and its **terminal point**, or **head**, is at *B*. The head of the arrow indicates direction, and we refer to this vector as the vector from *A* to *B*.

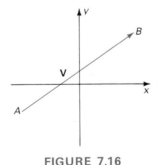

FIGURE 7.16

To denote a vector, we may use either boldface letters or letters with arrows. For example, the vector in Fig. 7.16 may be denoted by **AB**, **V**, $\overrightarrow{AB}$, or $\vec{V}$. The magnitude of a vector **V** can be denoted by |**V**| or by ordinary type, *V*. Thus if **A** = 3 m/s northward, then |**A**| = *A* = 3 m/s. The magnitude of a vector is always nonnegative. We shall feel free to use any of the above notations to indicate a vector or its magnitude, as convenient.

Let us be specific about when two vectors are *equal*.

Two vectors are **equivalent** or **equal** if and only if they have the same magnitude (length) and the same direction.

The vectors **A**, **B**, and **C** in Fig. 7.17 are equivalent; they have the same length and direction. They differ, however, in their initial and terminal points. The notion of

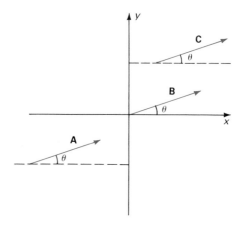

FIGURE 7.17

equivalent vectors is similar to the notion of equivalent rational numbers. That is, just as we can replace the number $\frac{2}{3}$ by $\frac{4}{6}$ without affecting a calculation, so can we represent a vector by any equivalent vector. That is, we are free to move a vector provided we do not change its magnitude and direction. More specifically, at times we shall find it convenient to represent a vector by an equivalent vector with initial point at the origin of a rectangular coordinate system. This type of vector is said to be in **standard position**. Vector **B** in Fig. 7.17 is in standard position.

Given a vector **A**, then the product $n\mathbf{A}$, where n is a number, is a vector with magnitude $|n|$ times that of **A** and with the following direction:

1. The same as **A** if $n > 0$.
2. Opposite that of **A** if $n < 0$.
3. Anywhere if $n = 0$. In this case the product is called the **zero vector**.

The product $(-1)\mathbf{A}$ is simply written $-\mathbf{A}$ and is called the **negative** of **A**; it has the same magnitude as **A** but points in the opposite direction.

Example 1

Given the vector **F** with magnitude 1 N in the direction due northeast, Fig. 7.18 shows **F**, $-\mathbf{F}$, 2**F**, and $-3\mathbf{F}$.

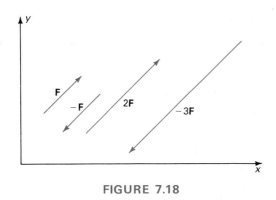

FIGURE 7.18

In Fig. 7.19(a), two successive displacements from A to B and then from B to C are represented by the vectors $\mathbf{d}_1$ and $\mathbf{d}_2$, respectively. The resulting effect, or net

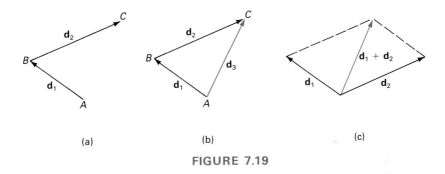

(a)   (b)   (c)

FIGURE 7.19

displacement, is the same as the displacement from A to C. This is represented in Fig. 7.19(b) by the vector $\mathbf{d}_3$ from the tail of $\mathbf{d}_1$ to the head of $\mathbf{d}_2$. We call $\mathbf{d}_3$ the **vector sum** or **resultant** of $\mathbf{d}_1$ and $\mathbf{d}_2$. Symbolically, we write

$$\mathbf{d}_1 + \mathbf{d}_2 = \mathbf{d}_3. \tag{1}$$

We point out that the plus sign in Eq. 1 does not imply ordinary addition of numbers. The vector addition in Eq. 1 implies all the geometric considerations necessary to account for the directions of the quantities involved.

The previous sum $\mathbf{d}_1 + \mathbf{d}_2$ can be expressed another way. If we place $\mathbf{d}_1$ and $\mathbf{d}_2$ so that their tails are at a common point, then [see Fig. 7.19(c)] the resultant is the vector from the common point to the opposite corner of the parallelogram having adjacent sides of $\mathbf{d}_1$ and $\mathbf{d}_2$. That is, the resultant is the directed diagonal of this parallelogram. This is a common interpretation for the sum of two vectors.

Two fundamental properties of vector addition are given by the commutative and associative laws:

Commutative law: $\mathbf{A} + \mathbf{B} = \mathbf{B} + \mathbf{A}$.

Associative law: $\mathbf{A} + (\mathbf{B} + \mathbf{C}) = (\mathbf{A} + \mathbf{B}) + \mathbf{C}$.

These laws imply that regardless of the order in which vectors are added and regardless of how they are grouped, their vector sum is the same. Given the vectors $\mathbf{A}$, $\mathbf{B}$, and $\mathbf{C}$ in Fig. 7.20(a), the commutative law is illustrated in Fig. 7.20(b) and the associative law in Fig. 7.20(c).

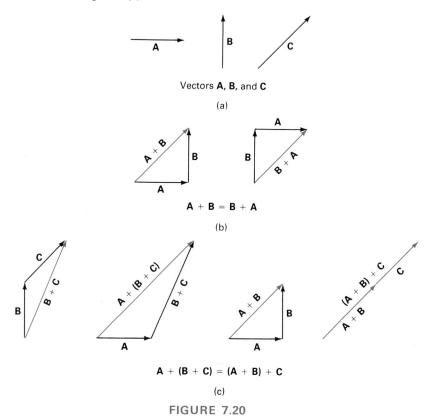

Vectors **A**, **B**, and **C**

(a)

A + **B** = **B** + **A**

(b)

A + (**B** + **C**) = (**A** + **B**) + **C**

(c)

FIGURE 7.20

To add vectors $\mathbf{A}, \mathbf{B}, \mathbf{C}, \mathbf{D}, \ldots, \mathbf{N}$ geometrically, we may follow these steps:

1. Draw **A** to scale on a diagram.
2. Draw **B** with its tail at the head of **A**.
3. Draw **C** with its tail at the head of **B**.
4. Continue in this manner until **N** has been drawn.
5. The resultant is the vector from the tail of **A** to the head of **N**.

Example 2

A newspaper carrier covers a route by traveling 3 blocks west, 3 blocks north, and then 7 blocks east. Determine the displacement of the carrier at the end of the route.

Solution   The displacement of the carrier in moving from one point to another is the directed straight-line distance from the first point to the second point. The carrier undergoes three separate displacements:

$$\mathbf{d}_1 = 3 \text{ blocks, west;}$$

$$\mathbf{d}_2 = 3 \text{ blocks, north;}$$

$$\mathbf{d}_3 = 7 \text{ blocks, east.}$$

The displacement of the carrier at the end of the route is the resultant displacement $\mathbf{d}$, where

$$\mathbf{d} = \mathbf{d}_1 + \mathbf{d}_2 + \mathbf{d}_3.$$

Following the procedure given above for adding vectors, we chose a scale of 1 unit = 1 block and constructed the diagram shown in Fig. 7.21. Note that $\mathbf{d}_1$ is placed in standard position.

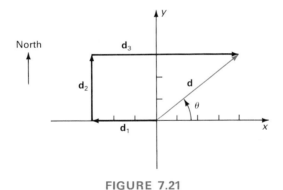

FIGURE 7.21

The resultant displacement is represented by the vector drawn from the tail of $\mathbf{d}_1$ to the head of $\mathbf{d}_3$ and is shown by the colored arrow. Measuring its length, we find that it is 5 units long, which represents a distance of 5 blocks according to our scale. The angle θ is measured with a protractor and is found to be approximately 36.9°. Thus the displacement of the carrier at the end of the route is 5 blocks, 36.9° north of east.

As illustrated by the answer in Example 2, a nonzero vector $\mathbf{V}$ in standard position can be specified by giving its magnitude V and the angle θ that $\mathbf{V}$ makes with the positive x-axis (Fig. 7.22). That is, θ specifies the direction of $\mathbf{V}$. We say that V and θ are **polar coordinates** for $\mathbf{V}$. *We shall usually choose θ between 0° and 360°.*

Another way of specifying a vector $\mathbf{V}$ in standard position is by giving the rectangular coordinates V_x and V_y of the head of $\mathbf{V}$, as shown in Fig. 7.23. That is, $\mathbf{V}$ is specified by the point (V_x, V_y). The numbers V_x and V_y are called the **horizontal** and

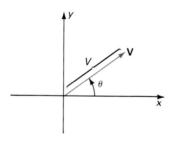

FIGURE 7.22

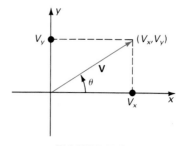

FIGURE 7.23

vertical (scalar) **components** of **V**, respectively. Observe that if θ is measured as in Fig. 7.23, then $\cos \theta = V_x/V$, from which

$$V_x = V \cos \theta,$$

and $\sin \theta = V_y/V$, from which

$$V_y = V \sin \theta.$$

We also have

$$V = \sqrt{V_x^2 + V_y^2}$$

and

$$\tan \theta = \frac{V_y}{V_x}.$$

Note that V_x and V_y can be positive or negative, depending on the quadrant in which **V** lies. If **V** is horizontal, then $V_y = 0$ and θ is either 0° or 180°. For a vertical vector, $V_x = 0$ and θ is either 90° or 270°.

Example 3

Find the components of **V** given that $V = 20$ and $\theta = 140°$.

Solution An appropriate diagram is given in Fig. 7.24. The components are

$$V_x = V \cos 140° = 20(-0.7660) = -15.32$$
$$V_y = V \sin 140° = 20(0.6428) = 12.86.$$

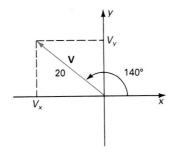

FIGURE 7.24

Example 4

Find the magnitude V and direction θ of **V** if $V_x = -4$ and $V_y = -3$.

Solution We have

$$V = \sqrt{V_x^2 + V_y^2} = \sqrt{(-4)^2 + (-3)^2} = \sqrt{25} = 5.$$

Because V_x and V_y are negative, the head of **V** is in the third quadrant (Fig. 7.25). Also,

$$\tan \theta = \frac{V_y}{V_x} = \frac{-3}{-4} = 0.75.$$

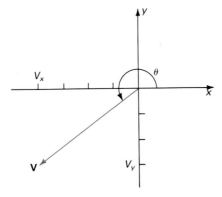

FIGURE 7.25

We find the reference angle for θ to be $36.9°$, so

$$\theta = 180° + 36.9° = 216.9°.$$

Therefore, $V = 5$ and $\theta = 216.9°$.

Whenever a given vector $\mathbf{V}$ is the resultant of two vectors, the two vectors are called **vector components** of $\mathbf{V}$, and $\mathbf{V}$ is said to be *resolved* into these two components. This notion is especially useful when the components are mutually perpendicular. For example, in Fig. 7.26 the vector $\mathbf{V}$ is resolved into a horizontal vector component

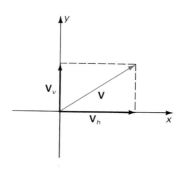

FIGURE 7.26

$\mathbf{V}_h$ and a vertical vector component $\mathbf{V}_v$. Because $\mathbf{V} = \mathbf{V}_h + \mathbf{V}_v$, $\mathbf{V}$ can be replaced by its vector components. Figure 7.27 shows other vectors $\mathbf{V}$ resolved into horizontal and vertical vector components. We point out that any vector can be replaced by its vector components because their sum is the given vector.

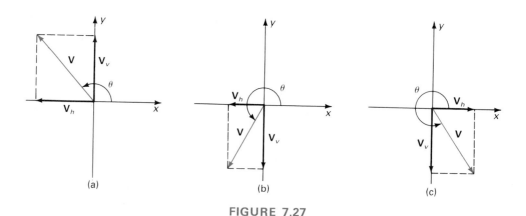

FIGURE 7.27

Example 5

A straight portion of a hill is inclined at an angle of 38° with the horizontal. If the speed of a sled down this hill is 18.3 m/s, find the magnitudes of the horizontal and vertical vector components of the sled's velocity. See Fig. 7.28(a).

(a)   (b)

FIGURE 7.28

Solution   The velocity vector **V** is shown in Fig. 7.28(b). The magnitudes of the vector components V_v and V_h can easily be found from the right triangle shown.

$$\sin 38° = \frac{\text{opp}}{\text{hyp}} = \frac{V_v}{18.3}$$

$$V_v = 18.3 \sin 38° = 18.3(0.6157)$$

$$= 11.3 \text{ m/s.}$$

$$\cos 38° = \frac{V_h}{18.3}$$

$$V_h = 18.3 \cos 38° = 18.3(0.7880)$$

$$= 14.4 \text{ m/s.}$$

Example 6

The weight **W** of an object on an inclined plane can be resolved into two vector components: one, **W**$_1$, parallel to the plane, and one, **W**$_2$, perpendicular to the plane (Fig. 7.29). Suppose a 150-N weight rests on an incline that rises 4 m vertically to every 3 m horizontally (Fig. 7.30). Find the magnitudes of the vector components of the weight.

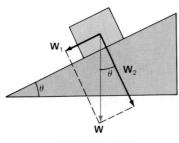

FIGURE 7.29

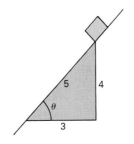

FIGURE 7.30

Solution In Fig. 7.29 note that there are two similar right triangles with acute angle θ. From the smaller one we have

$$\sin \theta = \frac{\text{opp}}{\text{hyp}} = \frac{W_1}{W}.$$

Solving for W_1 and substituting the data from Fig. 7.30 give

$$W_1 = W \sin \theta = 150\left(\frac{4}{5}\right) = 120 \text{ N}.$$

Similarly,

$$W_2 = W \cos \theta = 150\left(\frac{3}{5}\right) = 90 \text{ N}.$$

Problem Set 7.2

1. A car travels 5.0 km east, 3.0 km south, and then 2.0 km west. Graphically estimate the resultant displacement of the car. (See Example 2.)
2. A helicopter flies 240 km, 45° north of east, and then flies 100 km due south. Graphically estimate the displacement of the helicopter. (See Example 2.)

In Problems 3–12, find the horizontal and vertical components of the vector with the given polar coordinates.

3. $V = 150$, $\theta = 35°$.
4. $V = 10$, $\theta = 15.2°$
5. $V = 10.1$, $\theta = 273°$.
6. $V = 0.100$, $\theta = 226°$.
7. $V = 32$, $\theta = 184°$.
8. $V = 50$, $\theta = 316°$.
9. $V = 200$, $\theta = 101°$.
10. $V = 3000$, $\theta = 137°$.
11. $V = 150$, $\theta = 306°$.
12. $V = 720$, $\theta = 67°$.

In Problems 13–20, the components of a vector **V** *in standard position are given. Find V and θ.*

13. $V_x = 5$, $V_y = 7$.
14. $V_x = -3.4$, $V_y = 7.9$.
15. $V_x = 6.1$, $V_y = -3.2$.
16. $V_x = 23.5$, $V_y = 15.8$.
17. $V_x = -2.1$, $V_y = -4.8$.
18. $V_x = 0$, $V_y = 17$.
19. $V_x = -3$, $V_y = 0$.
20. $V_x = 0$, $V_y = -4$.

21. A baseball is thrown into the air with a speed of 32 m/s directed at an angle of 32° with the horizontal. Find the magnitudes of the horizontal and vertical vector components of the ball's initial velocity. (See Example 5.)
22. An airplane flies 168 km in the direction 64° west of south. How far south and how far west has it traveled?

23. An automobile travels down a hill with an acceleration of 5 m/s². The hill makes an angle of 40° with the horizontal. Find the magnitudes of the horizontal and vertical vector components of the acceleration.

24. A home-run ball moving at 42.6 m/s is caught by a fan in the bleachers. At the instant the ball was caught, it was moving at an angle of 27.5° below the horizontal. Find the magnitudes of the horizontal and vertical vector components of the ball's velocity as it was caught.

25. A child pulls a sled by exerting a 30-N force at an angle of 20° above the horizontal. Find (a) the magnitude of the horizontal vector component of the force which tends to move the sled along the ground, and (b) the magnitude of the vertical vector component of the force that tends to lift the sled.

26. A person *pushes* a lawn mower by applying a force of 10 N along its handle. If the handle is inclined at an angle of 42° with the horizontal, find (a) the magnitude of the horizontal vector component of the applied force that tends to move the lawn mower along the ground, and (b) the magnitude of the downward vertical vector component of the applied force.

27. A block weighing 65 N is placed on a 55° inclined plane. Find the magnitudes of the vector components of the weight that are (a) parallel and (b) perpendicular to the inclined surface. (See Example 6.)

28. Consider two displacements: one of magnitude 3 m and one of magnitude 4 m. Show geometrically how these displacement vectors can give a resultant displacement of magnitude (a) 7 m, (b) 1 m, and (c) 5 m.

7.3   ADDITION OF VECTORS—ANALYTICAL METHOD

The geometric method of vector addition that was given in the previous section is useful in only the simplest cases. In more complicated situations, we use an analytic method involving the use of components.

If **R** is the resultant (or vector sum) of two or more vectors in standard position, then the following can be shown geometrically:

1. R_x, the horizontal component of **R**, is the sum of the horizontal components of the vectors in the sum.

2. R_y, the vertical component of **R**, is the sum of the vertical components of the vectors in the sum.

These facts are illustrated in Fig. 7.31, where the horizontal component of **A** + **B** is $A_x + B_x$ and the vertical component of **A** + **B** is $A_y + B_y$.

As a result of the previous facts, we can perform the following steps to determine the resultant of two or more vectors.

1. Find the horizontal and vertical components of each of the given vectors.

2. Determine the sum R_x of the horizontal components.

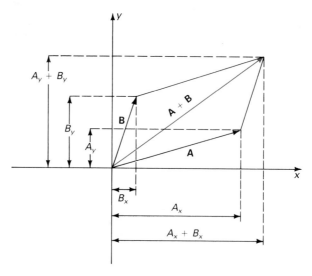

$A_y + B_y$

B_y

B

A_y

$A + B$

A

B_x

A_x

$A_x + B_x$

FIGURE 7.31

3. Determine the sum R_y of the vertical components.
4. Use the Pythagorean theorem and basic trigonometric relations to find the magnitude and direction of the resultant vector **R**. That is,

$$R = \sqrt{R_x^2 + R_y^2},$$

$$\tan \theta = \frac{R_y}{R_x}, \qquad 0° \le \theta < 360°.$$

If $R_x = 0$, then $\theta = 90°$ if $R_y > 0$ and $\theta = 270°$ if $R_y < 0$. If $R_y = 0$, then $\theta = 0°$ if $R_x > 0$ and $\theta = 180°$ if $R_x < 0$.

Example 1

Find the resultant of **A**, **B**, and **C** if their polar coordinates are:

$$A = 30 \text{ N}, \qquad \theta_A = \ 30°;$$
$$B = 10 \text{ N}, \qquad \theta_B = 270°;$$
$$C = 10 \text{ N}, \qquad \theta_C = 135°.$$

These vectors are shown in Fig. 7.32.

Solution The components of **A** are

$$A_x = A \cos 30° = (30)(0.8660) = 25.98 \text{ N},$$
$$A_y = A \sin 30° = (30)(0.5000) = 15 \text{ N}.$$

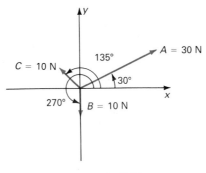

FIGURE 7.32

The components of **B** are

$$B_x = B \cos 270° = (10)(0) = 0 \text{ N,}$$
$$B_y = B \sin 270° = (10)(-1) = -10 \text{ N.}$$

For vector **C**, the reference angle is 45°:

$$C_x = C \cos 135° = C(-\cos 45°) = (10)(-0.7071) = -7.071 \text{ N,}$$
$$C_y = C \sin 135° = C \sin 45° = (10)(0.7071) = 7.071 \text{ N.}$$

The components R_x and R_y of the resultant **R** are

$$R_x = A_x + B_x + C_x = 25.98 + 0 - 7.071 = 18.91 \text{ N,}$$
$$R_y = A_y + B_y + C_y = 15 - 10 + 7.071 = 12.07 \text{ N.}$$

Because these components are positive, they correspond to points on the positive x- and y-axes. Thus **R** lies in the first quadrant (Fig. 7.33). By the Pythagorean theorem,

$$R = \sqrt{R_x^2 + R_y^2}$$
$$= \sqrt{(18.91)^2 + (12.07)^2} = 22.4 \text{ N.}$$

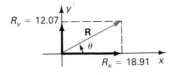

FIGURE 7.33

To determine the direction of θ, we have

$$\tan \theta = \frac{R_y}{R_x} = \frac{12.07}{18.91} = 0.6383$$

$$\theta = 32.6°.$$

Thus the resultant force **R**, which can replace **A**, **B**, and **C**, has magnitude $R = 22.4$ N and direction $\theta = 32.6°$.

Example 2

A boat in still water leaves a dock and travels 200 km at an angle of 23° east of north. It then travels 100 km due east. How far is the boat from the dock?

Solution The displacements of the boat are shown in Fig. 7.34(a). Putting the displacement vectors in standard position gives Fig. 7.34(b), where $A = 200$ km and $B = 100$ km. For the resultant displacement **d**, we have

$$d_x = A_x + B_x = 200 \cos 67° + 100 \cos 0°$$

$$= 200(0.39073) + 100(1) = 178.15 \text{ km},$$

$$d_y = A_y + B_y = 200 \sin 67° + 100 \sin 0°$$

$$= 200(0.92050) + 100(0) = 184.10 \text{ km}$$

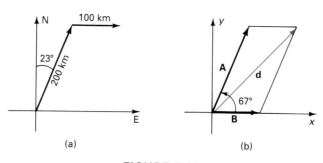

(a) (b)

FIGURE 7.34

The distance of the boat from the dock is given by the magnitude of **d**:

$$d = \sqrt{d_x^2 + d_y^2} = \sqrt{178.15^2 + 184.10^2}$$

$$= 256.2 \text{ km}.$$

Example 3

A plane is traveling due east at an airspeed of 400 km/h. If a wind of 100 km/h is blowing due southeast, find the magnitude and direction of the plane's resultant velocity.

Solution The resultant velocity **V** of the plane is the sum of the velocities 400 km/h east and 100 km/h southeast. The vector sum is shown in Fig. 7.35. The components of the wind velocity (**W**) are

$$W_x = 100 \cos 315° = 70.711 \text{ km/h},$$

$$W_y = 100 \sin 315° = -70.711 \text{ km/h}.$$

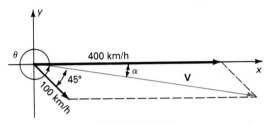

FIGURE 7.35

(Note the use of 315°, not 45°.) The components of the air velocity (**A**) of the plane are $A_x = 400$ km/h and $A_y = 0$ km/h. Thus the components of **V** are

$$V_x = W_x + A_x = 70.711 + 400 = 470.711 \text{ km/h},$$

$$V_y = W_y + A_y = -70.711 + 0 = -70.711 \text{ km/h},$$

so

$$V = \sqrt{V_x^2 + V_y^2} = \sqrt{470.711^2 + (-70.711)^2}$$

$$= 476.0 \text{ km/h}.$$

Also,

$$\tan \theta = \frac{V_y}{V_x} = \frac{-70.711}{470.711}.$$

Because $V_x > 0$ and $V_y < 0$, θ is a fourth-quadrant angle. The reference angle α is 8.5°, so the direction of **V** is

$$\theta = 351.5°.$$

Thus the plane's velocity is 476.0 km/h at 351.5°, which can be given as 8.5° south of east or 81.5° east of south.

Example 4

A block of weight W newtons rests on a plane inclined at an angle of 30° with the horizontal (Fig. 7.36). If the frictional force is 3 N and the block remains at rest, find the weight of the block. Assume all forces act at the center of the block.

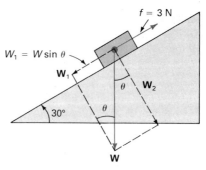

FIGURE 7.36

Solution   The weight **W** of the block is the force due to gravity and is directed vertically downward. The frictional force **f**, which always opposes motion or impending motion, is directed up the plane. Because the block does not move, *the net force up the plane must equal the net force down the plane*. The weight can be resolved into vector components **W**$_1$ (parallel to the plane) and **W**$_2$ (perpendicular to the plane). The frictional force **f** is the only force upward, and the vector component **W**$_1$ of the weight is the only force downward. Using geometry we can show that $\theta = 30°$. Hence $\sin 30° = W_1/W$ or $W_1 = W \sin 30°$. Thus

$$W_1 = f$$

$$W \sin 30° = f$$

$$W = \frac{f}{\sin 30°} = \frac{3}{\frac{1}{2}} = 6 \text{ N}.$$

Example 5

An object is dropped from an airplane flying horizontally at a speed of 78 m/s (Fig. 7.37). The vertical component of its velocity as a function of time t (in seconds) is given by $V_y = -9.8t$ m/s. What is the magnitude of the velocity of the object after 5 s, and at what angle with the horizontal is the object moving at that time?

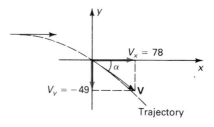

FIGURE 7.37

Solution   It is shown in physics that when air resistance is negligible, the horizontal component of the velocity **V** of such a free-falling object is constant. At every point along the object's flight path, V_x is the same as its initial value, namely, 78 m/s. At $t = 5$ s, the vertical component of **V** is $V_y = -9.8(5) = -49$ m/s. Thus the magnitude of the velocity is

$$V = \sqrt{78^2 + (-49)^2} = 92.1.$$

To find the angle α, which is measured below the horizontal, we have

$$\tan \alpha = \frac{|V_y|}{V_x} = \frac{49}{78}$$

$$\alpha = 32.1°.$$

Thus, at $t = 5\,\text{s}$, the object's velocity is 92.1 m/s directed at an angle of 32.1° below the horizontal.

Problem Set 7.3

In Problems 1–10, determine the magnitude R and direction θ of the resultant of the given vectors.

1. $A = 220$, $\theta_A = 225°$
   $B = 100$, $\theta_B = 16°$.

2. $A = 50$, $\theta_A = 30°$
   $B = 120$, $\theta_B = 30°$
   $C = 100$, $\theta_C = 90°$.

3. $A = 200$, $\theta_A = 210°$
   $B = 300$, $\theta_B = 45°$
   $C = 400$, $\theta_C = 120°$.

4. $A = 500$, $\theta_A = 0°$
   $B = 300$, $\theta_B = 180°$
   $C = 200$, $\theta_C = 315°$.

5. $A = 300$, $\theta_A = 80°$
   $B = 10$, $\theta_B = 270°$
   $C = 100$, $\theta_C = 30°$.

6. $A = 200$, $\theta_A = 200°$
   $B = 50$, $\theta_B = 50°$
   $C = 400$, $\theta_C = 400°$.

7. $A = 310$, $\theta_A = 330°$
   $B = 260$, $\theta_B = 150°$
   $C = 550$, $\theta_C = 120°$.

8. $A = 10$, $\theta_A = 0°$
   $B = 5$, $\theta_B = 226°$
   $C = 10$, $\theta_C = 180°$.

9. $A = 80$, $\theta_A = 30°$
   $B = 50$, $\theta_B = 50°$
   $C = 60$, $\theta_C = 60°$
   $D = 30$, $\theta_D = 80$.

10. $A = 10$, $\theta_A = 10°$
   $B = 20$, $\theta_B = 20°$
   $C = 30$, $\theta_C = 30°$
   $D = 40$, $\theta_D = 40°$
   $E = 50$, $\theta_E = 50°$.

In Problems 11 and 12, find the resultant of the forces shown.

11. See Fig. 7.38(a).

12. See Fig. 7.38(b).

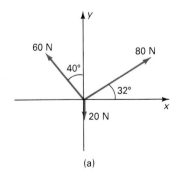

(a)

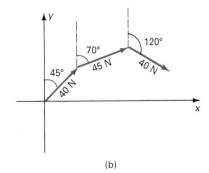

(b)

FIGURE 7.38

13. A plane in calm air leaves an airport and flies 300 km at an angle of 32° west of north. It then flies 200 km due west. How far from the airport is the plane? (See Example 2.)

14. Given a displacement of 300 m, 30° north of east, followed by a displacement of 400 m due north, find the magnitude and direction of the resultant displacement.

15. A plane is traveling eastward at an airspeed of 480 km/h. If a wind of 90 km/h is blowing northward, find the magnitude and direction of the plane's resultant velocity. (See Example 3.)

16. Find the magnitude and direction of the plane's resultant velocity in Problem 15 if the wind is blowing due northwest.

17. In still water a boat has an average speed of 41.3 km/h. If the boat heads due west in a river that flows due south at 14.1 km/h, find the resultant velocity of the boat. (See Example 3.)

18. Find the magnitude and direction of the resultant velocity in Problem 17 if the boat heads 22.3° south of west.

19. The angle between two displacements having magnitudes 20 km and 30 km is 42°. Find the magnitude of the resultant displacement. *Hint*: Place one of the displacements along the *x*-axis.

20. Two vectors **A** and **B** are added. Suppose $A = 10$ m, $B = 14$ m, and the magnitude of the resultant is 10 m. If the resultant **R** is perpendicular to **A**, determine the angle between **A** and **B**. *Hint*: Place **A** along the *x*-axis and **R** along the *y*-axis.

21. A block having a weight of 15 N rests on a 35.7° inclined plane. If the block does not move, determine the frictional force. (See Example 4.)

22. The resultant of two mutually perpendicular forces has magnitude 20 N. If one of the forces makes an angle of 28° with the resultant, what is the magnitude of that force?

23. An object is dropped from a plane that is traveling at a speed of 100 m/s at a direction of 30° above the horizontal. The vertical component of the object's velocity is given by $V_y = 50 - 9.8t$, where *t* is time in seconds. In the absence of air resistance, what is the speed of the object after 2 s, and in what direction with respect to the horizontal is the object moving at that time. (See Example 5.)

24. An object is dropped from a plane flying horizontally at a speed of 110 m/s.
    (a) After 6 s, what is the magnitude and direction of the object's velocity if its vertical component is given by $V_y = -9.8t$, where *t* is in seconds? (See Example 5.)
    (b) How long after being released will the object have a velocity that is directed at an angle of 45° below the horizontal?

25. In calm air the speed of a plane is 200 m/s. If a wind is blowing due east at 20 m/s, in what direction should the plane head if it is to fly due south?

7.4   REVIEW

Important Terms and Symbols

Section 7.1   Reference angle.

Section 7.2   Scalar quantity, vector quantity, vector (**V**), initial point (tail), terminal point (head), equivalent vectors, vector in standard position, zero vector, negative of vector, vector sum, resultant, polar coordinates for a vector, horizontal scalar component (V_x) of a vector, vertical scalar component (V_y) of a vector, horizontal vector component of a vector ($\mathbf{V}_h$), vertical vector component of a vector ($\mathbf{V}_v$).

Formula Summary

POSITIVE TRIGONOMETRIC FUNCTIONS

Quadrant I	all
Quadrant II	sin, csc
Quadrant III	tan, cot
Quadrant IV	cos, sec

REFERENCE ANGLES FOR θ, WHERE $0° < \theta < 360°$

Quadrant of θ	Reference Angle
I	θ
II	$180° - \theta$
III	$\theta - 180°$
IV	$360° - \theta$

RULE FOR REFERENCE ANGLES

To find the value of a trigonometric function of an angle θ, find the value of the same trigonometric function of the reference angle of θ and attach the proper sign to the result. This sign depends on the function involved and the quadrant of θ.

Trig function of $\theta = \pm$ trig function of reference angle.

$\uparrow$

Sign depends on trig function and quadrant of θ

Trigonometric Functions for Quadrantal Angles				
	$0°$ (or 0)	$90°$ $\left(or\ \dfrac{\pi}{2}\right)$	$180°$ (or π)	$270°$ $\left(or\ \dfrac{3\pi}{2}\right)$
sin	0	1	0	-1
cos	1	0	-1	0
tan	0	$-$	0	$-$
cot	$-$	0	$-$	0
sec	1	$-$	-1	$-$
csc	$-$	1	$-$	-1

SCALAR COMPONENTS OF VECTOR

$$V_x = V \cos \theta$$

$$V_y = V \sin \theta$$

$$V = \sqrt{V_x^2 + V_y^2}$$

$$\tan \theta = \frac{V_y}{V_x}$$

Review Questions

1. Which is larger, cos 0° or cos 90°? _____ .
2. Which of the following statements is (are) true? _____ .
   (a) cos $(-10°) = \cos 10°$.   (b) sin $(-20°) = -\sin 20°$.
   (c) sin $100° = -\sin 80°$.   (d) 20° is the reference angle for 160°.
   (e) sin $90° = \sin 270°$.   (f) sin $20° = \sin 380°$.
3. The only trigonometric functions that have positive values for an angle in Quadrant III are the _____ functions.
4. The value of sin 280° is equal to the negative of the sine of what acute angle? _____ .
5. If $0° \le \theta < 360°$, then sin $\theta = 0$ for what two values of θ? _____ .
6. A vector quantity has both magnitude and _____ .
7. In Fig. 7.39, what angle does the resultant of the given vectors make with the positive x-axis?

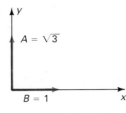

FIGURE 7.39

8. True or false: If sin A = sin B, then A must be equal to B. ___(a)___ . If A and B are coterminal
angles, then sin A = sin B. ___(b)___ .
9. If sin θ is negative and tan θ is positive, then θ lies in the _____ quadrant.

Answers to Review Questions

1. cos 0°.   2. a, b, d, and f.   3. Tangent and cotangent.   4. 80°.   5. 0°, 180°.
6. Direction.   7. 60°.   8. (a) False, (b) True.   9. Third.

Review Problems

In Problems **1–4**, *the given point lies on the terminal side of an angle θ in standard position. Find
the values of the six trigonometric functions of θ.*

1. $(1, -6)$.   2. $(-2, 0)$.   3. $(-2, -3\sqrt{5})$.   4. $(\sqrt{11}, -5)$.

5. If $\sin \theta = \dfrac{1}{5}$ and $\cos \theta = -\dfrac{2\sqrt{6}}{5}$, find the other trigonometric values of θ.

6. If $\tan \theta = -\dfrac{3}{5}$ and $\sec \theta = -\dfrac{\sqrt{34}}{5}$, find the other trigonometric values of θ.

7. If $\cos \theta = \dfrac{3}{7}$ and $\tan \theta$ is negative, find the other trigonometric values of θ.

8. If $\tan \theta = -\dfrac{\sqrt{51}}{7}$ and $\sin \theta$ is positive, find the other trigonometric values of θ.

In Problems **9–20**, *find the given values without the use of a calculator.*

9. $\cos \dfrac{2\pi}{3}$.   10. $\tan \dfrac{5\pi}{6}$.   11. $\sec \pi$.   12. $\sin (-120°)$.

13. $\csc 135°$.   14. $\cot 270°$.   15. $\tan 210°$.   16. $\sec 405°$.

17. $\sin (-180°)$.   18. $\cos \dfrac{11\pi}{6}$.   19. $\csc \dfrac{4\pi}{3}$.   20. $\cot \left(-\dfrac{3\pi}{2}\right)$.

In Problems **21** *and* **22**, *find the magnitude and direction of the resultant of the given vectors.*

21. $A = 100$, $\theta_A = 320°$
    $B = 120$, $\theta_B = 40°$
    $C = 50$, $\theta_C = 55°$.

22. $A = 20$, $\theta_A = 220°$
    $B = 30$, $\theta_B = 310°$
    $C = 10$, $\theta_C = 50°$.

23. An alternating emf ε is produced by rotating a coil in a magnetic field. For a particular coil, the emf (in volts) is given by $\varepsilon = 110 \sin \theta$, where θ is the angle through which the coil has turned. Find ε when θ is (a) 120°, (b) 598°, and (c) 8.5 rad.

24. A balloon, which rises 10 m/s, is released when there is a wind of 4 m/s blowing due east. Find the magnitude and direction of the resultant velocity of the balloon.

25. An airplane flies 400 km in the direction 30° west of south. Find the horizontal and vertical components of the displacement **d** of the airplane.

8

Straight Lines

8.1 SLOPE OF A LINE

Many physical relationships in science and engineering can be represented by straight lines. One feature of a straight line is its *steepness*. By steepness we mean the extent to which a line deviates from the horizontal. For example, in Fig. 8.1 line L_1 has more upward slant when compared to the *x*-axis than does line L_2. That is, L_1 rises faster as it goes from left to right than does L_2. In this sense L_1 is steeper than L_2.

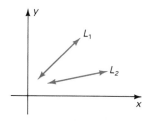

FIGURE 8.1

To measure the steepness of a line, we use the notion of *slope*. Basically, this involves examining two points on the line to see how the *y*-values change as the *x*-values change. For example, consider the line in Fig. 8.2. As we move along the line from $(2, 1)$ to $(4, 5)$, the *x*-coordinate increases from 2 to 4 (an increase of two units) and the *y*-coordinate increases from 1 to 5 (an increase of four units). To determine

291

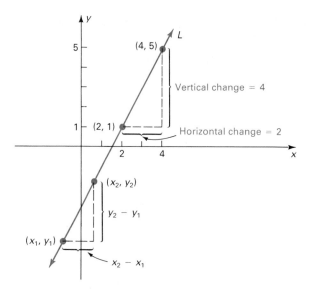

FIGURE 8.2

how y changes for each one-unit increase in x, we divide the change in y (vertical change) by the change in x (horizontal change). This gives the **average rate of change** of y with respect to x. Denoting the change in x by the symbol Δx (read *delta x*) and the change in y by Δy, we have

$$\frac{\Delta y}{\Delta x} = \frac{\text{change in } y}{\text{change in } x} = \frac{5-1}{4-2} = \frac{4}{2} = \frac{2}{1} = 2.$$

This means that for each one-unit increase in x, there is a two-unit *increase* in y. Thus the line must *rise* from left to right. Choosing two other different points on this line would give the same rate of change. For if (x_1, y_1) and (x_2, y_2) were such points, then the right triangles shown in Fig. 8.2 are similar. Thus the ratios of the lengths of their corresponding sides are equal. Therefore

$$\frac{4}{2} = \frac{y_2 - y_1}{x_2 - x_1},$$

so the rate of change is always constant (2). We call 2 the *slope* of the line.

SLOPE OF A STRAIGHT LINE

Let (x_1, y_1) and (x_2, y_2) be two points on a line, where $x_1 \neq x_2$. The **slope** of the line is the number m given by

$$m = \frac{y_2 - y_1}{x_2 - x_1} \left(\frac{\text{vertical change}}{\text{horizontal change}} \right).$$

In the slope formula, the order of the subscripts in the denominator *must* be the same as in the numerator. It is *incorrect* to write $m = (y_2 - y_1)/(x_1 - x_2)$, but correct to write $m = (y_1 - y_2)/(x_1 - x_2)$, which is obtained by multiplying the fraction in the box by $(-1)/(-1)$.

   *Slope is **not defined** for a vertical line.* The reason is that any two points on such a line must have $x_1 = x_2$ [see Fig. 8.3(a)], so the denominator in the slope formula would be zero. On the other hand, *a horizontal line does have a slope.* For such a line [see Fig. 8.3(b)], any two points must have $y_1 = y_2$. Thus the numerator in the slope formula is zero, so $m = 0$.

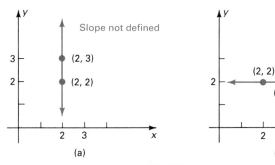

FIGURE 8.3

Example 1

Find the slope of the line through $(3, 4)$ and $(5, 1)$.

Solution If we choose $(3, 4)$ as (x_1, y_1) and $(5, 1)$ as (x_2, y_2), then

$$m = \frac{y_2 - y_1}{x_2 - x_1} = \frac{1 - 4}{5 - 3} = \frac{-3}{2} = -\frac{3}{2}.$$

Here the slope is negative, $-\frac{3}{2}$. This means that for each one-unit increase in x, there corresponds a *decrease* in y of $\frac{3}{2}$. Because of this decrease, the line *falls* from left to right (see Fig. 8.4). If we choose $(5, 1)$ as (x_1, y_1) and $(3, 4)$ as (x_2, y_2), then

$$m = \frac{y_2 - y_1}{x_2 - x_1} = \frac{4 - 1}{3 - 5} = \frac{3}{-2} = -\frac{3}{2},$$

as before.

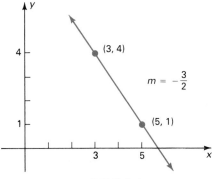

FIGURE 8.4

In summary, we have the following.

Zero slope: horizontal line.
Undefined slope: vertical line.
Positive slope: line rises from left to right.
Negative slope: line falls from left to right.

Now you can see why the slope is a measure of the steepness of a line. If two lines have slopes of 4 and 2, respectively, then in the first case we go up four units for a one-unit change in x to the right, whereas in the second case we would go up two units for the same change in x (see Fig. 8.5). In this sense, the first line is steeper than the

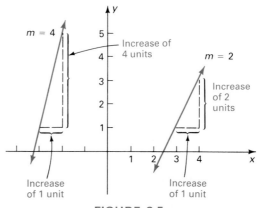

FIGURE 8.5

second. Similarly, if two lines had slopes of -3 and $-\frac{2}{3}$, respectively, then in the first case we would go down three units while moving one unit to the right, whereas in the second case we would go down two-thirds unit (see Fig. 8.6). Thus there is more downward slant in the first line than in the second.

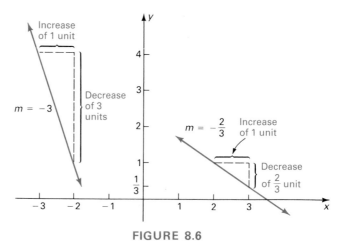

FIGURE 8.6

Figure 8.7 shows lines with different slopes. It makes it clear that **the closer the slope is to 0, the more nearly horizontal is the line. The greater the absolute value of the slope, the more nearly vertical is the line.**

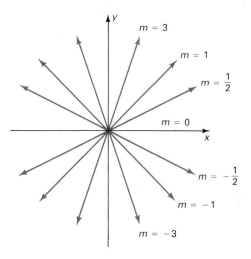

FIGURE 8.7

When the graph of a relationship between physical quantities is a straight line, its slope may have special physical significance, as Example 2 shows.

Example 2

Under the condition of constant acceleration, the graph of the velocity v of an object as a function of time t will always be a straight line. The acceleration of the object is the rate of change of velocity with respect to time. That is, it is the slope of the line. Suppose $v = 11$ m/s when $t = 2$ s, and $v = 18$ m/s when $t = 4$. Find and interpret the slope.

Solution A graph of the relationship appears in Fig. 8.8. In the slope formula we replace the y's by v's and the x's by t's. Letting $(2, 11) = (t_1, v_1)$ and $(4, 18) = (t_2, v_2)$, we have

$$m = \frac{\Delta v}{\Delta t} = \frac{v_2 - v_1}{t_2 - t_1} = \frac{18 \text{ m/s} - 11 \text{ m/s}}{4 \text{ s} - 2 \text{ s}} = \frac{7 \text{ m/s}}{2 \text{ s}}$$

$$= 3.5 \text{ m/s}^2.$$

Notice that the slope has units attached to it, namely, m/s², which are the units of an acceleration. The positive slope of 3.5 m/s² means that for each increase in time of 1 s, there corresponds an *increase* in velocity of 3.5 m/s.

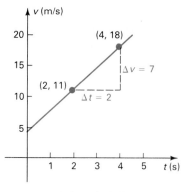

FIGURE 8.8

In summary, keep these points in mind:

1. Slope is a measure of steepness.
2. Slope indicates a rate of change.
3. Slope may have units attached to it.

Problem Set 8.1

In Problems **1-8**, *find the slope of the line through the given points.*

1. $(1, 2)$, $(4, 8)$.

2. $(-1, 9)$, $(1, 5)$.

3. $(6, -3)$, $(-7, 5)$.

4. $(2, -4)$, $(3, -4)$.

5. $(-2, 4)$, $(-2, 8)$.

6. $(0, -6)$, $(3, 0)$.

7. $(5, -2)$, $(4, -2)$.

8. $(1, -6)$, $(1, 0)$.

9. The slope of the line joining the points $(2, 5)$ and $(3, k)$ is 4. Find k.

10. The slope of the line joining the points $(1, k)$ and $(3, 3k)$ is 5. Find k.

11. When a small ball is thrown straight downward from the roof of a building, it falls with constant acceleration. Suppose the ball has a velocity of 24.6 m/s after 2 s of motion, and a velocity of 54.0 m/s after 5 s of motion. Find the acceleration of the ball. (See Example 2.)

12. The velocity of a certain object as a function of time is given by $v = 3t + 6$, where t is in seconds and v is in meters per second. The graph of this equation is a straight line. Find the acceleration of the object. (See Example 2.)   *Hint*: First find two points on the line.

13. For a metallic conductor, the graph of the voltage V, in volts, as a function of current i, in amperes, is a straight line. The slope of this line gives the resistance R of the conductor. If $i = 4$ A, then $V = 2$ V; if $i = 12$ A, then $V = 6$ V. Find the resistance (in ohms).

14. For an object moving in a straight line without accelerating, the graph of the displacement s, in meters, as a function of time t, in seconds, is a straight line. The slope of this line gives the velocity of the object, in meters per second. Find the velocity if $s = 13$ m when $t = 2$ s, and $s = 65$ m when $t = 10$ s.

15. In an experiment, a steel rod is found to have an initial length of 1.0000 m at a temperature of 0°C and a length of 1.0012 m at a temperature of 100°C. The graph of the length L as a function of temperature T is a straight line. The slope m of this line is related to α, the coefficient of linear expansion of steel. Specifically, $m = L_0\alpha$, where L_0 is the original length of the rod, in meters. From the given data, determine α.

16. One form of a capacitor, a device commonly used in electronics, consists of two parallel metal plates. When these plates are given equal and opposite electric charges, the graph of the potential difference V, in volts, as a function of the separation s of the plates, in meters, is a straight line. The slope of this line gives the strength of the electric field (in volts per meter) between the plates. For a particular capacitor, it is found that $V = 0.09$ V when $s = 0.001$ m, and $V = 0.45$ V when $s = 0.005$ m. Find the electric field strength between the plates of this capacitor.

17. An object has a constant acceleration of -6 m/s^2. Its velocity at time $t = 2$ s is 4 m/s. Find the velocity at $t = 5$ s. (See Example 2.)

8.2   EQUATIONS OF LINES

If we know the slope of a line and a point on the line, we can find an equation whose graph is that line. Suppose that line L has slope m and passes through the point (x_1, y_1). If (x, y) is any other point on L (see Fig. 8.9), then applying the slope formula

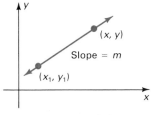

FIGURE 8.9

to (x_1, y_1) and (x, y) must result in m.

$$\frac{y - y_1}{x - x_1} = m,$$

from which

$$y - y_1 = m(x - x_1). \tag{1}$$

That is, the coordinates of every point on L satisfy Eq. 1. It can also be shown that every point with coordinates that satisfy Eq. 1 must lie on L. Thus we say that Eq. 1 is an equation for L. More specifically:

POINT-SLOPE FORM

The **point-slope form** of an equation for the line through the point (x_1, y_1) and having slope m is

$$y - y_1 = m(x - x_1).$$

Example 1

Find an equation of the line that has slope 2 and passes through the point $(2, -3)$. Sketch the graph of the line.

Solution   Using the preceding point-slope form, we set $m = 2$ and $(x_1, y_1) = (2, -3)$.

$$y - y_1 = m(x - x_1)$$
$$y - (-3) = 2(x - 2)$$
$$y + 3 = 2x - 4,$$

which simplifies to

$$y = 2x - 7.$$

To sketch the line, only two points need be plotted, because two points determine a straight line. If $x = 0$, then from the last equation we get $y = -7$. See Fig. 8.10.

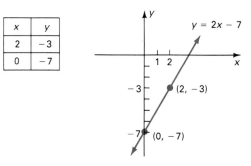

x	y
2	-3
0	-7

FIGURE 8.10

An equation of the line through two points can be found. First we determine the slope of the line. Then we use a point-slope form with either point as (x_1, y_1). This procedure is shown in the next example.

Example 2

Find an equation of the line through the points $(4, 1)$ and $(-5, 4)$.

Solution   We first find the slope. Let $(4, 1) = (x_1, y_1)$ and $(-5, 4) = (x_2, y_2)$. Then

$$m = \frac{y_2 - y_1}{x_2 - x_1} = \frac{4 - 1}{-5 - 4} = \frac{3}{-9} = -\frac{1}{3}.$$

Using a point-slope form with $(4, 1)$, we have

$$y - y_1 = m(x - x_1)$$

$$y - 1 = -\frac{1}{3}(x - 4)$$

$$y - 1 = -\frac{1}{3}x + \frac{4}{3}.$$

This can be written as

$$y = -\frac{1}{3}x + \frac{7}{3}.$$

Had we chosen the point $(-5, 4)$ as (x_1, y_1), the same result would have been obtained:

$$y - 4 = -\frac{1}{3}[x - (-5)]$$

$$y - 4 = -\frac{1}{3}x - \frac{5}{3}$$

$$y = -\frac{1}{3}x + \frac{7}{3}.$$

In Fig. 8.11 the line has slope m and intersects the y-axis at the point $(0, b)$. The

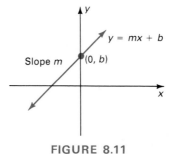

FIGURE 8.11

y-coordinate b is called the **y-intercept** of the line. Using the point-slope form, we have

$$y - y_1 = m(x - x_1)$$
$$y - b = m(x - 0)$$
$$y - b = mx$$
$$y = mx + b,$$

which is called the *slope-intercept form* of an equation for the line.

The **slope-intercept form** of an equation for the line with slope m and y-intercept b is

$$y = mx + b.$$

Example 3

a. An equation of the line with slope -2 and y-intercept -3 is

$$y = mx + b$$
$$y = -2x + (-3)$$
$$y = -2x - 3.$$

b. The equation $y - \frac{8}{9}x + 7 = 0$ can be written $y = \frac{8}{9}x - 7$, which has the form $y = mx + b$, where $m = \frac{8}{9}$ and $b = -7$ (not 7). Thus its graph is a line with slope $\frac{8}{9}$ and y-intercept -7.

One comment is worthy of emphasis. The function $f(x) = ax + b$, where $a \neq 0$, is called a **linear function**. If we replace $f(x)$ by y, then $y = ax + b$, which is an equation of a line with slope a and y-intercept b. Thus *the graph of a linear function is a straight line*. For example, the graph of the linear function $f(x) = 3x + 5$ is a straight line with slope 3 and y-intercept 5. We remark that the y-intercept of the graph of a linear function is the value of the dependent variable when the independent variable is zero.

Example 4

When an object moves in a straight line with constant acceleration, its velocity v is a linear function of time t. The slope of the graph of this function is the acceleration of the object. Suppose $v = 3$ m/s when $t = 2$ s, and the acceleration is $\frac{1}{4}$ m/s^2.

a. Determine v as a linear function of t.

Solution Because v is a linear function of t, it has the form $v = at + b$. The slope is $\frac{1}{4}$, so $a = \frac{1}{4}$. Thus

$$v = \frac{1}{4}t + b.$$

To determine b, we use the fact that $v = 3$ when $t = 2$.

$$3 = \frac{1}{4}(2) + b$$

$$\frac{5}{2} = b.$$

Thus $v = \frac{1}{4}t + \frac{5}{2}$, which expresses v as a linear function of t.

b. Find the initial velocity of the object (the velocity when $t = 0$).

Solution   The v-intercept $\frac{5}{2}$ m/s gives the initial velocity of the object.

We now turn to equations of horizontal and vertical lines. In Fig. 8.12 the vertical line through the point $(5, 3)$ consists of all points (x, y) for which $x = 5$. Thus an equation of the line is $x = 5$. More generally, we have the following.

VERTICAL LINE

An equation of the *vertical* line through the point (a, b) is

$$x = a.$$

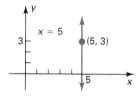

FIGURE 8.12

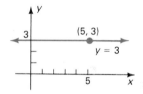

FIGURE 8.13

Similarly, in Fig. 8.13, an equation for the horizontal line through $(5, 3)$ is $y = 3$. In general, we have the following.

HORIZONTAL LINE

An equation of the *horizontal* line through the point (a, b) is

$$y = b.$$

An equation of the x-axis is $y = 0$, and an equation of the y-axis is $x = 0$.

It turns out that every straight line is the graph of an equation that can be put in the form

$$Ax + By + C = 0,$$

where A, B, and C are constants and A and B are not both zero. We call this form a **general linear equation** (or an *equation of the first degree*) **in the variables x and y**, and x and y are said to be **linearly related**. For example, we can get a general linear equation for the line $y = 7x - 2$:

$$y = 7x - 2$$
$$-7x + y + 2 = 0$$
$$(-7)x + (1)y + (2) = 0 \quad [A = -7, B = 1, C = 2].$$

On the other hand, the graph of every general linear equation is a straight line. For example, by solving the equation

$$3x + 4y - 5 = 0$$

for y, we can get a slope-intercept form:

$$3x + 4y - 5 = 0$$
$$4y = -3x + 5$$
$$y = -\frac{3}{4}x + \frac{5}{4}.$$

This is the slope-intercept form of a line with slope $-\frac{3}{4}$ and y-intercept $\frac{5}{4}$.

Example 5

Sketch the graph of $2x - 3y + 6 = 0$.

Solution **Method 1:** Because this is a general linear equation, its graph is a straight line. Thus we need only to get two points on the graph. If $x = 0$, then $y = 2$. This gives the y-intercept 2. Now, if $y = 0$, then $x = -3$. Thus $(-3, 0)$ lies on the line. The graph is given in Fig. 8.14(a). The point $(-3, 0)$ is the point where the line intersects the x-axis. The number -3 is called the **x-intercept** of the line.

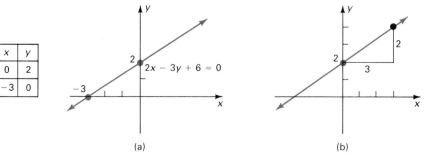

x	y
0	2
-3	0

2x − 3y + 6 = 0

(a)

(b)

FIGURE 8.14

Method 2: Here we use the slope and y-intercept to sketch the line. First we solve for y to get the slope-intercept form.

$$2x - 3y + 6 = 0$$

$$-3y = -2x - 6$$

$$y = \frac{2}{3}x + 2.$$

Hence the line intersects the y-axis at 2 and has slope $\frac{2}{3}$. Thus as x increases by 3 units, y increases by 2 units. This means that if we begin at the point $(0, 2)$, move 3 units to the right, and then move 2 units upward, we obtain a second point on the line [see Fig. 8.14(b)]. Using these two points, we sketch the graph.

Example 6

Fahrenheit temperature F and Celsius temperature C are linearly related. Use the facts that $32°F = 0°C$ and $212°F = 100°C$ to find an equation that relates F and C. (Express C in terms of F.) Also, find C when $F = 50$.

Solution   Because F and C are linearly related, the graph of the equation is a straight line. In Fig. 8.15 we used F for the horizontal axis and C for the vertical. (We could just as well have reversed our choices.) When $F = 32$, then $C = 0$, so the point $(32, 0)$ is on the line. Likewise, $(212, 100)$ is on it.

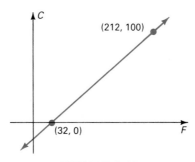

FIGURE 8.15

The slope m of the line is given by

$$m = \frac{C_2 - C_1}{F_2 - F_1} = \frac{100 - 0}{212 - 32} = \frac{100}{180} = \frac{5}{9}.$$

Using the point-slope form with the point $(32, 0)$, we get

$$C - C_1 = m(F - F_1)$$

$$C - 0 = \frac{5}{9}(F - 32)$$

$$C = \frac{5}{9}(F - 32).$$

Here C is in terms of F. We can use this equation to find C when $F = 50$:

$$C = \frac{5}{9}(F - 32) = \frac{5}{9}(50 - 32) = \frac{5}{9}(18) = 10.$$

Thus $50°F = 10°C$.

For your convenience, Table 8.1 gives a summary of the various forms of equations of straight lines.

TABLE 8.1 Forms of Equations of Lines

Point-slope form	$y - y_1 = m(x - x_1)$
Slope-intercept form	$y = mx + b$
Vertical line	$x = a$
Horizontal line	$y = b$
General linear form	$Ax + By + C = 0$

Problem Set 8.2

In Problems **1–8**, *find an equation of the line that has the given properties.*

1. Passes through $(1, 2)$ and has slope 6.
2. Passes through $(2, 4)$ and has slope 3.
3. Passes through $(5, -2)$ and has slope $-\frac{1}{5}$.
4. Passes through $(3, 5)$ and has slope 0.
5. Passes through $(3, -6)$ and has slope $\frac{1}{3}$.
6. Has slope $\frac{3}{5}$ and passes through $(0, 2)$.
7. Has slope -5 and passes through origin.
8. Has slope $\frac{1}{2}$ and passes through $(-\frac{1}{2}, -\frac{3}{2})$.

In Problems **9–16**, *find an equation of the line passing through the given points.*

9. $(2, 4)$, $(8, 7)$.
10. $(0, 0)$, $(5, -5)$.
11. $(2, -5)$, $(3, 4)$.
12. $(-2, 5)$, $(-2, 3)$.
13. $(-2, 5)$, $(3, 5)$.
14. $(2, 3)$, $(-2, -4)$.
15. $(-1, -1)$, $(-3, -3)$.
16. $(2, -7)$, $(3, -7)$.

In Problems **17–26**, *find an equation of the line that has the given properties.*

17. Has slope 2 and y-intercept 4.
18. Has slope 7 and y-intercept -5.
19. Has slope $-\frac{1}{2}$ and y-intercept -3.
20. Has slope 0 and y-intercept $-\frac{1}{2}$.
21. Is horizontal and passes through $(2, 4)$.

22. Is vertical and passes through $(-3, 2)$.

23. Passes through $(2, -3)$ and is vertical.

24. Passes through the origin and is horizontal.

25. Passes through $(2, 4)$ and has y-intercept 3.

26. Passes through $(-3, 2)$ and has x-intercept -3.

In Problems 27–38, write each line in slope-intercept form and find the slope m and y-intercept b. Sketch each line. For vertical lines, give the graph only.

27. $y = 2x - 1$.   **28.** $y = -3x + 2$.

29. $2y = -6x + 4$.   **30.** $y = x$.

31. $y - 4x = 0$.   **32.** $x = 2y + 1$.

33. $x + 2y - 3 = 0$.   **34.** $3x - 8 = 8y$.

35. $y = 1$.   **36.** $y - 7 = 3(x - 4)$.

37. $x = -3$.   **38.** $x = 0$.

In Problems 39–48, find a general linear form $(Ax + By + C = 0)$ of the given equation. Express your answer so that all coefficients are integers.

39. $y = x - 3$.   **40.** $x = y$.

41. $x = -2y + 4$.   **42.** $3x + 2y = 6$.

43. $4x + 9y - 5 = 0$.   **44.** $2(x - 3) - 4(y + 2) = 8$.

45. $\dfrac{3}{4}x = \dfrac{7}{3}y + \dfrac{1}{4}$.   **46.** $\dfrac{y}{-2} + \dfrac{x}{3} = 1$.

47. $\dfrac{x}{2} - \dfrac{y}{3} = -4$.   **48.** $y = \dfrac{1}{300}x + 8$.

In Problems 49–52, find the slope of the graph of the given linear function.

49. $f(x) = x + 2$.   **50.** $f(x) = 2x + 3$.

51. $f(x) = -3x$.   **52.** $f(x) = x$.

53. A straight line has slope 2 and y-intercept 1. Does the point $(-1, -1)$ lie on the line?

54. A straight line has slope -3 and passes through the point $(4, -1)$. Find the abscissa of the point on the line that has an ordinate of -2.

55. Suppose that s and t are linearly related such that $s = 40$ when $t = 12$, and $s = 25$ when $t = 18$.
(a) Find an equation that relates s and t.
(b) Find s when $t = 24$.

56. The force F exerted on a spring and the stretch S of the spring that the force produces are linearly related. When $F = 1$ N, then $S = 0.4$ cm. When $F = 4$ N, then $S = 1.6$ cm.
(a) Find an equation relating F and S.
(b) Find the stretch when a force of 2 N is exerted.

57. In a circuit the voltage V (in volts) and current i (in amperes) are linearly related. When $i = 4$, then $V = 2$; when $i = 12$, then $V = 6$.
(a) Find an equation relating V and i.
(b) Find the voltage when the current is 10.

58. A ball is thrown straight up in the air with an initial velocity of 50 m/s. The ball's velocity at the end of t seconds is given by $v(t) = 50 - 9.8t$. Sketch this linear function for $0 \le t \le 5$.

59. The pressure P of a fixed volume of gas, in centimeters of mercury, is linearly related to temperature T, in degrees Celsius. In an experiment with dry air, it was found that $P = 90$ when $T = 40$, and that $P = 100$ when $T = 80$. Express P as a function of T.

60. When a graph of the terminal potential difference V, in volts, of a Daniell cell is plotted as a function of the current i, in amperes, delivered to an external resistor, a straight line is obtained. The slope of this line is the negative of the internal resistance of the cell. For a particular cell with an internal resistance of $0.06\ \Omega$, it was found that $V = 0.6$ V when $i = 0.12$ A. Express V as a function of i.

61. A formula used in hydraulics is

$$Q = 3.340b^3 + 1.8704b^2x,$$

where b is a constant.
(a) Is the graph of this equation a straight line?
(b) If so, what is its slope when $b = 1$?

62. The graph in Fig. 8.16 was constructed by observing the first 10 s of motion of a racing car.
(a) Given that the slope of each part of the graph is the acceleration, determine the acceleration of the racing car for each of the time intervals $0 < t < 5$ and $5 < t < 10$.
(b) Find an equation for the velocity v of the car as a function of time t for each of these same intervals.

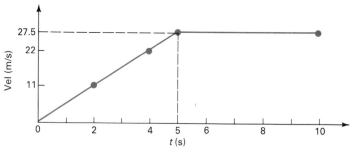

FIGURE 8.16

63. When the acceleration is constant, the velocity of an object varies linearly with time. Under this condition the velocity v of a certain object is 20 m/s at $t = 1$ s, and 2 s *later* its velocity is 24 m/s.
(a) Express the velocity of the object as a function of time.
(b) Did the object start from rest at $t = 0$?

64. Small oscillations of a simple pendulum can be used to determine a local value of g, the acceleration due to gravity. When the *square* of the period T, the time in seconds for one complete oscillation, is plotted as a function of the pendulum length l in meters, a straight line is obtained. The slope of the line gives the value of $4\pi^2/g$. Given that $T = 1.418$ when $l = 0.5$ and $T = 1.553$ when $l = 0.6$, (a) determine the slope, and (b) find the value of g (to one decimal place).

65. For reasons of comparison, a teacher wants to rescale the scores on a set of test papers so that the maximum score is still 100 but the mean (average) is 80 instead of 56.
   (a) Find a linear equation that will do this. *Hint*: You want 56 to become an 80 and 100 to remain 100. Consider the points $(56, 80)$ and $(100, 100)$ and, more generally, (x, y) where x is the old score and y is the new score. Find the slope and use a point-slope form. Express y in terms of x.
   (b) If 60 on the new scale is the lowest passing score, what was the lowest passing score on the original scale?

66. Biologists have found that the number of chirps made per minute by crickets of a certain species is related to the temperature. The relationship is very close to being linear. At 68°F, those crickets chirp about 124 times a minute. At 80°F, they chirp about 172 times a minute.
   (a) Find an equation that gives Fahrenheit temperature t in terms of the number of chirps c per minute.
   (b) If you count chirps for only 15 s, how can you quickly estimate the temperature?

8.3 PARALLEL AND PERPENDICULAR LINES

We can determine if one line is parallel or perpendicular to another line by comparing their slopes.

PARALLEL LINES

Two lines with slopes m_1 and m_2 are **parallel** if and only if $m_1 = m_2$. Two vertical lines are parallel.

PERPENDICULAR LINES

Two lines with slopes m_1 and m_2 are **perpendicular** if and only if $m_1 = -\dfrac{1}{m_2}$. A vertical line and a horizontal line are perpendicular.

Thus two lines are perpendicular when the slope of one line is the negative reciprocal of the slope of the other.*

Example 1

a. Show that the lines $y = 2x - 3$ and $y = 2x + 2$ are parallel.

   Solution   Both lines have slope 2, so they are parallel. See Fig. 8.17(a).

b. Show that the line $y = \frac{1}{2}x + 3$ is perpendicular to the line $y = -2x + 7$.

---

* This is shown in Sec. 17.2.

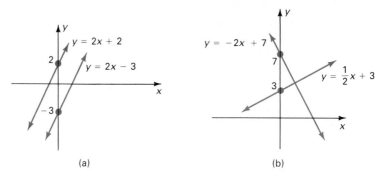

(a)   (b)

FIGURE 8.17

Solution The line $y = \frac{1}{2}x + 3$ has slope $m_1 = \frac{1}{2}$, whereas $y = -2x + 7$ has slope $m_2 = -2$. Because $\frac{1}{2}$ is the negative reciprocal of -2 (that is, $\frac{1}{2} = -\frac{1}{-2}$), the lines are perpendicular. See Fig. 8.17(b).

Example 2

Find equations of the lines that pass through $(3, -2)$ and are parallel and perpendicular to the line $y = 3x + 1$. See Fig. 8.18.

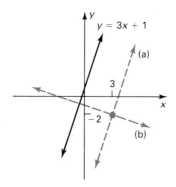

FIGURE 8.18

Solution The slope of $y = 3x + 1$ is 3. Thus the line through $(3, -2)$ that is *parallel* to $y = 3x + 1$ also has slope 3. Using the point-slope form, we have

$$y - (-2) = 3(x - 3)$$

$$y + 2 = 3x - 9$$

$$y = 3x - 11.$$

The slope of a line *perpendicular* to $y = 3x + 1$ must be $-\frac{1}{3}$ ($=$ the negative reciprocal of 3). Using the point-slope form, we have

$$y - (-2) = -\frac{1}{3}(x - 3)$$

$$y + 2 = -\frac{1}{3}x + 1$$

$$y = -\frac{1}{3}x - 1.$$

Problem Set 8.3

In Problems 1–10, determine whether the lines are parallel, perpendicular, or neither.

1. $y = 7x + 2$, $y = 7x - 3$.
2. $y = 4x + 3$, $y = 5 + 4x$.
3. $y = 5x + 2$, $-5x + y - 3 = 0$.
4. $y = x$, $y = -x$.
5. $x + 2y + 1 = 0$, $y = 2x$.
6. $x + 2y = 0$, $x + y - 4 = 0$.
7. $y = 3$, $y = -\frac{1}{3}$.
8. $x = 3$, $x = -4$.
9. $3x + y = 4$, $3x - y + 1 = 0$.
10. $x - 1 = 0$, $y = 0$.

In Problems 11–20, find an equation of the line satisfying the given conditions. Give the answer in slope-intercept form if possible.

11. Passing through $(-1, 3)$ and parallel to $y = 4x - 5$.
12. Passing through $(2, -8)$ and parallel to $x = -4$.
13. Passing through $(2, 1)$ and parallel to $y = 2$.
14. Passing through $(3, -4)$ and parallel to $y = 3 + 2x$.
15. Perpendicular to $y = 3x - 5$ and passing through $(3, 4)$.
16. Perpendicular to $y = -4$ and passing through $(1, 1)$.
17. Passing through $(7, 4)$ and perpendicular to $y = -4$.
18. Passing through $(-5, 4)$ and perpendicular to $2y = -x + 1$.
19. Passing through $(-7, -5)$ and parallel to $2x + 3y + 6 = 0$.
20. Passing through $(-2, 1)$ and parallel to the y-axis.
21. A coordinate map of a college campus gives the coordinates (x, y) of three major buildings as follows: computation center, $(3.5, -1)$; engineering lab, $(0.5, 0)$; and library, $(-1, -4.5)$. Find the equations (in slope-intercept form) of the straight-line paths connecting (a) the engineering lab with the computation center, and (b) the engineering lab with the library. Show that these two paths are perpendicular to each other.

8.4   LINEAR INTERPOLATION

For a laboratory experiment in optics, suppose that you need the value of the index of refraction (n) of pure water at $31°C$. In a standard reference table, you find the information shown in Table 8.2.

TABLE 8.2

T (°C)	n	T (°C)	n
20	1.33299	30	1.33192
22	1.33281	32	1.33164
24	1.33262	34	1.33136
26	1.33241	36	1.33107
28	1.33219	38	1.33079

Here you want an n-value for a T-value of 31, which is between two T-values in the table. This type of situation occurs often in scientific and technological work. A useful method of estimating the value of n is called **linear interpolation**. As its name implies, linear interpolation assumes that for a small change in T, the variable n is linearly related to T. In other words, we assume that for small changes in T, the graph of n as a function of T is a straight line and, therefore, n changes *uniformly*. Although this assumption is not necessarily true, the method of linear interpolation usually gives good estimates. The method is described in the following examples.

Example 1

Estimate the index of refraction of water at 31°C.

Solution   The value $T = 31$°C lies between the consecutive entries 30°C and 32°C in Table 8.2. Since 31°C is half of the way from 30°C to 32°C, we assume that the value of n at 31°C is also half of the way from the n-value for 30°C to that for 32°. These are 1.33192 and 1.33164, respectively, which differ by 0.00028. Note that n *decreases* in value. Thus one-half of the decrease, or $\frac{1}{2}(0.00028) = 0.00014$, must be *subtracted* from 1.33192. Therefore, at 31°C we estimate that $n = 1.33192 - 0.00014 = 1.33178$. These results can be shown in a tabular arrangement:

$$2\left[1\begin{bmatrix} \begin{array}{cc} T\,(°C) & n \\ 30 & 1.33192 \\ 31 & ? \\ 32 & 1.33164 \end{array} \end{bmatrix}d\right]0.00028 \text{ decrease}$$

$$d = \frac{1}{2}(0.00028) = 0.00014,$$

$$n = 1.33192 - 0.00014 = 1.33178.$$

We remark that the difference d can also be obtained by solving the proportion

$$\frac{1}{2} = \frac{d}{0.00028}.$$

Example 2

Use Table 8.2 and linear interpolation to estimate the index of refraction of water at 24.3°C.

Solution Following the procedure used in Example 1, we write the needed information as follows:

$$2\begin{bmatrix} 0.3\begin{bmatrix} \begin{array}{ccc} T\,(^\circ C) & & n \\ 24 & & 1.33262 \\ 24.3 & & ? \end{array}\end{bmatrix}d \\ 26 \qquad 1.33241 \end{bmatrix}0.00021\ \text{decrease}$$

Assuming that n for $T = 24.3°C$ is $0.3/2$ of the way from 1.33262 to 1.33241, we compute $0.3/2$ of the decrease.

$$d = \frac{0.3}{2}\,(0.00021) = 0.00003,$$

where we have rounded to five decimal places. Thus at 24.3°C,

$$n = 1.33262 - 0.00003 = 1.33259.$$

Our next example of linear interpolation deals with the total pressure (in millimeters of mercury) of saturated steam at various temperatures, as given in Table 8.3.

TABLE 8.3

T (°C)	P (mm Hg)	T (°C)	P (mm Hg)	T (°C)	P (mm Hg)
0	4.579	5	6.541	10	9.205
1	4.924	6	7.011	11	9.840
2	5.290	7	7.511	12	10.513
3	5.681	8	8.042	13	11.226
4	6.097	9	8.606	14	11.980

Example 3

The total pressure of saturated steam is found to be 7.821 millimeters of mercury. Estimate the temperature of the steam by using linear interpolation. Give your answer to one decimal place.

Solution The given value of pressure, 7.821, is between the consecutive entries 7.511 and 8.042, which correspond to 7°C and 8°C, respectively. Following the same basic technique used in Examples 1 and 2, we tabulate the needed information.

$$\text{1 increase}\begin{bmatrix} d\begin{bmatrix} \begin{array}{cc} T\,(^\circ C) & P\,(\text{mm Hg}) \\ 7 & 7.511 \\ ? & 7.821 \end{array}\end{bmatrix}0.310 \\ 8 \qquad 8.042 \end{bmatrix}0.531$$

The given pressure is 0.310/0.531 of the way from the pressure for 7°C to that for 8°C. So the desired temperature is assumed to be 0.310/0.531 of the way from 7°C to 8°C, which differ by 1°C:

$$d = \left(\frac{0.310}{0.531}\right)(1) = 0.6,$$

when rounded to one decimal place. Because the temperature *increases* as P changes from 7.511 to 8.042, the temperature of the saturated steam is estimated to be 7°C + 0.6°C = 7.6°C.

Problem Set 8.4

In Problems **1–8**, *use Table 8.2 and the method of linear interpolation.*

1. Find n at 20.4°C.
2. Find n at 24.7°C.
3. Find n at 37.3°C.
4. Find n at 31.8°C.
5. If n = 1.33253, find T to one decimal place.
6. If n = 1.33147, find T to one decimal place.
7. If n = 1.33114, find T to one decimal place.
8. If n = 1.33210, find T to one decimal place.

In Problems **9–12**, *use Table 8.3 and the method of linear interpolation.*

9. Find P at 7.3°C.
10. Find P at 11.4°C.
11. If P = 5.436, find T to one decimal place.
12. If P = 9.962, find T to one decimal place.
13. The table below gives some numbers (N) and the square of each number (N^2). Use linear interpolation to find an approximate value of the square of 90.3. By how much is your estimate in error?

N	90	91	92	93
N^2	8100	8281	8464	8649

14. Use the table in Problem 13 and linear interpolation to estimate the number having a square of 8300. Express your answer to one decimal place. Use a calculator to check the accuracy of this interpolation method.
15. When x milliliters of a 0.2-molar HCl solution are added to 25 mL of a 0.2-molar KCl solution, the pH of the resulting solution is as given in the following table.

x	33.6	26.6	20.7	16.2
pH	1.30	1.40	1.50	1.60

Using linear interpolation, estimate how many milliliters of the HCl solution are needed to give a pH of 1.43.

16. In Problem 15, if 23.2 mL of the HCl solution are used, what is the approximate pH of the resulting solution?

17. The surface tension γ (the Greek lowercase letter *gamma*) of a 34% solution (by volume) of alcohol in water varies with temperature as follows:

$T(°C)$	20	40	50
$\gamma\ (10^{-2}\ \text{N/m})$	3.324	3.158	3.070

Use linear interpolation to estimate the temperature at which the surface tension is 3.094×10^{-2} N/m.

8.5 REVIEW

Important Terms and Symbols

Section 8.1	Slope, average rate of change, Δx (delta x).
Section 8.2	Point-slope form, y-intercept, slope-intercept form, linear function, general linear equation, linearly related, x-intercept.
Section 8.3	Parallel lines, perpendicular lines.
Section 8.4	Linear interpolation.

Formula and Rules Summary

SLOPE FORMULA

$$m = \frac{y_2 - y_1}{x_2 - x_1}$$

ORIENTATION OF LINE

Zero slope: horizontal line.
Undefined slope: vertical line.
Positive slope: line rises from left to right.
Negative slope: line falls from left to right.

LINEAR FUNCTION

The graph of $f(x) = ax + b$ is a straight line.

Review Questions

1. A linear equation in x and y is one that can be written in the general form _____.
2. The line in Fig. 8.19 has a __(positive) (negative)__ slope.

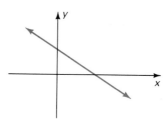

FIGURE 8.19

3. The graph of $x = 7$ is a line parallel to the _____ axis.
4. If the points $(4, 5)$ and $(2, 1)$ lie on the graph of a straight line, then the line has a slope of _____.
5. The y-intercepts of the lines $y = x$ and $y = 2x$ are __(a)__ and __(b)__, respectively.
6. The point-slope form of an equation of the line through $(2, -3)$ with slope 5 is _____.
7. The slope of the line $y = 2x - 1$ is _____.
8. An equation of the vertical line passing through $(5, -3)$ is _____.
9. A straight line for which the slope is not defined has an equation of the form __$(x=a)(y=b)$__.
10. The slope of the line $2y = 3x + 2$ is _____.
11. True or false: The value obtained for the slope of a straight line depends on which two points on the line are used for its computation. _____
12. A line for which the slope is zero is __(parallel) (perpendicular)__ to the y-axis.
13. If the line $y = kx + 4$ is parallel to the line $y = 7x + 16$, then $k = $ _____.
14. The graph of a linear function is a _____.

Answers to Review Questions

1. $Ax + By + C = 0$, where A and B are not both zero.   2. Negative.   3. y.
4. 2.   5. (a) 0,   (b) 0.   6. $y + 3 = 5(x - 2)$.   7. 2.   8. $x = 5$.   9. $x = a$.
10. $\frac{3}{2}$.   11. False.   12. Perpendicular.   13. 7.   14. Straight line.

Review Problems

*In Problems **1–4**, find the slope of the line passing through the given points.*

1. $(2, 3)$, $(-1, 4)$.   2. $(1, -1)$, $(2, 3)$.
3. $(2, 1)$, $(5, 1)$.   4. $(-2, 3)$, $(3, -2)$.

*In Problems **5–18**, find an equation of the line satisfying the given conditions. Give the answer in slope-intercept form if possible.*

5. Passes through $(2, -3)$ and has slope -2.
6. Passes through $(-6, 2)$ and has slope $\frac{1}{3}$.
7. Passes through $(-2, 3)$ and $(4, 5)$.
8. Passes through $(1, -1)$ and $(3, 0)$.
9. Passes through $(-2, 2)$ and has slope 0.
10. Passes through $(1, 2)$ and is vertical.
11. Passes through the origin and is vertical.
12. Passes through $(4, -1)$ and is parallel to the x-axis.
13. Has slope 3 and y-intercept -4.
14. Has slope -1 and passes through $(0, 2)$.
15. Passes through $(1, 2)$ and is perpendicular to $-3y + 5x = 7$.
16. Passes through $(-2, 4)$ and is horizontal.
17. Parallel to $y = 3 - 5x$ and passes through $(1, 2)$.
18. Has slope 1 and passes through $(1, 0)$.

*In Problems **19–24**, determine if the lines are parallel, perpendicular, or neither.*

19. $x + 4y + 2 = 0$, $8x - 2y - 2 = 0$.   20. $y - 2 = 2(x - 1)$, $2x + 4y - 3 = 0$.
21. $x - 3 = 2(y + 4)$, $y = 4x + 2$.   22. $3x + 5y + 4 = 0$, $6x + 10y = 0$.
23. $y = \frac{1}{2}x + 5$, $2x = 4y - 3$.   24. $y = 7x$, $y = 7$.

*In Problems **25** and **26**, write the given line in slope-intercept form and a general linear form. Find the slope.*

25. $3x - 2y = 4$.   26. $x = -3y + 4$.

27. Suppose f is a linear function such that $f(1) = 5$, and $f(x)$ decreases by 4 units for every 3-unit increase in x. Find $f(x)$.

28. Determine whether the point $(0, -7)$ lies on the graph of the straight line passing through $(1, -3)$ and $(4, 9)$.

29. Suppose s and t are linearly related so that $s = 1$ when $t = 2$, and $s = 2$ when $t = 1$. Find a general linear form of an equation that relates s and t. Find s when $t = 3$.

Given the corresponding values of D, S, and C in Table 8.4, use linear interpolation in Problems 30–35.

30. Find S and C for $D = 23.02$.

31. Find S and C for $D = 23.36$.

32. Find D if $S = 0.3912$.

33. Find D if $S = 0.3969$.

34. Find D if $C = 0.9187$.

35. Find D if $C = 0.9200$.

TABLE 8.4

D	S	C
23.0	0.3907	0.9205
23.1	0.3923	0.9198
23.2	0.3939	0.9191
23.3	0.3955	0.9184
23.4	0.3971	0.9178

36. A small aircraft weighs 17,800 N at takeoff. If the plane burns 1200 N of fuel each hour, express the weight W of the plane as a function of the time of flight t, in hours.

37. The hydrostatic pressure at a point 2 m below the surface of a body of water is 19,600 N/m². The pressure is 49,000 N/m² at a depth of 5 m. The graph of the pressure p as a function of depth d is a straight line. Find the slope of this line, which is the weight density of the water.

Systems of Equations

9.1 SYSTEMS OF LINEAR EQUATIONS IN TWO VARIABLES

When a physical situation must be described mathematically, it is not unusual for a *set* of equations to arise. For example, suppose an airplane travels 1440 km in 3 h with the aid of a tail wind, but takes 3 h 36 min for the return trip, in which the pilot flies against the same wind. What is the average speed of the airplane in still air, and what is the speed of the wind?

Suppose we let x be the (average) speed of the airplane in still air and y be the speed of the wind, where both speeds are in kilometers per hour. Then, with the wind the speed of the airplane is $x + y$, and against the wind its speed is $x - y$. The distance in both cases is 1440 km and

$$(\text{Rate}) \, (\text{time}) = \text{distance}.$$

Thus we have

$$(x + y)(3) = 1440$$

and (because 3 h 36 min $= 3\frac{3}{5}$ h $= \frac{18}{5}$ h)

$$(x - y)(\tfrac{18}{5}) = 1440.$$

317

Equivalently, we have the following set of equations:

$$\begin{cases} 3x + 3y = 1440 & \text{(1)} \\ \frac{18}{5}x - \frac{18}{5}y = 1440, & \text{(2)} \end{cases}$$

which is called a **system** of two linear equations in the variables (or unknowns) x and y. The brace indicates that each equation is to be considered in conjunction with the other. A **solution of the system** consists of values of x and y that satisfy *both* equations *simultaneously*.

Because Eqs. 1 and 2 are linear, their graphs are straight lines, call them L_1 and L_2. Now, the coordinates of any point on a line satisfy the equation of that line; that is, they make the equation true. Thus the coordinates of any point of intersection of L_1 and L_2 will satisfy both equations. This means that a point of intersection gives a solution of the system.

If L_1 and L_2 are drawn on the same plane, they will appear in one of three ways.

1. L_1 and L_2 may intersect at exactly one point, (x_0, y_0). See Fig. 9.1(a). Thus the system has the solution $x = x_0$ and $y = y_0$.
2. L_1 and L_2 may be parallel and have no points in common [see Fig. 9.1(b)]. Thus there is no solution.
3. L_1 and L_2 may be the same line [see Fig. 9.1(c)]. Thus the coordinates of *any* point on the line is a solution of the system, so there are infinitely many solutions. In this case the equations for L_1 and L_2 must be equivalent.

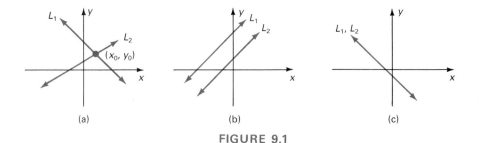

FIGURE 9.1

If a solution to a system exists, the system is said to be **consistent**. Otherwise, the system is **inconsistent**. For example, Figs. 9.1(a) and (c) represent consistent systems, but Fig. 9.1(b) represents an inconsistent system. A system of linear equations with exactly one solution is said to be **independent**. If more than one solution exists, the system is **dependent**. The system represented in Fig. 9.1(a) is independent, but that in Fig. 9.1(c) is dependent.

Returning to our original system concerning the airplane, we shall graphically solve

$$\begin{cases} 3x + 3y = 1440 \\ \frac{18}{5}x - \frac{18}{5}y = 1440. \end{cases}$$

To sketch the graph of the first equation (or line), we note that if $x = 0$, then $3y = 1440$, so $y = 480$. If $y = 0$, then $x = 480$. Thus the x- and y-intercepts are both 480. Using these points we graph the line in Fig. 9.2. For the second line, the x- and

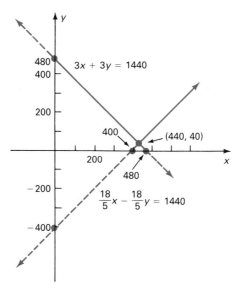

$$3x + 3y = 1440$$

$$(440, 40)$$

$$\frac{18}{5}x - \frac{18}{5}y = 1440$$

FIGURE 9.2

y-intercepts are found to be 400 and -400, respectively. These intercepts are then used to sketch the graph. Figure 9.2 shows the graph of the system. Because x and y are speeds, the lines have no physical meaning for $x < 0$ or $y < 0$. We estimate the point of intersection to be $x = 440$, $y = 40$. By substituting these values into *both* of the given equations, we find that the system is satisfied. For example, the first equation becomes $3(440) + 3(40) = 1440$, which is true. Thus the speed of the airplane in still air is 440 km/h and the speed of the wind is 40 km/h.

Example 1

Graphically solve the linear system

$$\begin{cases} 4T_1 + 2T_2 = 8 \\ 2T_1 + T_2 = -3. \end{cases}$$

Solution   Choosing T_1 for the horizontal axis and T_2 for the vertical axis, we obtain Fig. 9.3. Note that the lines are different but seem to be parallel, which implies the system has no solution. To confirm this we can analyze an equivalent system (that is, one with identical solutions) in which both lines are written in slope-intercept form:

$$\begin{cases} T_2 = -2T_1 + 4 \\ T_2 = -2T_1 - 3. \end{cases}$$

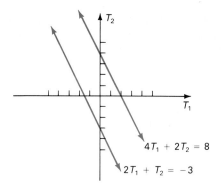

FIGURE 9.3

Because the lines have the same slope ($m = -2$) but different T_2-intercepts (4 and -3), the system is indeed represented by different parallel lines. Thus no solution exists and the system is inconsistent.

Example 2

Graphically solve the system

$$\begin{cases} 2y = 4x + 2 & (3) \\ y = 2x + 1. & (4) \end{cases}$$

Solution   Graphing both equations gives the same line, shown in Fig. 9.4. Observe that if both sides of Eq. 4 are multiplied by 2, the result is Eq. 3. Thus the equations are equivalent. Therefore, the coordinates of any point on the line $y = 2x + 1$ is a solution, so there are infinitely many solutions. For example, one solution is $x = 0$, $y = 1$. The system is consistent and dependent.

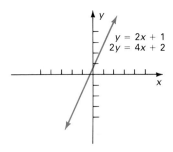

FIGURE 9.4

In the next section we shall focus our attention on algebraic methods of solving a system of equations. The graphical approach is not precise and should be used only when other methods are not practical.

Problem Set 9.1

Solve the following systems by graphing.

1. $\begin{cases} x - 2y = -3 \\ 3x + y = -2. \end{cases}$

2. $\begin{cases} y = x - \frac{1}{2} \\ 2y = -1 + x. \end{cases}$

3. $\begin{cases} y + 2x = 2 \\ y - 5 = -2x. \end{cases}$

4. $\begin{cases} 2x - y = -11 \\ y + 5x = -7. \end{cases}$

5. $\begin{cases} 2x + y = 4 \\ 3y - 2 = 4x. \end{cases}$

6. $\begin{cases} \frac{3}{2}y = 2x - 1 \\ y + \frac{2}{3} = \frac{4}{3}x. \end{cases}$

7. $\begin{cases} y = 6x - 3 \\ 18x = 9 + 3y. \end{cases}$

8. $\begin{cases} 3F_1 - 2F_2 = 1 \\ 2F_1 - 3F_2 = -6. \end{cases}$

9. $\begin{cases} 2i_2 = -i_1 + 4 \\ i_2 + 5 = 2i_1. \end{cases}$

10. $\begin{cases} 3(x + y) = 4(x - y) + 3 \\ 2(x - y) = 2y + 1. \end{cases}$

9.2 METHODS OF ELIMINATION

We now turn to algebraic methods of solving a system of two linear equations. These methods involve using algebraic operations to obtain an equation with only one variable in it. The other variable is *eliminated*. After solving that equation, we can easily find the value of the eliminated variable.

To illustrate, we shall solve the system

$$\begin{cases} 2x - 3y = -12 & (1) \\ 3x + y = -7 & (2) \end{cases}$$

by obtaining an equation in which y does not appear. That is, we shall eliminate y. First, we find an equivalent system in which the coefficients of the y-terms are numerically the same but differ in sign. To do this we can multiply Eq. 2 by 3 (that is, multiply both sides of Eq. 2 by 3).

$$\begin{cases} 2x - 3y = -12 & (3) \\ 9x + 3y = -21. & (4) \end{cases}$$

Because the left and right sides of Eq. 3 are equal, each side can be *added* to the corresponding side of Eq. 4. This gives

$$11x = -33,$$

which has only one variable, as planned. Solving this gives

$$x = -3.$$

To find y, we replace x by -3 in either one of the *original* equations, such as Eq. 1, and then solve for y.

$$2(-3) - 3y = -12$$
$$-6 - 3y = -12$$
$$-3y = -6$$
$$y = 2.$$

The solution is $x = -3$ and $y = 2$. We can check our answer by substituting $x = -3$ and $y = 2$ into *both* of the original equations. In Eq. 1 we have $2(-3) - 3(2) = -12$, or $-12 = -12$. In Eq. 2 we get $3(-3) + 2 = -7$, or $-7 = -7$. Thus the solution is

$$x = -3 \quad \text{and} \quad y = 2.$$

The method we used is called **elimination by addition**. Although we chose to eliminate y, we could eliminate x as follows. Multiplying Eq. 1 by 3 and Eq. 2 by -2 gives

$$\begin{cases} 6x - 9y = -36 & \text{(5)} \\ -6x - 2y = 14. & \text{(6)} \end{cases}$$

Adding Eq. 5 to Eq. 6 (that is, adding each side of Eq. 5 to the corresponding side of Eq. 6) gives

$$-11y = -22$$
$$y = 2.$$

Finally, replacing y in Eq. 1 by 2 gives $x = -3$, as expected. Figure 9.5 shows the graph of the system.

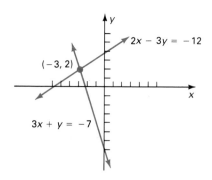

FIGURE 9.5

Example 1

Use elimination by addition to solve the system

$$\begin{cases} y - 2x = 4 \\ 4x - 2y = -5. \end{cases}$$

Solution Aligning the x- and y-terms for convenience gives

$$\begin{cases} -2x + y = 4 & \text{(7)} \\ 4x - 2y = -5. & \text{(8)} \end{cases}$$

To eliminate y we multiply Eq. 7 by 2.

$$\begin{cases} -4x + 2y = 8 & \text{(9)} \\ 4x - 2y = -5. & \text{(10)} \end{cases}$$

Adding Eq. 9 to Eq. 10 gives

$$0 = 3. \tag{11}$$

Because Eq. 11 is *never* true, there is **no solution** to the system. There is a reason for this. By using the slope-intercept form, we can write the original system as

$$\begin{cases} y = 2x + 4 \\ y = 2x + \frac{5}{2}. \end{cases}$$

These equations represent straight lines with slopes of 2 but different y-intercepts, 4 and $\frac{5}{2}$. That is, their graphs are different parallel lines (see Fig. 9.6). The system is inconsistent.

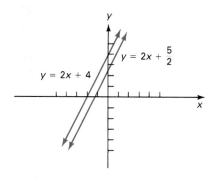

FIGURE 9.6

Example 2

By using elimination by addition, solve the system

$$\begin{cases} 2x + y = 1 & \text{(12)} \\ 4x + 2y = 2. & \text{(13)} \end{cases}$$

Solution We shall eliminate y. Multiplying Eq. 12 by -2 gives

$$\begin{cases} -4x - 2y = -2 & (14) \\ 4x + 2y = 2. & (15) \end{cases}$$

Adding Eq. 14 to Eq. 15, we have

$$0 = 0,$$

which is *always* true. We might have expected this result, since each term in Eq. 13 is two times the corresponding term in Eq. 12. Equation 13 is said to be a *multiple* of Eq. 12. Because of this, the graphs of Eqs. 12 and 13 are the same line (see Fig. 9.7). The coordinates of any point on the line $2x + y = 1$ is a solution, so there are infinitely many solutions. Because we can write that line as $y = 1 - 2x$, we can easily find some specific solutions. If $x = 0$, then $y = 1$. Thus $x = 0$, $y = 1$ is one solution. Similarly, some others are $x = 1$, $y = -1$ and $x = -3$, $y = 7$. The given system is consistent and dependent.

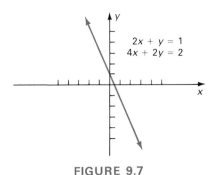

FIGURE 9.7

We can solve the system

$$\begin{cases} 2i_1 + 3i_2 = -10 & (16) \\ 3i_1 - 2i_2 = -2 & (17) \end{cases}$$

by a method other than elimination by addition. We first choose one of the equations and solve it for one variable in terms of the other. For example, solving Eq. 17 for i_1 in terms of i_2 gives

$$3i_1 - 2i_2 = -2$$

$$3i_1 = 2i_2 - 2$$

$$i_1 = \frac{2i_2 - 2}{3}. \qquad (18)$$

Next we *substitute* the right side of Eq. 18 for i_1 in Eq. 16:

$$2\left(\frac{2i_2 - 2}{3}\right) + 3i_2 = -10. \qquad (19)$$

Thus i_1 has been eliminated. (Note that we did not substitute the right side of Eq. 18 for i_1 in Eq. 17 because Eq. 18 was derived from Eq. 17.) Solving Eq. 19, we have

$$\frac{4}{3}i_2 - \frac{4}{3} + 3i_2 = -10$$

$$4i_2 - 4 + 9i_2 = -30 \qquad \text{[clearing fractions]}$$

$$13i_2 = -26$$

$$i_2 = -2.$$

Replacing i_2 in Eq. 18 by -2 gives

$$i_1 = \frac{2(-2) - 2}{3} = \frac{-6}{3} = -2.$$

Thus the solution is $i_1 = -2$, $i_2 = -2$, which you should verify. The method we used is called **elimination by substitution**.

Problem Set 9.2

In Problems **1–4**, *use elimination by addition to solve the systems.*

1. $\begin{cases} x + 2y = 1 \\ 3x + y = -2. \end{cases}$

2. $\begin{cases} 2x + 3y = -10 \\ 3x - 2y = -2. \end{cases}$

3. $\begin{cases} 4T_1 - 3T_2 = 6 \\ 3T_1 + 2T_2 = 13. \end{cases}$

4. $\begin{cases} 5i_1 + 3i_2 = 10 \\ 3i_2 + 4i_2 = 6. \end{cases}$

In Problems **5–8**, *use elimination by substitution to solve the systems.*

5. $\begin{cases} 4x + y = 6 \\ 3x + 2y = 2. \end{cases}$

6. $\begin{cases} x + 5y = -14 \\ -2x - 7y = 16. \end{cases}$

7. $\begin{cases} 5x + 7y = 13 \\ -x + 5y = 7. \end{cases}$

8. $\begin{cases} 4x - 3y = 13 \\ 3x + y = 0. \end{cases}$

In Problems **9–20**, *solve the systems by addition or substitution.*

9. $\begin{cases} 3F_1 + F_2 = 7 \\ 2F_1 + 2F_2 = -2. \end{cases}$

10. $\begin{cases} 2x - y = -11 \\ y + 5x = -7. \end{cases}$

11. $\begin{cases} 3x - 4y - 13 = 0 \\ 2x + 3y - 3 = 0. \end{cases}$

12. $\begin{cases} 5x - 3y = 2 \\ -10x + 6y = 4. \end{cases}$

13. $\begin{cases} 2i_2 = 36 - 5i_1 \\ 8i_1 = 3i_2 - 54. \end{cases}$

14. $\begin{cases} T_1 = 3 - T_2 \\ 3T_1 + 2T_2 = 19. \end{cases}$

15. $\begin{cases} 4x + 12y = 6 \\ 2x + 6y = 3. \end{cases}$

16. $\begin{cases} 2u - v - 1 = 0 \\ -u + 2v - 7 = 0. \end{cases}$

17. $\begin{cases} 2x + y = 4 \\ 10y - 41 = -20x. \end{cases}$

18. $\begin{cases} 3y = 4x - 2 \\ y + \frac{2}{3} = \frac{4}{3}x. \end{cases}$

19. $\begin{cases} \frac{2}{3}x + \frac{1}{2}y = 2 \\ \frac{3}{8}x + \frac{5}{6}y = -\frac{11}{2}. \end{cases}$

20. $\begin{cases} 3(x + y) = 4(x - y) + 3 \\ 2(x - y) = 2y + 1. \end{cases}$

21. A 49-N block is to be moved without acceleration along a horizontal surface by means of a force F parallel to the surface. The coefficient of kinetic friction between the block and the surface is 0.3. For this situation, Newton's laws lead to the system

$$\begin{cases} F - 0.3N = 0 \\ N - 49 = 0, \end{cases}$$

where N is the normal force exerted on the block by the surface, and F and N are in newtons. Find F and N.

22. If the block in Problem 21 is to accelerate at 2 m/s^2, the appropriate system of equations is

$$\begin{cases} F - 0.3N = \dfrac{49}{9.8}(2) \\ N - 49 = 0. \end{cases}$$

Find F and N.

23. In a given electrical circuit, two currents i_1 and i_2, in amperes, can be found by solving the system

$$\begin{cases} 3i_1 + 5i_2 = -3 \\ 4i_1 + 2i_2 = 4. \end{cases}$$

Find i_1 and i_2.

24. Masses of 7 kg and 9 kg are attached to a cord passing over a light frictionless pulley. When the masses are released, the acceleration a of each mass (in meters per second squared) and the tension T in the cord (in newtons) are related by the system

$$\begin{cases} T - 68.6 = 7a \\ 88.2 - T = 9a. \end{cases}$$

Find T and a.

25. A 1-kg ball moving at 12 m/s has a head-on collision with a 2-kg ball moving in the opposite direction at 24 m/s. If the coefficient of restitution for the collision is $\frac{2}{3}$, then the velocities v_1 and v_2 of the balls after the collision are related by the system

$$\begin{cases} v_1 + 2v_2 = -36 \\ v_2 - v_1 = 24, \end{cases}$$

where v_1 and v_2 are measured in meters per second. Find v_1 and v_2.

26. If the collision described in Problem 25 were perfectly elastic, then the system of equations would be

$$\begin{cases} v_1 + 2v_2 = -36 \\ v_2 - v_1 = 36. \end{cases}$$

Solve for v_1 and v_2.

27. A child pulls a 5-kg sled across the snow with constant velocity. The rope used to pull the sled makes an angle of $36.9°$ with the horizontal. The relationship between the force F exerted by the rope and the normal force N exerted by the ground, both measured in newtons, is given by the system

$$\begin{cases} F \sin 36.9° + N - 49 = 0 \\ F \cos 36.9° - 0.3N \quad = 0. \end{cases}$$

Find F and N.

28. If an athlete puts the shot, then (if we neglect air resistance) the horizontal distance x and height y of the shot after t seconds, until the shot hits the ground, are given by

$$x = (V \cos \alpha)t.$$

$$y = -\frac{1}{2} gt^2 + (V \sin \alpha)t + h,$$

where

h = the height at which the shot is released,

V = the speed at which the shot is released,

α = the angle with the horizontal at which the shot is released, and

g = the acceleration due to gravity.

In August, 1975, Marianne Adam (E. Germany) put the shot 21.6 m. If $\alpha = 44°$, $g = 9.8$ m/s^2, and $h = 1.6$ m, what was the speed of the shot when she released it?

9.3 SYSTEMS OF LINEAR EQUATIONS IN THREE VARIABLES

An equation of the form $Ax + By + Cz = D$, where A, B, C, and D are constants and A, B, and C are not all zero, is called a general **linear equation in the variables x, y, and z**. To solve a system of three linear equations in three unknowns, we use the methods of elimination discussed in the previous section.

One approach is to select two *pairs* of the given equations and eliminate the *same* variable from each pair. The resulting two equations, being in the same two variables, are then solved by elimination. Finally, we substitute these two values in any of the original equations to find the value of the third variable. This technique can be extended to a system of n linear equations in n variables by reducing the system to $n - 1$ equations in $n - 1$ variables, which is then reduced, and so on. An alternative method will be given in Example 2.

Example 1

Solve the system

$$\begin{cases} 4x - y - 3z = 1 & (1) \\ 2x + y + 2z = 5 & (2) \\ 8x + y - z = 5. & (3) \end{cases}$$

Solution   Here we have a system of three linear equations in three variables. As our first pair of equations, we select Eqs. 1 and 2:

$$\begin{cases} 4x - y - 3z = 1 \\ 2x + y + 2z = 5. \end{cases}$$

Adding these to eliminate y gives

$$6x - z = 6. \qquad (4)$$

As our second pair of equations we shall select Eqs. 2 and 3:

$$\begin{cases} 2x + y + 2z = 5 \\ 8x + y - z = 5. \end{cases}$$

We must also eliminate y from these. Multiplying the second equation by -1 gives

$$\begin{cases} 2x + y + 2z = 5 \\ -8x - y + z = -5. \end{cases}$$

Adding these, we have

$$-6x + 3z = 0. \qquad (5)$$

A new system of two equations in two variables is formed by Eqs. 4 and 5:

$$\begin{cases} 6x - z = 6 & (4) \\ -6x + 3z = 0. & (5) \end{cases}$$

Adding Eq. 4 to Eq. 5 and solving for z gives

$$2z = 6$$

$$z = 3.$$

Substituting 3 for z in Eq. 4 gives

$$6x - 3 = 6$$

$$6x = 9$$

$$x = \frac{3}{2}.$$

Substituting 3 for z and $\frac{3}{2}$ for x into Eq. 1, we can find y.

$$4x - y - 3z = 1$$
$$4(\tfrac{3}{2}) - y - 3(3) = 1$$
$$-y - 3 = 1$$
$$-y = 4$$
$$y = -4.$$

The solution is $x = \frac{3}{2}$, $y = -4$, and $z = 3$.

Example 2

Solve the system

$$\begin{cases} 2x + y - z = -2 & (6) \\[2mm] x - 2y = \dfrac{13}{2} & (7) \\[2mm] 3x + 2y - 2z = -\dfrac{9}{2}. & (8) \end{cases}$$

Solution   Since Eq. 7 can be written $x - 2y + 0z = \frac{13}{2}$, we can view Eqs. 6, 7, and 8 as a system of three linear equations in the variables x, y, and z. We shall solve this system by a different approach than that of the previous example. From Eq. 7, $x = 2y + \frac{13}{2}$. By substituting for x in Eqs. 6 and 8, we obtain

$$\begin{cases} 2\left(2y + \dfrac{13}{2}\right) + y - z = -2 \\[3mm] 3\left(2y + \dfrac{13}{2}\right) + 2y - 2z = -\dfrac{9}{2}, \end{cases}$$

or, more simply,

$$\begin{cases} 5y - z = -15 & (9) \\[2mm] 4y - z = -12. & (10) \end{cases}$$

We now solve the system formed by Eqs. 9 and 10. Multiplying Eq. 9 by -1 and adding the result to Eq. 10 gives $-y = 3$, or $y = -3$. Substituting -3 for y in Eq. 9 gives $5(-3) - z = -15$, from which $z = 0$. Substituting these values into Eq. 7 gives $x - 2(-3) = \frac{13}{2}$, from which $x = \frac{1}{2}$. Thus the solution of the original system is $x = \frac{1}{2}$, $y = -3$, and $z = 0$.

Problem Set 9.3

In Problems **1–10**, *solve the systems.*

1. $\begin{cases} x + y + z = 6 \\ x - y + z = 2 \\ 2x - y + 3z = 6. \end{cases}$

2. $\begin{cases} 2x - y + 3z = 12 \\ x + 2y - 3z = -10 \\ x + y - z = -3. \end{cases}$

3. $\begin{cases} x - z = 14 \\ y + z = 21 \\ x - y + z = -10. \end{cases}$

4. $\begin{cases} x + y = -6 \\ z = 4 \\ -x + y + 2z = 16. \end{cases}$

5. $\begin{cases} 2x + y + 6z = 3 \\ x - y + 4z = 1 \\ 3x + 2y - 2z = 2. \end{cases}$

6. $\begin{cases} 5x - 7y + 4z = 2 \\ 3x + 2y - 2z = 3 \\ 2x - y + 3z = 4. \end{cases}$

7. $\begin{cases} 2x - 3y + z = -2 \\ 3x + 3y - z = 2 \\ x - 6y + 3z = -2. \end{cases}$

8. $\begin{cases} x + y + z = -1 \\ 3x + y + z = 1 \\ 4x - 2y + 2z = 0. \end{cases}$

9. $\begin{cases} x + y + 5z = 6 \\ x + 2y + w = 4 \\ 2y + z + w = 6 \\ 3x - 4z = 2. \end{cases}$

10. $\begin{cases} x - y + 3z + w = -14 \\ x + 2y - 3w = 12 \\ 2x + 3y + 6z + w = 1 \\ x + y + z + w = 6. \end{cases}$

11. The boom in Fig. 9.8 has weight **W** and is acted upon by the forces shown, in newtons. The force exerted by the string is **T**, and **V** and **H** are forces exerted by the wall. In mechanics it is shown that for the boom to be in equilibrium,

$$\begin{cases} H - T \cos 45° = 0 \\ V + T \sin 45° - W = 0 \\ W(\tfrac{1}{2}) - T \sin 45° = 0. \end{cases}$$

If $W = 100$ N, find the exact values for H, T, and V.

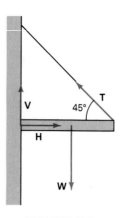

FIGURE 9.8

12. If the weight of the boom in Problem 11 is doubled, what are the exact values for H, T, and V?

13. In the analysis of direct-current circuits, Kirchhoff's laws are often used. Application of these laws to the circuit of Fig. 9.9 gives the following system:

$$\begin{cases} R_1 i_1 + R_2 i_1 + R_3 i_3 + \mathcal{E}_1 = 0 \\ \qquad\quad R_3 i_3 + R_4 i_2 + \mathcal{E}_2 = 0 \\ \qquad\qquad\qquad i_1 + i_2 = i_3, \end{cases}$$

where the R's are resistances, the i's are currents, and the $\mathcal{E}$'s are potential differences. Given that $R_1 = 2\,\Omega$, $R_2 = 3\,\Omega$, $R_3 = 5\,\Omega$, $R_4 = 3\,\Omega$, $\mathcal{E}_1 = 15$ V, and $\mathcal{E}_2 = 2$ V, solve the system of equations for i_1, i_2, and i_3. This gives the current, in amperes, in each branch of the circuit.

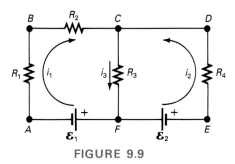

FIGURE 9.9

14. In the circuit of Problem 13, if $R_1 = 1\,\Omega$, $R_2 = 1\,\Omega$, $R_3 = 4\,\Omega$, $R_4 = 6\,\Omega$, $\mathcal{E}_1 = 2$ V, and $\mathcal{E}_2 = 60$ V, find i_1, i_2, and i_3.

15. In the circuit of Problem 13, with $\mathcal{E}_1 = \mathcal{E}_2$ a student measured and found that $i_1 = i_2$. Show that, as a consequence,

$$R_1 + R_2 - R_4 = 0.$$

You may assume that $i_1 \neq 0$.

9.4 DETERMINANTS OF ORDER 2

There is a method other than elimination to use in solving systems of linear equations. This method involves *determinants*. In this section you will learn what determinants are and how to use them to solve linear systems.

We begin by using elimination to solve a system of two linear equations in two variables of the general form $ax + by = c$, namely,

$$\begin{cases} a_1 x + b_1 y = c_1 & \text{(1)} \\ a_2 x + b_2 y = c_2. & \text{(2)} \end{cases}$$

To eliminate y we first multiply Eq. 1 by b_2 and Eq. 2 by $-b_1$:

$$\begin{cases} a_1 b_2 x + b_1 b_2 y = b_2 c_1 \\ -a_2 b_1 x - b_1 b_2 y = -b_1 c_2. \end{cases}$$

Adding these equations, we have

$$a_1 b_2 x - a_2 b_1 x = b_2 c_1 - b_1 c_2.$$

The left side can be written as the product $x(a_1 b_2 - a_2 b_1)$.

$$x(a_1 b_2 - a_2 b_1) = b_2 c_1 - b_1 c_2.$$

Solving for x gives

$$x = \frac{b_2 c_1 - b_1 c_2}{a_1 b_2 - a_2 b_1}, \tag{3}$$

provided that the denominator $a_1 b_2 - a_2 b_1$ is not zero. Similarly, we can show that

$$y = \frac{a_1 c_2 - a_2 c_1}{a_1 b_2 - a_2 b_1}. \tag{4}$$

Note that the denominators in Eqs. 3 and 4 are the same number, $a_1 b_2 - a_2 b_1$. This number will be denoted by a *square array of numbers* enclosed by vertical bars:

$$\begin{vmatrix} a_1 & b_1 \\ a_2 & b_2 \end{vmatrix} = a_1 b_2 - a_2 b_1.$$

In this form the number $a_1 b_2 - a_2 b_1$ is called a **determinant**. Since there are two rows and two columns, we say that this determinant has **order** 2, or is a 2 × 2 (read 2 *by* 2) determinant. The rows and columns are identified in the more general array:

$$\begin{array}{l} \text{Row 1} \rightarrow \\ \text{Row 2} \rightarrow \end{array} \begin{vmatrix} a & b \\ c & d \end{vmatrix} = ad - bc.$$

$$\begin{array}{cc} \uparrow & \uparrow \\ \text{Col. 1} & \text{Col. 2} \end{array}$$

Notice that ad is the product of the entries on the diagonal that goes from the upper left corner to the lower right corner, whereas bc is the product of the entries on the other diagonal. The value of the determinant is the difference of these products.

$$\begin{vmatrix} a & b \\ c & d \end{vmatrix} = ad - bc.$$

In the future we shall find it convenient to simply call the above array a determinant, although what is meant is the *number* represented by the array.

Example 1

a. $\begin{vmatrix} 2 & 1 \\ 5 & 7 \end{vmatrix} = (2)(7) - (1)(5) = 14 - 5 = 9.$

b. $\begin{vmatrix} -6 & -2 \\ 4 & 3 \end{vmatrix} = (-6)(3) - (-2)(4) = -18 + 8 = -10.$

c. $\begin{vmatrix} 2 & 0 \\ 0 & 2 \end{vmatrix} = (2)(2) - (0)(0) = 4.$

Returning to the given system of equations, we can now write Eqs. 3 and 4, which form the solution of the system, as

$$x = \frac{\begin{vmatrix} c_1 & b_1 \\ c_2 & b_2 \end{vmatrix}}{\begin{vmatrix} a_1 & b_1 \\ a_2 & b_2 \end{vmatrix}} \quad \text{and} \quad y = \frac{\begin{vmatrix} a_1 & c_1 \\ a_2 & c_2 \end{vmatrix}}{\begin{vmatrix} a_1 & b_1 \\ a_2 & b_2 \end{vmatrix}}.$$

Using these formulas is a convenient way to solve the original system,

$$\begin{cases} a_1 x + b_1 y = c_1 \\ a_2 x + b_2 y = c_2. \end{cases}$$

The determinants in the formulas can be easily remembered. The first column of the determinant in the denominators consists of the coefficients of the x-terms in the system; the second column consists of the coefficients of the y-terms. The numerator for x is the same as its denominator, except that the "x-column" is replaced by the constant terms of the system. Similarly, the numerator for y can be found by replacing the "y-column" of the denominator by the column of constant terms. Usually we use the letter D to stand for the determinants in the denominators.

This method of using determinants to solve a system of linear equations is called *Cramer's rule*.

CRAMER'S RULE

The system

$$\begin{cases} a_1 x + b_1 y = c_1 \\ a_2 x + b_2 y = c_2 \end{cases}$$

has the solution

$$x = \frac{\begin{vmatrix} c_1 & b_1 \\ c_2 & b_2 \end{vmatrix}}{D}, \qquad y = \frac{\begin{vmatrix} a_1 & c_1 \\ a_2 & c_2 \end{vmatrix}}{D},$$

where $D = \begin{vmatrix} a_1 & b_1 \\ a_2 & b_2 \end{vmatrix}$. We assume that $D \neq 0$.

You cannot use Cramer's rule to solve a system if $D = 0$. In that situation, use the method of elimination. It can be shown that the system will have either no solution or infinitely many solutions.

Example 2

Use Cramer's rule to solve

$$\begin{cases} 2x + y + 5 = 0 \\ 3y + x = 6. \end{cases}$$

Solution   First, we rewrite the system so that it has the form given in Cramer's rule.

$$\begin{cases} 2x + y = -5 \\ x + 3y = 6. \end{cases}$$

In both equations the x- and y-terms are aligned and the constant terms are on the right sides. Now we evaluate D.

$$D = \begin{vmatrix} 2 & 1 \\ 1 & 3 \end{vmatrix} = (2)(3) - (1)(1) = 6 - 1 = 5.$$

By Cramer's rule,

$$x = \frac{\begin{vmatrix} -5 & 1 \\ 6 & 3 \end{vmatrix}}{D} = \frac{(-5)(3) - (1)(6)}{5} = \frac{-15 - 6}{5} = -\frac{21}{5}$$

and

$$y = \frac{\begin{vmatrix} 2 & -5 \\ 1 & 6 \end{vmatrix}}{D} = \frac{(2)(6) - (-5)(1)}{5} = \frac{12 + 5}{5} = \frac{17}{5}.$$

Example 3

Solve the following system by using Cramer's rule.

$$\begin{cases} 0.24i_1 - 0.34i_2 = 0.28 \\ 0.11i_1 + 0.21i_2 = 0.86. \end{cases}$$

Solution   We have

$$D = \begin{vmatrix} 0.24 & -0.34 \\ 0.11 & 0.21 \end{vmatrix} = (0.24)(0.21) - (-0.34)(0.11)$$

$$= 0.0878.$$

Thus

$$i_1 = \frac{\begin{vmatrix} 0.28 & -0.34 \\ 0.86 & 0.21 \end{vmatrix}}{D} = \frac{(0.28)(0.21) - (-0.34)(0.86)}{0.0878}$$

$$= \frac{0.3512}{0.0878} = 4$$

and

$$i_2 = \frac{\begin{vmatrix} 0.24 & 0.28 \\ 0.11 & 0.86 \end{vmatrix}}{D} = \frac{(0.24)(0.86) - (0.28)(0.11)}{0.0878}$$

$$= \frac{0.1756}{0.0878} = 2.$$

The solution is $i_1 = 4$ and $i_2 = 2$.

Problem Set 9.4

In Problems **1–10**, *evaluate.*

1. $\begin{vmatrix} 2 & 1 \\ 3 & 2 \end{vmatrix}.$

2. $\begin{vmatrix} 1 & 2 \\ 3 & 4 \end{vmatrix}.$

3. $\begin{vmatrix} 2 & -6 \\ 5 & 3 \end{vmatrix}.$

4. $\begin{vmatrix} 3 & 2 \\ -5 & -4 \end{vmatrix}.$

5. $\begin{vmatrix} \frac{1}{2} & \frac{1}{3} \\ \frac{1}{4} & -4 \end{vmatrix}.$

6. $\begin{vmatrix} -2 & -3 \\ -4 & -6 \end{vmatrix}.$

7. $\begin{vmatrix} -3 & 1 \\ -a & b \end{vmatrix}.$

8. $\begin{vmatrix} -2 & -a \\ -a & 2 \end{vmatrix}.$

9. $\dfrac{\begin{vmatrix} 1 & 2 \\ 3 & 4 \end{vmatrix}}{\begin{vmatrix} 2 & 1 \\ 5 & 6 \end{vmatrix}}.$

10. $\dfrac{\begin{vmatrix} 6 & 2 \\ 1 & 5 \end{vmatrix}}{\begin{vmatrix} 2 & -6 \\ 5 & 3 \end{vmatrix}}.$

11. Solve for k if

$$\begin{vmatrix} 2 & 3 \\ 4 & k \end{vmatrix} = 12.$$

12. Solve for k if

$$\begin{vmatrix} k & k \\ 6 & 9 \end{vmatrix} = -4.$$

In Problems 13–24, use Cramer's rule to solve the system.

13. $\begin{cases} 2x - y = 4 \\ 3x + y = 5. \end{cases}$

14. $\begin{cases} 3x + y = 6 \\ 7x - 2y = 5. \end{cases}$

15. $\begin{cases} \frac{3}{2}x - \frac{1}{4}z = 1 \\ \frac{1}{3}x + \frac{1}{2}z = 2. \end{cases}$

16. $\begin{cases} s - \frac{1}{4}t = 1 \\ s + t = -4. \end{cases}$

17. $\begin{cases} 0.14i_1 - 0.31i_2 = 1.97 \\ 0.34i_1 + 0.52i_2 = -1.58. \end{cases}$

18. $\begin{cases} 0.6x - 0.7y = 0.33 \\ 2.1x - 0.9y = 0.69. \end{cases}$

19. $\begin{cases} x - 2y = 4 \\ x - 6 = -y. \end{cases}$

20. $\begin{cases} x + 4 = 3x \\ 3x + 2y = 8. \end{cases}$

21. $\begin{cases} -2x = 4 - 3y \\ y = 6x - 1. \end{cases}$

22. $\begin{cases} x + 2y - 6 = 0 \\ y - 1 = 3x. \end{cases}$

23. $\begin{cases} 2 - u = t \\ 3 + t = -u. \end{cases}$

24. $\begin{cases} \frac{3}{2}w + \frac{1}{4}z = 1 \\ \frac{1}{3}w = 2 - \frac{1}{2}z. \end{cases}$

25. Show that Cramer's rule does *not* apply to

$$\begin{cases} x + y = 2 \\ 3 + x = -y, \end{cases}$$

but that from geometrical considerations there is no solution.

9.5 DETERMINANTS OF ORDER 3

The value of the **third-order determinant**

$$\begin{matrix} & \text{Col. 1} & \text{Col. 2} & \text{Col. 3} \\ & \downarrow & \downarrow & \downarrow \\ \text{Row 1} \rightarrow & a_1 & b_1 & c_1 \\ \text{Row 2} \rightarrow & a_2 & b_2 & c_2 \\ \text{Row 3} \rightarrow & a_3 & b_3 & c_3 \end{matrix}$$

is defined in the following manner. With a given entry in the array, we associate the second-order determinant obtained by crossing out the row and column in which the entry lies. Hence, for a_2 we cross out the second row, which consists of the entries a_2, b_2, c_2, and the first column, which consists of the entries a_1, a_2, a_3:

$$\begin{vmatrix} \cancel{a_1} & b_1 & c_1 \\ \cancel{a_2} & \cancel{b_2} & \cancel{c_2} \\ \cancel{a_3} & b_3 & c_3 \end{vmatrix}.$$

This leaves the determinant

$$\begin{vmatrix} b_1 & c_1 \\ b_3 & c_3 \end{vmatrix},$$

which is called the **minor** of the entry a_2. With each entry is also associated the number

$$(-1)^{i+j},$$

where i is the number of the row and j is the number of the column in which the entry lies. Because a_2 lies in row 2 and column 1, we associate $(-1)^{2+1} = (-1)^3 = -1$. The **cofactor** of a_2 is the product of this number and the minor of a_2:

$$-1 \cdot \begin{vmatrix} b_1 & c_1 \\ b_3 & c_3 \end{vmatrix}.$$

Similarly, the cofactor of b_2 is

$$(-1)^{2+2} \begin{vmatrix} a_1 & c_1 \\ a_3 & c_3 \end{vmatrix}.$$

To find the value of a 3×3 determinant, we use the following rule.

To evaluate a **determinant of order 3**, choose *any* row (or column) and multiply each entry in that row (or column) by its cofactor. The *sum* of these products is the value of the determinant.

Example 1

Find the value of $\begin{vmatrix} 2 & -1 & 3 \\ 4 & 0 & -5 \\ -2 & 1 & 1 \end{vmatrix}$.

Solution We shall apply the previous rule to the first row. (This is called *expanding along the first row*.)

$$\begin{vmatrix} 2 & -1 & 3 \\ 4 & 0 & -5 \\ -2 & 1 & 1 \end{vmatrix} = 2\begin{pmatrix} \text{cofactor} \\ \text{of } 2 \end{pmatrix} + (-1)\begin{pmatrix} \text{cofactor} \\ \text{of } -1 \end{pmatrix} + 3\begin{pmatrix} \text{cofactor} \\ \text{of } 3 \end{pmatrix}$$

$$= 2(-1)^{1+1}\begin{vmatrix} 0 & -5 \\ 1 & 1 \end{vmatrix} + (-1)(-1)^{1+2}\begin{vmatrix} 4 & -5 \\ -2 & 1 \end{vmatrix} + 3(-1)^{1+3}\begin{vmatrix} 4 & 0 \\ -2 & 1 \end{vmatrix}$$

$$= 2(1)[0 - (-5)] + (-1)(-1)[4 - 10] + 3(1)[4 - 0]$$

$$= 2(1)(5) + (-1)(-1)(-6) + 3(1)(4)$$

$$= 10 - 6 + 12 = 16.$$

If we had expanded along the second column, we would have obtained

$$\begin{vmatrix} 2 & -1 & 3 \\ 4 & 0 & -5 \\ -2 & 1 & 1 \end{vmatrix} = (-1)\begin{pmatrix} \text{cofactor} \\ \text{of} -1 \end{pmatrix} + 0\begin{pmatrix} \text{cofactor} \\ \text{of } 0 \end{pmatrix} + (1)\begin{pmatrix} \text{cofactor} \\ \text{of } 1 \end{pmatrix}$$

$$= (-1)(-1)^{1+2}\begin{vmatrix} 4 & -5 \\ -2 & 1 \end{vmatrix} + 0 + (1)(-1)^{3+2}\begin{vmatrix} 2 & 3 \\ 4 & -5 \end{vmatrix}$$

$$= (-1)(-1)[4 - 10] + 0 + (1)(-1)[-10 - 12]$$

$$= (-1)(-1)(-6) + 0 + (1)(-1)(-22)$$

$$= -6 + 0 + 22 = 16,$$

as before. The computation is easier here, since we did not have to find the cofactor of 0. In general it's a good idea to expand along the row or column with the greatest number of zeros.

Example 2

a. Evaluate $\begin{vmatrix} 12 & -1 & 3 \\ -3 & 1 & -1 \\ -10 & 2 & -3 \end{vmatrix}$.

Solution   Expanding along the first row gives

$$\begin{vmatrix} 12 & -1 & 3 \\ -3 & 1 & -1 \\ -10 & 2 & -3 \end{vmatrix}$$

$$= 12(-1)^{1+1}\begin{vmatrix} 1 & -1 \\ 2 & -3 \end{vmatrix} + (-1)(-1)^{1+2}\begin{vmatrix} -3 & -1 \\ -10 & -3 \end{vmatrix} + 3(-1)^{1+3}\begin{vmatrix} -3 & 1 \\ -10 & 2 \end{vmatrix}$$

$$= 12(1)(-1) + (-1)(-1)(-1) + 3(1)(4) = -1.$$

b. Evaluate $\begin{vmatrix} 0 & 1 & 1 \\ 2 & 3 & 2 \\ 0 & -1 & 3 \end{vmatrix}$.

Solution   Expanding along column 1 for conveneince, we have

$$\begin{vmatrix} 0 & 1 & 1 \\ 2 & 3 & 2 \\ 0 & -1 & 3 \end{vmatrix} = 0 + 2(-1)^{2+1}\begin{vmatrix} 1 & 1 \\ -1 & 3 \end{vmatrix} + 0 = 2(-1)(4) = -8.$$

Cramer's rule can be extended to handle systems of three linear equations in three unknowns.

CRAMER'S RULE

The system

$$\begin{cases} a_1x + b_1y + c_1z = d_1 \\ a_2x + b_2y + c_2z = d_2 \\ a_3x + b_3y + c_3z = d_3 \end{cases}$$

has the solution

$$x = \frac{D_x}{D}, \qquad y = \frac{D_y}{D}, \qquad z = \frac{D_z}{D},$$

where

$$D = \begin{vmatrix} a_1 & b_1 & c_1 \\ a_2 & b_2 & c_2 \\ a_3 & b_3 & c_3 \end{vmatrix}$$

and

$$D_x = \begin{vmatrix} d_1 & b_1 & c_1 \\ d_2 & b_2 & c_2 \\ d_3 & b_3 & c_3 \end{vmatrix}, \qquad D_y = \begin{vmatrix} a_1 & d_1 & c_1 \\ a_2 & d_2 & c_2 \\ a_3 & d_3 & c_3 \end{vmatrix}, \qquad D_z = \begin{vmatrix} a_1 & b_1 & d_1 \\ a_2 & b_2 & d_2 \\ a_3 & b_3 & d_3 \end{vmatrix}.$$

We assume that $D \neq 0$.

Again, there is a pattern for the determinants. For D, notice that columns 1, 2, and 3 consist of the coefficients of the x-, y-, and z-terms, respectively, in the given system. The numerator D_x is found by replacing the x-column of D by the constant terms. The numerators D_y and D_z are found similarly.

Example 3

Solve by using Cramer's rule:

$$\begin{cases} 2x + y + z = 0 \\ 4x + 3y + 2z = 2 \\ 2x - y - 3z = 0. \end{cases}$$

Solution   This system is in the form stated in Cramer's rule. We have

$$D = \begin{vmatrix} 2 & 1 & 1 \\ 4 & 3 & 2 \\ 2 & -1 & -3 \end{vmatrix},$$

$$D_x = \begin{vmatrix} 0 & 1 & 1 \\ 2 & 3 & 2 \\ 0 & -1 & -3 \end{vmatrix}, \quad D_y = \begin{vmatrix} 2 & 0 & 1 \\ 4 & 2 & 2 \\ 2 & 0 & -3 \end{vmatrix}, \quad D_z = \begin{vmatrix} 2 & 1 & 0 \\ 4 & 3 & 2 \\ 2 & -1 & 0 \end{vmatrix}.$$

We shall evaluate D by expanding along the first row.

$$D = 2(-1)^{1+1} \begin{vmatrix} 3 & 2 \\ -1 & -3 \end{vmatrix} + (1)(-1)^{1+2} \begin{vmatrix} 4 & 2 \\ 2 & -3 \end{vmatrix} + (1)(-1)^{1+3} \begin{vmatrix} 4 & 3 \\ 2 & -1 \end{vmatrix}$$

$$= 2(1)(-7) + (1)(-1)(-16) + (1)(1)(-10)$$

$$= -14 + 16 - 10 = -8.$$

For D_x, we expand along column 1.

$$D_x = 0 + 2(-1)^{2+1} \begin{vmatrix} 1 & 1 \\ -1 & -3 \end{vmatrix} + 0$$

$$= 2(-1)(-2) = 4.$$

For D_y, we expand along column 2.

$$D_y = 0 + 2(-1)^{2+2} \begin{vmatrix} 2 & 1 \\ 2 & -3 \end{vmatrix} + 0$$

$$= 2(1)(-8) = -16.$$

For D_z, we expand along column 3.

$$D_z = 0 + 2(-1)^{2+3} \begin{vmatrix} 2 & 1 \\ 2 & -1 \end{vmatrix} + 0$$

$$= 2(-1)(-4) = 8.$$

Thus

$$x = \frac{D_x}{D} = \frac{4}{-8} = -\frac{1}{2},$$

$$y = \frac{D_y}{D} = \frac{-16}{-8} = 2,$$

$$z = \frac{D_z}{D} = \frac{8}{-8} = -1.$$

Finally, we remark that a determinant of order n, where $n > 3$, can also be evaluated by the method of minors and cofactors.

Problem Set 9.5

In Problems 1–5, evaluate the determinants.

1. $\begin{vmatrix} 2 & 1 & 3 \\ 2 & 0 & 1 \\ -4 & 0 & 6 \end{vmatrix}$.

2. $\begin{vmatrix} 3 & 2 & 1 \\ 1 & -2 & 3 \\ -1 & 3 & 2 \end{vmatrix}$.

3. $\begin{vmatrix} 1 & 2 & -3 \\ 4 & 5 & 4 \\ 3 & -2 & 1 \end{vmatrix}$.

4. $\begin{vmatrix} 1 & 0 & -1 \\ 0 & 1 & 0 \\ 1 & -1 & 1 \end{vmatrix}$.

5. $\begin{vmatrix} 2 & 1 & 5 \\ -3 & 4 & -1 \\ 0 & 6 & -1 \end{vmatrix}$.

6. A *nomogram* is one type of graphical process that can be used to solve certain types of equations. One nomogram yields the basic (determinant) equation

$$\begin{vmatrix} \frac{1}{2}f_1 & \frac{1}{2} & 1 \\ -f_2 & 1 & 1 \\ -3f_1 & 0 & 1 \end{vmatrix} = 0.$$

What is the relation between f_1 and f_2?

In Problems 7–16, solve the systems by using Cramer's rule.

7. $\begin{cases} x + y + z = 6 \\ x - y + z = 2 \\ 2x - y + 3z = 6. \end{cases}$

8. $\begin{cases} 2x - y + 3z = 12 \\ x + y - z = -3 \\ x + 2y - 3z = -10. \end{cases}$

9. $\begin{cases} 2i_1 - 3i_2 + 4i_3 = 0 \\ i_1 + i_2 - 3i_3 = 4 \\ 3i_1 + 2i_2 - i_3 = 0. \end{cases}$

10. $\begin{cases} 3r - t = 7 \\ 4r - s + 3t = 9 \\ 3s + 2t = 15. \end{cases}$

11. $\begin{cases} 2x - 3y + z = -2 \\ x - 6y + 3z = -2 \\ 3x + 3y - 2z = 2. \end{cases}$

12. $\begin{cases} T_1 - T_3 = 14 \\ T_2 + T_3 = 21 \\ T_1 - T_2 + T_3 = -10. \end{cases}$

13. $\begin{cases} 2a + b + 6c = 3 \\ a - b + 4c = 1 \\ 3a + 2b - 2c = 2. \end{cases}$

14. $\begin{cases} x + y + z = -1 \\ 3x + y + z = 1 \\ 4x - 2y + 2z = 0. \end{cases}$

15. Solve for y:
$$\begin{cases} 5x - 7y + 4z = 2 \\ 3x + 2y - 2z = 3 \\ 2x - y + 3z = 4. \end{cases}$$

16. Solve for z:
$$\begin{cases} 3x - 2y + z = 0 \\ -2x + y - 2z = 5 \\ \frac{3}{2}x + \frac{4}{5}y + 4z = 10. \end{cases}$$

17. Loads weighing 200 N and 300 N are suspended from the pulley system shown in Fig. 9.10. One pulley is fixed in position and the other is free to move up and down. If the

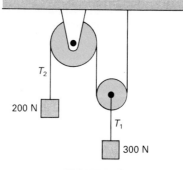

FIGURE 9.10

pulleys are frictionless and without mass, the tensions T_1 and T_2 in the cords and the acceleration a of the 200-N load are found by solving the system

$$\begin{cases} T_1 = 2T_2 \\ T_1 - 300 = 15.3a \\ 200 - T_2 = 20.4a, \end{cases}$$

where T_1 and T_2 are in newtons and a is in meters per second squared. Find the approximate values of T_1, T_2, and a.

9.6   PROPERTIES OF DETERMINANTS (OPTIONAL)

We can often simplify the task of evaluating a determinant by using various properties, some of which we now list.

---

1. **If each of the entries in a row (or column) of a determinant is zero, then the value of the determinant is zero.**

Thus

$$\begin{vmatrix} 6 & 2 & 5 \\ 7 & 1 & 4 \\ 0 & 0 & 0 \end{vmatrix} = 0.$$

---

2. **If two rows (or columns) of a determinant are identical or proportional, then the value of the determinant is zero.**

Thus

$$\begin{vmatrix} 2 & 5 & 2 & 1 \\ 2 & 6 & 2 & 3 \\ 2 & 4 & 2 & 1 \\ 6 & 5 & 6 & 1 \end{vmatrix} = 0 \qquad \text{[because column 1 = column 3]}$$

and

$$\begin{vmatrix} 1 & 2 & 3 \\ 7 & 4 & 2 \\ 2 & 4 & 6 \end{vmatrix} = 0 \qquad \text{[because row 3 is 2 times row 1].}$$

3. **If all the entries below (or above) the main diagonal entries (from the upper left corner to the lower right corner) of a determinant are zero, then the value of the determinant is equal to the product of the main diagonal entries.**

Thus

$$\begin{vmatrix} 2 & 6 & 1 & 0 \\ 0 & 5 & 7 & 6 \\ 0 & 0 & -2 & 5 \\ 0 & 0 & 0 & 1 \end{vmatrix} = (2)(5)(-2)(1) = -20.$$

4. **The value of a determinant is unchanged if a multiple of one row (or column) is added to another row (or column).**

Thus if we add -4 times row 2 to row 3 below, then

$$\begin{vmatrix} 1 & 3 & 5 \\ 0 & 2 & 3 \\ 0 & 8 & -1 \end{vmatrix} = \begin{vmatrix} 1 & 3 & 5 \\ 0 & 2 & 3 \\ 0 & 0 & -13 \end{vmatrix} = (1)(2)(-13) = -26 \qquad \text{[Property 3].}$$

5. **If two rows (or two columns) of a determinant are interchanged, then the value of the determinant is multiplied by -1.**

Thus by interchanging rows 2 and 4 below and using Property 3, we have

$$\begin{vmatrix} 2 & 2 & 1 & 6 \\ 0 & 0 & 0 & 1 \\ 0 & 0 & 2 & 0 \\ 0 & 1 & -3 & 4 \end{vmatrix} = - \begin{vmatrix} 2 & 2 & 1 & 6 \\ 0 & 1 & -3 & 4 \\ 0 & 0 & 2 & 0 \\ 0 & 0 & 0 & 1 \end{vmatrix} = -(4) = -4.$$

6. **If each entry of a row (or column) of a determinant is multiplied by the same constant k, then the value of the determinant is multiplied by k.**

Thus

$$\begin{vmatrix} 2 \cdot 3 & 2 \cdot 5 & 2 \cdot 7 \\ 5 & 2 & 1 \\ 6 & 4 & 3 \end{vmatrix} = 2 \begin{vmatrix} 3 & 5 & 7 \\ 5 & 2 & 1 \\ 6 & 4 & 3 \end{vmatrix}.$$

Example 1

Evaluate $\begin{vmatrix} 1 & 1 & 0 & 5 \\ 1 & 2 & 1 & 0 \\ 0 & 2 & 1 & 1 \\ 3 & 0 & 0 & -4 \end{vmatrix}.$

Solution We shall express the determinant in a form in which all entries below the main diagonal are zero. Then, by Property 3, we shall take the product of the entries of the main diagonal.

$$\begin{vmatrix} 1 & 1 & 0 & 5 \\ 1 & 2 & 1 & 0 \\ 0 & 2 & 1 & 1 \\ 3 & 0 & 0 & -4 \end{vmatrix} = \begin{vmatrix} 1 & 1 & 0 & 5 \\ 0 & 1 & 1 & -5 \\ 0 & 2 & 1 & 1 \\ 0 & -3 & 0 & -19 \end{vmatrix}$$ [by adding -1 times row 1 to row 2; adding -3 times row 1 to row 4]

$$= \begin{vmatrix} 1 & 1 & 0 & 5 \\ 0 & 1 & 1 & -5 \\ 0 & 0 & -1 & 11 \\ 0 & 0 & 3 & -34 \end{vmatrix}$$ [by adding -2 times row 2 to row 3; adding 3 times row 2 to row 4]

$$= \begin{vmatrix} 1 & 1 & 0 & 5 \\ 0 & 1 & 1 & -5 \\ 0 & 0 & -1 & 11 \\ 0 & 0 & 0 & -1 \end{vmatrix}$$ [by adding 3 times row 3 to row 4]

$$= (1)(1)(-1)(-1) = 1.$$

Problem Set 9.6

Evaluate the determinants in Problems 1–9 by using properties of determinants.

1. $\begin{vmatrix} -2 & 5 & 7 \\ 0 & 0 & 0 \\ 5 & 9 & 4 \end{vmatrix}.$

2. $\begin{vmatrix} -3 & -5 & 6 \\ 0 & 4 & 6 \\ 0 & 0 & 2 \end{vmatrix}.$

3. $\begin{vmatrix} 2 & 1 & -6 \\ -1 & 7 & 3 \\ 3 & 2 & -9 \end{vmatrix}$.

4. $\begin{vmatrix} 4 & 2 & -4 \\ 2 & 4 & 6 \\ 4 & 2 & -4 \end{vmatrix}$.

5. $\begin{vmatrix} 1 & 7 & -3 & 8 \\ 0 & 1 & -5 & 4 \\ 0 & 0 & 1 & 7 \\ 0 & 0 & 0 & 1 \end{vmatrix}$.

6. $\begin{vmatrix} 1 & 2 & -3 & 4 \\ 3 & -1 & 2 & 4 \\ -2 & -4 & 6 & -8 \\ 0 & 3 & -1 & 2 \end{vmatrix}$.

7. $\begin{vmatrix} 1 & 0 & 0 & 0 \\ 0 & -2 & 0 & 0 \\ 0 & 0 & 4 & 0 \\ 0 & 0 & 0 & -3 \end{vmatrix}$.

8. $\begin{vmatrix} 7 & 6 & 0 & 5 \\ -3 & 2 & 0 & 1 \\ 4 & -3 & 0 & 2 \\ 1 & 0 & 0 & 6 \end{vmatrix}$.

9. $\begin{vmatrix} 1 & 0 & 3 & 2 \\ 4 & -1 & 0 & 1 \\ 2 & 1 & 0 & 3 \\ -1 & 2 & 3 & -1 \end{vmatrix}$.

10. Suppose the value of a determinant of order 4 is 12. What is the value of the determinant obtained by multiplying every entry in the given determinant by 2?

9.7 WORD PROBLEMS

Each word problem in this section gives rise to a system of linear equations. After an appropriate system has been determined, you may use any method to solve it. The following examples illustrate some basic techniques.

Example 1

In a laboratory a student is to combine a 25% hydrogen peroxide solution (25% by volume is hydrogen peroxide) with a 40% hydrogen peroxide solution to obtain 2 L of a 30% solution. How many liters of each solution should the student mix?

Solution Let x be the number of liters of the 25% solution and y be the number of liters of the 40% solution that should be mixed (refer to Fig. 9.11). Then

$$x + y = 2. \tag{1}$$

In 2 L of a 30% solution there is $0.30(2) = 0.6$ L of hydrogen peroxide. This hydrogen peroxide comes from two sources: $0.25x$ liters of it come from the 25% solution, and $0.40y$ liters come from the 40% solution. Thus

$$0.25x + 0.40y = 0.6. \tag{2}$$

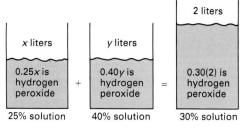

FIGURE 9.11

Equations 1 and 2 form a system. Solving Eq. 1 for x gives $x = 2 - y$. Substituting $2 - y$ for x in Eq. 2 gives

$$0.25(2 - y) + 0.40y = 0.6$$
$$0.5 - 0.25y + 0.40y = 0.6$$
$$0.15y = 0.1$$
$$y = \frac{0.1}{0.15} = \frac{10}{15} = \frac{2}{3}.$$

From Eq. 1,

$$x = 2 - y = 2 - \frac{2}{3} = \frac{4}{3}.$$

Thus $\frac{4}{3}$ L of the 25% solution and $\frac{2}{3}$ L of the 40% solution must be mixed.

Example 2

Suppose it is known that quantities x and y are linearly related by an equation of the form

$$y = mx + b.$$

If measurements are made and it is found that $y = 4$ when $x = 3$, and $y = 13$ when $x = 6$, find m and b.

Solution Substituting the data into the given equation, we obtain the system

$$\begin{cases} 4 = 3m + b \\ 13 = 6m + b. \end{cases}$$

Multiplying the first equation by -1 and adding the result to the second equation gives $9 = 3m$, from which $m = 3$. Substituting 3 for m in the first equation gives $4 = 9 + b$, from which $b = -5$. Thus $m = 3$ and $b = -5$.

Example 3

To determine the height h of the top of an antenna situated on a cliff (see Fig. 9.12), the angle of elevation of the top from a point on level ground is measured to be 35°. From a point 95 m farther away, the angle of elevation is 25°. Find h.

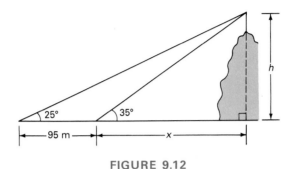

FIGURE 9.12

Solution   From the smaller right triangle in Fig. 9.12, we have

$$\tan 35° = \frac{h}{x}$$

$$x = \frac{h}{\tan 35°}. \tag{3}$$

From the larger right triangle,

$$\tan 25° = \frac{h}{95 + x}$$

$$h = (\tan 25°)(95 + x). \tag{4}$$

Equations 3 and 4 form a system of two linear equations in the unknowns h and x. Substituting $h/(\tan 35°)$ for x in Eq. 4 gives

$$h = (\tan 25°)\left(95 + \frac{h}{\tan 35°}\right)$$

$$h = 95 \tan 25° + \frac{h \tan 25°}{\tan 35°}$$

$$h = 44.299 + 0.66596h$$

$$0.33404h = 44.299$$

$$h = \frac{44.299}{0.33404} = 132.6 \text{ m.}$$

Example 4

In Fig. 9.13(a) a weight of 20 N is supported by two cables. Find the tensions T_1 and T_2 in the cables.

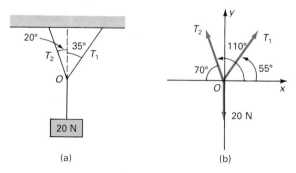

FIGURE 9.13

Solution   Because the system is in equilibrium (the weight is at rest), it is shown in physics that the sum of the forces acting at point O must be zero. As a result, at that point the sum of all horizontal components is zero and the sum of all vertical components is zero. From Fig. 9.13(b), this means that

Hor. $\quad \begin{cases} T_1 \cos 55° - T_2 \cos 70° + 20 \cos 270° = 0 \\ T_1 \sin 55° + T_2 \sin 70° + 20 \sin 270° = 0. \end{cases}$
Ver.

In setting up the first equation, for $T_2 \cos 110°$ we used the fact that $\cos 110° = -\cos 70°$. Similarly, in the second equation we used the fact that $\sin 110° = \sin 70°$. Thus we have

$$\begin{cases} 0.5736T_1 - 0.3420T_2 = 0 \\ 0.8192T_1 + 0.9397T_2 = 20. \end{cases}$$

By Cramer's rule,

$$T_1 = \frac{\begin{vmatrix} 0 & -0.3420 \\ 20 & 0.9397 \end{vmatrix}}{\begin{vmatrix} 0.5736 & -0.3420 \\ 0.8192 & 0.9397 \end{vmatrix}} = \frac{6.840}{0.8192} = 8.4 \text{ N}$$

and

$$T_2 = \frac{\begin{vmatrix} 0.5736 & 0 \\ 0.8192 & 20 \end{vmatrix}}{0.8192} = \frac{11.47}{0.8192} = 14.0 \text{ N}.$$

Problem Set 9.7

1. The perimeter of a rectangle is 26 m. The length exceeds the width by 3 m. Find the dimensions of the rectangle.

2. The length of a rectangle is twice the width. The perimeter is 60 m. Find the dimensions of the rectangle.

3. One of two complementary angles is three-fifths of the other one. Find the angles.

4. One of two supplementary angles is three-fifths of the other one. Find the angles.

5. A chemical manufacturer needs to fill an order for 700 L of a 24% acid solution (24% by volume is acid). Solutions of 20% and 30% are in stock. How many liters of each must be mixed to fill the order? (See Example 1.)

6. A chemical manufacturer needs to obtain 500 L of a 25% acid solution by mixing a 30% solution with an 18% solution. How many liters of each must be mixed?

7. A 10,000-L tank is to be filled with solvent from storage tanks A and B. Solvent from A is pumped at the rate of 20 L/min, and solvent from B is pumped at the rate of 30 L/min. Usually, both pumps operate at the same time. However, because of a blown fuse, the pump on A is delayed 10 min. How many liters from each storage tank will be used to fill the tank?

8. On a trip on a raft it took $\frac{3}{4}$ h to travel 20 km downstream. The return trip took $1\frac{1}{2}$ h. Find the speed of the raft in still water and the speed of the current.

9. Table 9.1 shows how alloys A, B, and C are composed (by mass). How much of A, B, and C must be mixed to produce 100 kg of an alloy that is 53% copper and 19% zinc?

TABLE 9.1

	A	B	C
Copper	50%	60%	40%
Zinc	30%	20%	
Nickel	20%	20%	60%

10. A company manufactures industrial control units. Their new models are the Argon I and the Argon II. Each Argon I unit requires 6 transistors and 3 integrated circuits. Each Argon II unit requires 10 transistors and 8 integrated circuits. The company receives a total of 760 transistors and 500 integrated circuits each day from its supplier. How many units of each model of the Argon can the company make each day? Assume that all the parts are used.

11. The graph of $y = ax^2 + bx + c$ passes through the points $(2, 0)$, $(0, 0)$, and $(-1, 3)$. Find a, b, and c. (See Example 2.)

12. Repeat Problem 11 if the graph passes through the points $(2, 5)$, $(-3, 5)$ and $(1, 1)$.

13. From a point on the ground, the angle of elevation of the top of a building is 35°. From a point 50 m farther away, the angle of elevation is 30°. Find the height of the building. (See Example 3.)

14. From a point on level ground, the angle of elevation of the top of a mountain is 58°. From a point 340 m farther away, the angle of elevation is 46°. Find the height of the mountain.

15. In Fig. 9.14, a weight of 15 N is supported by two cables. Find the tensions T_1 and T_2 in the cables. (See Example 4.)

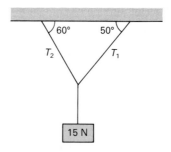

FIGURE 9.14

16. A very light boom is shown in Fig. 9.15. A 100-N load is suspended from its end. The forces acting at point O are the tension **T** in the horizontal cable, the 100-N force due to the load, and the force **F** due to the boom. Find T and F. (See Example 4.)

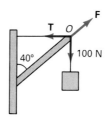

FIGURE 9.15

17. A 440-N weight is supported by three cables, as shown in Fig. 9.16. The tension in the vertical cable is 440 N and is due to the weight. Find the tensions T_1 and T_2 in the other two cables. (See Example 4.)

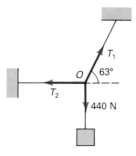

FIGURE 9.16

9.8 REVIEW

Important Terms

Section 9.1   System of equations, consistent system, inconsistent system, independent system, dependent system.

Section 9.2   Elimination by addition, elimination by substitution.

Section 9.3   Linear equation in three variables.

Section 9.4   Determinant, order of determinant, Cramer's rule.

Section 9.5   Minor, cofactor.

Formula Summary

Determinant

$$\begin{vmatrix} a_1 & b_1 \\ a_2 & b_2 \end{vmatrix} = a_1 b_2 - a_2 b_1.$$

Cramer's rule

A system of n linear equations in n unknowns has the solution

$$x = \frac{D_x}{D}, \qquad y = \frac{D_y}{D}, \qquad \text{and so on}$$

if $D \neq 0$. If $D = 0$, the system is inconsistent or dependent.

Review Questions

1. Two methods by which a system of two linear equations in two variables can be solved are by elimination by __(a)__ and elimination by __(b)__.

2. If a system of two linear equations is graphically represented by two different parallel lines, then what can be said about the solution of the system? _____ .

3. A system of two linear equations that has at least one solution is said to be _____ .

4. The equations in the system

$$\begin{cases} y = 2x - 4 \\ y = 3x - 4 \end{cases}$$

represent lines with different __(slopes)(y-intercepts)__ and hence must have exactly one solution.

5. To solve the system

$$\begin{cases} 3x - 4y = 10 \\ 8x + 5y = -40 \end{cases}$$

by eliminating y by addition, we can first multiply the top equation by 5 and the bottom equation by _____ .

6. The minor of the entry 3 in

$$\begin{vmatrix} 0 & 1 & 3 \\ 2 & 4 & 0 \\ 1 & 5 & 2 \end{vmatrix}$$

is equal to __(a)__ and its cofactor is equal to __(b)__ .

7. Expanding

$$\begin{vmatrix} 1 & 0 & 0 \\ 1 & 1 & 1 \\ 2 & 1 & 2 \end{vmatrix}$$

along the first row gives a value of _____ .

Answers to Review Questions

1. (a) Addition, (b) Substitution.   2. No solution.   3. Consistent.   4. Slopes.   5. 4.
6. (a) 6, (b) 6.   7. 1.

Review Problems

In Problems 1-16, solve the system by elimination.

1. $\begin{cases} 2x - y = 6 \\ 3x + 2y = 5. \end{cases}$
2. $\begin{cases} 3x + 5y = -6 \\ 2x - 5y = 6. \end{cases}$
3. $\begin{cases} 5x + 2y = 36 \\ 8x - 3y = -54. \end{cases}$

4. $\begin{cases} x + y = 3 \\ 3x + 2y = 19. \end{cases}$
5. $\begin{cases} 3s + t - 4 = 0 \\ 12s + 4t - 2 = 0. \end{cases}$
6. $\begin{cases} u - 3v + 11 = 0 \\ 4u + 3v - 9 = 0. \end{cases}$

7. $\begin{cases} 3x + \frac{1}{2}y = 2 \\ \frac{1}{2}x - \frac{1}{4}y = 0. \end{cases}$
8. $\begin{cases} \frac{1}{3}x - \frac{1}{2}y = 4 \\ \frac{1}{4}x - \frac{3}{8}y = 3. \end{cases}$

9. $\begin{cases} 6x = 3 - 9y \\ 12y = 4 - 8x. \end{cases}$
10. $\begin{cases} 4x = 7 + 12y \\ 5x = 15y - 2. \end{cases}$

11. $\begin{cases} 2(x - y) + 6(x + y) = 13 \\ 3(x + y) - 4(x - y) = 10. \end{cases}$
12. $\begin{cases} x + \dfrac{2y + x}{6} = 14 \\[2mm] y + \dfrac{3x + y}{4} = 20. \end{cases}$

13. $\begin{cases} x - y = 2 \\ x + z = 1 \\ y - z = 3. \end{cases}$
14. $\begin{cases} 2x - 4z = 8 \\ x - 2y - 2z = 14 \\ 3x + y + z = 0. \end{cases}$

15. $\begin{cases} 4r - s + 2t = 2 \\ 8r - 3s + 4t = 1 \\ r + 2s + 2t = 8. \end{cases}$
16. $\begin{cases} 3u - 2v + w = -2 \\ 2u + v + w = 1 \\ u + 3v - w = 3. \end{cases}$

In Problems **17–24**, *find the value of the determinant.*

17. $\begin{vmatrix} 2 & -1 \\ 4 & 7 \end{vmatrix}$.

18. $\begin{vmatrix} 5 & 8 \\ 3 & 0 \end{vmatrix}$.

19. $\begin{vmatrix} -1 & -2 \\ -3 & 4 \end{vmatrix}$.

20. $\begin{vmatrix} -2 & -6 \\ 3 & 5 \end{vmatrix}$.

21. $\begin{vmatrix} 1 & 2 & -1 \\ 0 & 1 & 4 \\ 1 & 2 & 2 \end{vmatrix}$.

22. $\begin{vmatrix} 2 & 0 & 3 \\ 1 & 4 & 6 \\ -1 & 2 & -1 \end{vmatrix}$.

23. $\begin{vmatrix} 2 & 1 & -1 \\ 1 & 1 & 3 \\ -1 & 1 & -1 \end{vmatrix}$.

24. $\begin{vmatrix} 0 & 1 & 2 \\ -2 & 0 & 3 \\ 4 & 2 & 0 \end{vmatrix}$.

In Problems **25–28**, *solve the system by Cramer's rule.*

25. $\begin{cases} 3x - y = 1 \\ 2x + 3y = 8. \end{cases}$

26. $\begin{cases} 2x = 5y \\ 4x + 3y = 0. \end{cases}$

27. $\begin{cases} x + y + z = 0 \\ x \quad\quad - z = 0 \\ x - y + 2z = 5. \end{cases}$

28. $\begin{cases} 3x + y + 4z = 1 \\ x \quad\quad + z = 0 \\ 2y + z = 2. \end{cases}$

29. Two alloys of copper are to be mixed so that the result is 15 kg of a 45% alloy (by mass). One alloy is 20% copper, and the other is 50% copper. How many kilograms of each should be used?

30. A 30-N weight is supported by two cables, as shown in Fig. 9.17. Find the tensions T_1 and T_2 in the cables.

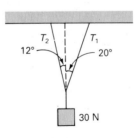

FIGURE 9.17

Factoring
and Fractions

10.1 SPECIAL PRODUCTS

In mathematics certain products occur so often that we find it worthwhile to memorize their patterns. But just "knowing" these *special products* is not enough. You have to be so familiar with them that you can recognize them in any form. Each special product is the result of applying the distributive law. This will be shown for the first special product, and you may check the others for yourself.

We begin with the square of a binomial:

$$(a + b)^2 = (a + b)(a + b)$$
$$= a(a + b) + b(a + b) = a^2 + ab + ba + b^2$$
$$(a + b)^2 = a^2 + 2ab + b^2.$$

That is, *the square of a binomial is equal to the square of the first term, plus twice the product of the terms, plus the square of the second term.* Similarly, we have $(a - b)^2 = a^2 - 2ab + b^2$. Thus we have our first two special products.

SQUARE OF A BINOMIAL

$$(a + b)^2 = a^2 + 2ab + b^2.$$
$$(a - b)^2 = a^2 - 2ab + b^2.$$

Note that the square of a binomial is a trinomial. We emphasize that

$$(a + b)^2 \neq a^2 + b^2 \quad \text{and} \quad (a - b)^2 \neq a^2 - b^2.$$

For example,

$$(1 + 2)^2 \neq 1^2 + 2^2 \quad \text{because} \quad 9 \neq 5.$$

Letters such as a and b that appear in the statements of special products are understood to denote any expression representing a number.

Example 1

Find the given products by using the rules for the square of a binomial.

a. $(x + 3)^2$.

Solution   Here x plays the role of a in $(a + b)^2$ and 3 plays the role of b.

$$(x + 3)^2 = x^2 + 2(x)(3) + 3^2 = x^2 + 6x + 9.$$

b. $(y - 4)^2 = y^2 - 2(y)(4) + 4^2 = y^2 - 8y + 16.$

c. $(2x + 4)^2$.

Solution   Here $2x$ plays the role of a.

$$(2x + 4)^2 = (2x)^2 + 2(2x)(4) + 4^2 = 4x^2 + 16x + 16.$$

d. $(x^2 - 3y^3)^2$.

Solution   Here x^2 is a and $3y^3$ is b.

$$(x^2 - 3y^3)^2 = (x^2)^2 - 2(x^2)(3y^3) + (3y^3)^2$$
$$= x^4 - 6x^2y^3 + 9y^6.$$

e. $-2b(-a + b)^2$.

Solution   You may immediately be tempted to use the distributive law. However, because of the exponent 2, the distributive law does not directly apply. It can be used *after* squaring the binomial $-a + b$.

$$-2b(-a + b)^2 = -2b[(-a)^2 + 2(-a)b + b^2]$$
$$= -2b[a^2 - 2ab + b^2]$$
$$= -2a^2b + 4ab^2 - 2b^3.$$

Note that the given problem could have been written as $-2b(b - a)^2$.

Formulas for the **cube of a binomial** are given in Problems 81 and 83 of Problem Set 10.1. The next special product shows how to find the product of the sum and difference of two terms.

PRODUCT OF THE SUM AND DIFFERENCE

$$(a + b)(a - b) = a^2 - b^2.$$

That is, *the product of the sum and difference of two terms is equal to the square of the first term, minus the square of the second term.* Note that the product is a binomial that is the difference of two squares.

Example 2

Find the given products.

a. $(x + 4)(x - 4) = x^2 - 4^2 = x^2 - 16.$

b. $(4x - 3y)(4x + 3y).$

> *Solution*   Here $4x$ plays the role of a, and $3y$ plays the role of b.
>
> $$(4x - 3y)(4x + 3y) = (4x)^2 - (3y)^2 = 16x^2 - 9y^2.$$

c. $(a^2 + 1)(a^2 - 1) = (a^2)^2 - 1^2 = a^4 - 1.$

d. $(y - \sqrt{2})(y + \sqrt{2}) = y^2 - (\sqrt{2})^2 = y^2 - 2.$

e. $(3x - 4)(-4 - 3x).$

> *Solution*   We can obtain the form $(a + b)(a - b)$ by writing $3x - 4$ as $-4 + 3x$.
>
> $$(3x - 4)(-4 - 3x) = (-4 + 3x)(-4 - 3x)$$
> $$= (-4)^2 - (3x)^2 = 16 - 9x^2.$$

The next special product involves multiplying two binomials with the same first terms.

$$(x + a)(x + b) = x^2 + (a + b)x + ab.$$

Notice that the product is a trinomial.

Example 3

Find the given products.

a. $(x + 7)(x + 4).$

Solution   Here 7 is *a* and 4 is *b*.

$$(x + 7)(x + 4) = x^2 + (7 + 4)x + (7)(4)$$
$$= x^2 + 11x + 28.$$

b. $(x - 2)(x + 6)$.

Solution   Here -2 is *a* because $x - 2 = x + (-2)$.

$$(x - 2)(x + 6) = x^2 + (-2 + 6)x + (-2)(6)$$
$$= x^2 + 4x - 12.$$

c. $(y + 1)(y - 5)$.

Solution   Here -5 is *b*.

$$(y + 1)(y - 5) = y^2 + (1 - 5)y + (1)(-5)$$
$$= y^2 - 4y - 5.$$

d. $(z - 1)(z - 2)$.

Solution   Here -1 is *a* and -2 is *b*.

$$(z - 1)(z - 2) = z^2 + (-1 - 2)z + (-1)(-2)$$
$$= z^2 - 3z + 2.$$

The products in Example 3 can also be found by the so-called FOIL method. The letters in FOIL stand for the words *first*, *outer*, *inner*, and *last*. These words describe a sequence of steps to follow. To illustrate this method, we shall redo Example 3(b): $(x - 2)(x + 6)$.

Step 1. Multiply the *first* terms in the binomials to obtain the first term of the result: $(x)(x) = x^2$.

First

$$(x - 2)(x + 6) = x^2$$

Step 2. Add the product of the *outer* terms to the product of the *inner* terms to get the middle term of the result: $6x + (-2x) = 4x$.

$$(x - 2)(x + 6) = x^2 + 4x$$

Inner

Outer

Step 3. Multiply the *last* terms to obtain the last term of the result: $(-2)(6) = -12$.

$$(x - 2)(x + 6) = x^2 + 4x - 12.$$

The FOIL method can also be used for multiplying two binomials with different first terms, as Example 4 shows.

Example 4

Find $(4x + 1)(5x - 3)$ by the FOIL method.

Solution

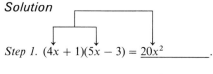

Step 1. $(4x + 1)(5x - 3) = \underline{20x^2}$.

Step 2. $(4x + 1)(5x - 3) = \underline{20x^2 - 7x}$.

Step 3. $(5x + 1)(5x - 3) = \underline{20x^2 - 7x - 3}$.

Example 5

Find each of the following by using the FOIL method.

a. $(x + 2)(x - 3) = x^2 - x - 6$.

b. $(2x - 1)(x - 3) = 2x^2 - 7x + 3$.

c. $(3y - 4)(2y + 3) = 6y^2 + y - 12$.

Example 6

Find each of the following by using special products.

a. $(x + 1)(x - 1) + (x + 2)^2 = [x^2 - 1] + [x^2 + 4x + 4]$
$$= 2x^2 + 4x + 3.$$

b. $(x^2 + 4)(x + 2)(x - 2) = (x^2 + 4)[(x + 2)(x - 2)]$
$$= (x^2 + 4)(x^2 - 4) = x^4 - 16.$$

c. $(x - 2)^2 - (3 - x)^2 = x^2 - 4x + 4 - (9 - 6x + x^2)$
$$= x^2 - 4x + 4 - 9 + 6x - x^2$$
$$= 2x - 5.$$

Note that parentheses were used to enclose the square of $3 - x$ because the minus sign in front of $(3 - x)^2$ applies to the entire square. That is,

$$(x - 2)^2 - (3 - x)^2 \neq x^2 - 4x + 4 - 9 - 6x + x^2.$$

d. $(x + 1)(x - 1)^2 = (x + 1)(x^2 - 2x + 1)$

$$= x(x^2 - 2x + 1) + 1(x^2 - 2x + 1)$$

$$= x^3 - 2x^2 + x + x^2 - 2x + 1$$

$$= x^3 - x^2 - x + 1.$$

Problem Set 10.1

*In Problems **1–62**, perform the operation by using special products.*

1. $(x + 4)^2$.
2. $(x + 6)^2$.
3. $(y + 10)^2$.
4. $(z + 5)^2$.
5. $(4x + 1)^2$.
6. $(2x + 10)^2$.
7. $(x + \frac{1}{2})^2$.
8. $(2 + x)^2$.
9. $(x - 6)^2$.
10. $(x - 2)^2$.
11. $(3x - 2)^2$.
12. $(3 - x)^2$.
13. $(2x - y)^2$.
14. $(2x - 3)^2$.
15. $(3x + 3)^2$.
16. $(2 - t)^2$.
17. $(2 - 4y)^2$.
18. $(3y - 4x)^2$.
19. $(x + 3)(x - 3)$.
20. $(x + 2)(x - 2)$.
21. $(x - 9)(x + 9)$.
22. $(7 - x)(7 + x)$.
23. $(1 + x)(1 - x)$.
24. $(4 + 2t)(4 - 2t)$.
25. $(3x^2 + 5)(3x^2 - 5)$.
26. $(2 - 3x)(2 + 3x)$.
27. $(2y + 3x)(2y - 3x)$.
28. $(y - 8)(y + 8)$.
29. $(6 - x)(x + 6)$.
30. $(x + yz)(x - yz)$.
31. $(x + 8)(x + 3)$.
32. $(x - 6)(x - 1)$.
33. $(x + 4)(x + 1)$.
34. $(x + 3)(x + 6)$.
35. $(x - 2)(x + 1)$.
36. $(x + 5)(x + 4)$.
37. $(t + 7)(t - 5)$.
38. $(x - 2)(x + 14)$.
39. $(x - 2)(x - 3)$.
40. $(x - 6)(x - 2)$.
41. $(x - 4)(x - 5)$.
42. $(x - 1)(x - 2)$.
43. $(y - 2)(y + 3)$.
44. $(t - 4)(t + 3)$.
45. $(x + 3)(2x - 4)$.
46. $(2x + 1)(2x + 2)$.
47. $(4x - 2)(2x - 1)$.
48. $(x - 3)(4x - 3)$.
49. $(2x + 1)(3x - 3)$.
50. $(2x + 1)(x - 2)$.
51. $(5x - 4)(2x - 1)$.
52. $(x + \sqrt{13})(x - \sqrt{13})$.
53. $(2 - 5t)(1 + 7t)$.
54. $(4 - t)(4 + 3t)$.
55. $(xy^2 + a)^2$.
56. $(x^3 - 1)(x^3 - 1)$.
57. $(x^2 - y^2)(x^2 + y^2)$.
58. $(x^2 - 2)^2$.
59. $(x - \sqrt{7})(x + \sqrt{7})$.
60. $(ab - c)(ab + c)$.
61. $(-2x + 1)^2$.
62. $(t - 2s)(t + 2s)$.

*In Problems **63–80**, perform the operations.*

63. $4(x + 3)^2$.
64. $x(x - 2)^2$.
65. $2x(y - 3)(y + 3)$.
66. $-2a(a + 4)^2$.
67. $(a^2b - 2m^2n)^2$.
68. $(x - 1)(x + 1)(x)$.

69. $(x + 2)(x - 2)(x^2 + 4)$.

70. $(x + 1)(x + 1)(x + 2)$.

71. $(a - b)^2 - (b - a)^2$.

72. $(x + y)^2 - (y - x)^2$.

73. $(2x - 3)^2 + (x + 1)(x + 2)$.

74. $(t + 3)(t - 3) - (t + 3)^2$.

75. $(x + 2)(x - 2)^2$.

76. $a(a - b)^2 - b(b - a)^2$.

77. $[4x(x - 3)]^2$.

78. $[2x^2(x - y)]^2$.

79. $3x(x + 2) - (\sqrt{3}x - 1)(\sqrt{3}x + 1)$.

80. $(5x^2 + 1)(\sqrt{5}x + 1)(\sqrt{5}x - 1)$.

81. Use the formula

$$(a + b)^3 = a^3 + 3a^2b + 3ab^2 + b^3$$

to find $(x + 2)^3$.

82. Use the formula in Problem 81 to find $(3x + 1)^3$.

83. Use the formula

$$(a - b)^3 = a^3 - 3a^2b + 3ab^2 - b^3$$

to find $(2x - 3)^3$.

84. Use the formula in Problem 83 to find $(3x^2 - 2y)^3$.

85. In the study of X-ray spectroscopy, it is shown that the frequency of the K_α line of any element is given by

$$\frac{3}{4} Rc(Z - 1)^2,$$

where c is the speed of light, Z is the atomic number of the element, and R is a constant called the Rydberg constant. Perform the indicated multiplication.

86. When a flexible chain l units long and of linear density w is released with a length c overhanging a smooth table, energy considerations result in the following expression for the square of the velocity of the chain as it leaves the table:

$$\frac{2g}{wl}\left[wc(l - c) + \frac{1}{2} w(l - c)^2 \right].$$

Perform all indicated operations and simplify.

87. The kinetic energy (KE) of an object of mass m moving with a speed v is given by

$$KE = \frac{1}{2} mv^2.$$

If the speed of an object at time t is given by $v = 2t + 1$, find an expression for the kinetic energy in terms of t and m. Perform the multiplications and simplify.

88. The speed v of a model rocket at time t is given by $v = t + 2$. The mass m of the rocket decreases as fuel is burned. If $m = 12 - 0.1t$, find the kinetic energy of the rocket in terms of t. (See Problem 87.) Perform the multiplications and simplify.

89. In studies of battery power, the expression

$$(R + r)^2(1) - R(2)(R + r)$$

arises. Perform the indicated operations and simplify.

90. For a body moving in a straight line with constant acceleration, it is shown in mechanics that

$$s = v_{ave} \cdot t, \tag{1}$$

$$v_{ave} = \frac{v_f + v_0}{2}, \tag{2}$$

and

$$t = \frac{v_f - v_0}{a}, \tag{3}$$

where s = displacement of the body, v_{ave} = average velocity, t = time, v_0 = initial velocity, v_f = final velocity, and a = acceleration. Show by substitution that

$$s = \frac{v_f^2 - v_0^2}{2a}.$$

10.2 FACTORING

The process of writing an expression as a product of its factors is called **factoring**. For example, because $2(x + 1) = 2x + 2$, then 2 and $x + 1$ are factors of $2x + 2$, and a *factored form* of $2x + 2$ is $2(x + 1)$. In later chapters, factoring will play an important role when we are working with fractions, solving equations, and investigating the behavior of certain expressions.

The simplest type of factoring is called **removing** (or *factoring out*) **a common factor**. If a factor is common to *each* term of a given expression, this factor can be *removed* by the distributive law:

> **COMMON FACTOR**
>
> $ab + ac = a(b + c).$

(Here, as with other factoring formulas, the letters a, b, and c can be replaced by any expression.) For example, given $6x + 9y$, note that 3 is a factor of both $6x$ and $9y$ because $6x = 3 \cdot 2x$ and $9y = 3 \cdot 3y$. Thus we can remove the factor 3 from both terms:

$$6x + 9y = 3(2x + 3y).$$

Thus $6x + 9y$ is factored. You can check the factored form by showing that the product $3(2x + 3y)$ is $6x + 9y$.

Another factored form for $6x + 9y$ is $6(x + \frac{3}{2}y)$, which you can check by multiplication. However, when factoring an expression that has integer coefficients, we shall usually choose factors whose terms also have integer coefficients. Thus we factor $6x + 9y$ as $3(2x + 3y)$, not as $6(x + \frac{3}{2}y)$. (Later in the book we may find it convenient to lift this restriction.)

Also, when factoring polynomials we shall choose factors that are themselves polynomials. For example, you can check that

$$6x + 9y = 3x^{-2}(2x^3 + 3x^2y).$$

But $3x^{-2}$ is not a polynomial, so we do not consider it as a factor of $6x + 9y$.

Furthermore, we usually factor a polynomial so that no factor can be written as a product of other factors except itself and 1 or -1. When this is done, we say that the polynomial is **completely factored**. For example, $x(x + 2)$ is completely factored, because x and $x + 2$ cannot be factored any further. To describe this we say that x and $x + 2$ are **prime polynomials**.

Let us now factor $2xy^2 + 8xy^4$. Observe that 2 and x are factors of both terms: $2xy^2 = 2x(y^2)$ and $8xy^4 = 2x(4y^4)$. Thus we have the factorization

$$2xy^2 + 8xy^4 = 2x(y^2 + 4y^4).$$

But this is not the completely factored form. The common factor y^2 can yet be removed from the factor $y^2 + 4y^4$ because $y^2 = y^2 \cdot 1$ and $4y^4 = y^2(4y^2)$. Thus we have

$$2xy^2 + 8xy^4 = 2xy^2(1 + 4y^2).$$

Although we must always factor *completely*, it does not necessarily involve more than one step, as in the previous case. If we had originally removed the factor $2xy^2$, then we would have the completely factored form in one step. In short, when removing a common factor, you should attempt to remove the *greatest* common factor.

Example 1

For each of the following, factor completely.

a. $x^5 - 2x^3 + x^2 = x^2(x^3 - 2x + 1)$.   *Do not forget the 1.*

b. $3a^2b^5 - 4a^3b^3 = a^2b^3(3b^2 - 4a)$.

c. $9x^2y + 3xy^2 - 6xy = 3xy(3x + y - 2)$.

Example 2

Completely factor $2(x - 1)^3(x + 2) + 3(x + 2)^2(x - 1)^2$.

Solution   Calculus students often come across an expression like this and must completely factor it. The expression consists of *two* terms:

$$2(x - 1)^3(x + 2) \quad \text{and} \quad 3(x + 2)^2(x - 1)^2.$$

The factors that are common to both terms are $(x - 1)^2$ and $x + 2$. Thus

$$2(x - 1)^3(x + 2) + 3(x + 2)^2(x - 1)^2 = (x - 1)^2(x + 2)[2(x - 1) + 3(x + 2)]$$
$$= (x - 1)^2(x + 2)[2x - 2 + 3x + 6]$$
$$= (x - 1)^2(x + 2)(5x + 4).$$

The special-product relationships given in the previous section form the basis of our remaining techniques of factoring. For example, because $(a + b)(a - b) = a^2 - b^2$, a difference of two squares can be factored into a sum and difference:

> **DIFFERENCE OF TWO SQUARES**
>
> $$a^2 - b^2 = (a + b)(a - b).$$

Example 3

Completely factor each of the following.

a. $x^2 - 4 = x^2 - 2^2 = (x + 2)(x - 2).$

b. $4y^2 - 1 = (2y)^2 - 1^2 = (2y + 1)(2y - 1).$

c. $9x^2 - 16y^2 = (3x)^2 - (4y)^2 = (3x + 4y)(3x - 4y).$

d. $x^4 - y^4 = (x^2)^2 - (y^2)^2 = (x^2 + y^2)(x^2 - y^2).$

Although $x^2 + y^2$ cannot be factored (it is prime), $x^2 - y^2$ can be factored into $(x + y)(x - y)$. Thus

$$x^4 - y^4 = (x^2 + y^2)(x + y)(x - y).$$

A second-degree trinomial with leading coefficient 1, such as $x^2 + 3x - 18$, can sometimes be factored as a product of two first-degree binomials of the form $x + a$ and $x + b$. Suppose

$$x^2 + 3x - 18 = (x + a)(x + b).$$

Then by the FOIL method of the previous section,

$$x^2 + 3x - 18 = x^2 + (a + b)x + ab.$$

By matching the coefficients on the right side with the corresponding ones on the left side, we must have

$$a + b = 3 \quad \text{and} \quad ab = -18.$$

There are several choices of a and b such that their product is -18. For example, two choices are 18 and -1, and -9 and 2. However, because the sum of a and b must be 3, we choose $a = 6$ and $b = -3$ (or vice versa). Thus

$$x^2 + 3x - 18 = (x + 6)(x - 3).$$

You can check this factorization by using the FOIL method on $(x + 6)(x - 3)$.

Example 4

Completely factor $x^2 - 7x + 12$.

Solution   We must find two numbers whose product is 12 and whose sum is -7. The numbers -3 and -4 work. Thus

$$x^2 - 7x + 12 = (x - 3)(x - 4).$$

Example 5

Completely factor the following trinomials.

a. $y^2 + 4y + 4 = (y + 2)(y + 2) = (y + 2)^2$, which is the square of a binomial.

b. $z^2 + 4z - 32 = (z - 4)(z + 8)$.

c. $z^2 - 4z - 32 = (z + 4)(z - 8)$. Compare this with part (b), where we chose -4 and $+8$ because we wanted the middle term to have a positive coefficient.

Sometimes a trinomial whose leading coefficient is not 1 can be factored as a product of two binomials. The method of factoring involves reversing the FOIL method and using trial and error. For example, to factor $6x^2 - 5x - 6$ we must have

$$6x^2 - 5x - 6 = (\_\_ + \_\_)(\_\_ + \_\_).$$

That is, the product of the first terms of the binomials must be $6x^2$, and the product of the last terms must be -6. One combination of binomials for which that is true is $(x + 3)(6x - 2)$. But here the sum of the products of the outer terms and inner terms, which is $16x$, is not equal to the middle term of $6x^2 - 5x - 6$, namely $-5x$. By trial and error we try other possible combinations to see if we can find one that works. Some possibilities are:

$$(3x - 1)(2x + 6) = 6x^2 + 16x - 6,$$

$$(x + 3)(6x - 2) = 6x^2 + 16x - 6,$$

$$(3x + 2)(2x - 3) = 6x^2 - 5x - 6.$$

The combination $(3x + 2)(2x - 3)$ works. Thus

$$6x^2 - 5x - 6 = (3x + 2)(2x - 3).$$

Check it. Here we factored a polynomial of degree two. This method may work for other types of trinomials, as Example 6(d) shows.

Example 6

Completely factor the following.

a. $2x^2 + 5x + 2 = (2x + 1)(x + 2).$

b. $8y^2 - 6y - 9 = (2y - 3)(4y + 3).$

c. $9y^2 - 12y + 4 = (3y - 2)(3y - 2) = (3y - 2)^2$, the square of a binomial.

d. $x^4 - 3x^2 + 2 = (x^2 - 1)(x^2 - 2) = (x + 1)(x - 1)(x^2 - 2)$. Note that we can write $x^2 - 2$ as $(x + \sqrt{2})(x - \sqrt{2})$, but since we require that factors have *integer* coefficients, we do not allow irrational coefficients. Thus we leave $x^2 - 2$ alone.

When factoring, you should *first remove any common factors*, as Example 7 shows.

Example 7

Completely factor the following.

a. $4x^3 - 36x = 4x(x^2 - 9) = 4x(x + 3)(x - 3).$

b. $8ay^2 + 8ay + 2a = 2a(4y^2 + 4y + 1)$
$$= 2a(2y + 1)(2y + 1) = 2a(2y + 1)^2.$$

c. $7a^5 + 7a^3 = 7a^3(a^2 + 1).$

d. $6x^5 - 4x^3 - 2x = 2x(3x^4 - 2x^2 - 1)$
$$= 2x(3x^2 + 1)(x^2 - 1)$$
$$= 2x(3x^2 + 1)(x + 1)(x - 1).$$

Example 8

The total work done on a body of mass m is given by

$$\tfrac{1}{2}mv_f^2 - \tfrac{1}{2}mv_0^2,$$

where v_0 and v_f are the initial and final speeds of the body, respectively. Completely factor this expression.

Solution

$$\tfrac{1}{2}mv_f^2 - \tfrac{1}{2}mv_0^2 = \tfrac{1}{2}m(v_f^2 - v_0^2) = \tfrac{1}{2}m(v_f + v_0)(v_f - v_0).$$

Problem Set 10.2

*In Problems **1-64**, factor completely.*

1. $8x + 8$.
2. $9x - 9$.
3. $10x - 5y + 25$.
4. $12x^2 + 24y - 4$.
5. $5cx + 9x$.
6. $16mx + 4m$.
7. $6xy + 3xz$.
8. $4xyz - 5yz$.
9. $2x^3 - x^2$.
10. $5x^8 - 4x^7$.
11. $2x^3y^3 + x^5y^5$.
12. $m^2y - y^2m$.
13. $4m^2x^3 - 8mx^4$.
14. $25a^5x^9 - 15a^4x^{10}$.
15. $9a^4y^3 + 3a^2y^5 - 6a^3y^4z$.
16. $by^5 - 2b^3y^4 - 8b^2y^2$.
17. $x^2 - 1$.
18. $x^2 - 49$.
19. $x^2 + 4x + 3$.
20. $x^2 + 5x + 6$.
21. $x^2 - 9x + 20$.
22. $x^2 + 3x - 10$.
23. $y^2 + 2y - 24$.
24. $y^2 - 10y + 9$.
25. $y^2 - 36$.
26. $4 - y^2$.
27. $x^2 + 12x + 36$.
28. $x^2 - 3x - 28$.
29. $x^2 - 4x - 32$.
30. $x^2 - 8x + 12$.
31. $y^2 - 10y + 25$.
32. $y^2 + 8y + 16$.
33. $3x^2 + 7x + 2$.
34. $5x^2 - 12x + 4$.
35. $2y^2 - 7y + 3$.
36. $7y^2 + 9y + 2$.
37. $16x^2 + 8x + 1$.
38. $4x^2 - 4x + 1$.
39. $9 - 4x^2y^2$.
40. $a^2b^2 - c^2d^2$.
41. $4y^2 + 7y - 2$.
42. $8y^2 + 2y - 3$.
43. $6x^2 - 11x - 10$.
44. $5x^2 + 14x - 3$.
45. $2x^2 + 4x - 6$.
46. $a^2x^2 + a^2x - 20a^2$.
47. $3x^3 + 18x^2 + 27x$.
48. $3x^4 - 15x^3 + 18x^2$.
49. $16s^2t^3 - 4s^2t$.
50. $a^2b^2 - a^4b^4$.
51. $4y^2 - 6y - 18$.
52. $30y^2 + 55y + 15$.
53. $(x + 3)^3(x - 1) + (x + 3)^2(x - 1)^2$.
54. $(x + 5)^2(x + 1)^3 + (x + 5)^3(x + 1)^2$.
55. $(x + 4)(2x + 1) + (x + 4)$.
56. $(x - 3)(2x + 3) - (2x + 3)(x + 5)$.
57. $x^4 - 16$.
58. $81x^4 - y^4$.
59. $y^8 - 1$.
60. $t^4 - 4$.
61. $x^4 + x^2 - 2$.
62. $x^4 - 5x^2 + 4$.
63. $x^5 - 2x^3 + x$.
64. $4x^3 - 6x^2 - 4x$.

65. The total surface area of a closed cylinder of radius r and altitude h is given by

$$2\pi r^2 + 2\pi rh.$$

Completely factor this expression.

66. The moment of inertia of a right circular cone of base radius r and altitude h about a certain axis is

$$\tfrac{1}{20}(\pi\rho r^4h) + \tfrac{1}{5}(\pi\rho r^2h^3),$$

where ρ (rho) is mass density. Completely factor this expression.

67. The effective emf (electromotive force) of a thermocouple is given by

$$c(T_1 - T_2) + k(T_1^2 - T_2^2),$$

where c and k are constants that depend on the metals used, and T_1 and T_2 are the temperatures of the hot and cold junctions, respectively. Completely factor this expression.

68. The power P supplied to a motor by a battery of emf V was found to be given by

$$P = \frac{mg}{BL}\left(V - \frac{Rmg}{BL}\right) + R\left(\frac{mg}{BL}\right)^2.$$

Completely factor and simplify this expression.

69. The moment of inertia of a right circular cylinder about a certain axis is found to be

$$\tfrac{1}{4}mR^2 + \tfrac{1}{12}mh^2.$$

Show that by factoring this expression properly, you can arrive at

$$\tfrac{1}{12}m(3R^2 + h^2),$$

which was found in a reference table.

70. When two bodies at different temperatures face each other, the net rate at which energy is radiated from the warmer body is given by

$$kT_2^4 - kT_1^4,$$

where k is a constant and T_1 and T_2 are the temperatures of the bodies. Completely factor this expression.

10.3 FACTORING, CONTINUED

Sometimes we can factor an expression by first grouping its terms in a suitable way. The next example shows how **factoring by grouping** may be done.

Example 1

Factor $x^2 + xy - 5x - 5y$ by grouping.

Solution   Grouping the first two terms and grouping the last two terms, we have

$$x^2 + xy - 5x - 5y$$
$$= (x^2 + xy) - (5x + 5y) \qquad \text{[note the signs for the second grouping]}$$
$$= x(x + y) - 5(x + y) \qquad \text{[factoring each group]}$$
$$= (x + y)(x - 5) \qquad \text{[removing the common factor } x + y \text{ from both terms].}$$

Another way to factor the given expression is to group the first and third terms and group the second and last terms:

$$x^2 + xy - 5x - 5y$$
$$= (x^2 - 5x) + (xy - 5y)$$
$$= x(x - 5) + y(x - 5) \qquad \text{[factoring each group]}$$
$$= (x - 5)(x + y) \qquad \text{[removing the common factor } x - 5 \text{ from both terms].}$$

Example 2

Factor $ax^2 - ay^2 + bx^2 - by^2$ by grouping.

Solution

$$ax^2 - ay^2 + bx^2 - by^2 = (ax^2 - ay^2) + (bx^2 - by^2)$$
$$= a(x^2 - y^2) + b(x^2 - y^2)$$
$$= (x^2 - y^2)(a + b)$$
$$= (x + y)(x - y)(a + b).$$

To factor the sum or difference of two cubes, use the following formulas.

SUM OR DIFFERENCE OF TWO CUBES

$$a^3 + b^3 = (a + b)(a^2 - ab + b^2).$$
$$a^3 - b^3 = (a - b)(a^2 + ab + b^2).$$

Example 3

Factor completely.

a. $x^3 + 8 = x^3 + 2^3 = (x + 2)[x^2 - (x)(2) + 2^2]$
$$= (x + 2)(x^2 - 2x + 4).$$

b. $2y^6 - 54 = 2(y^6 - 27)$ [removing common factor]
$$= 2[(y^2)^3 - 3^3]$$
$$= 2(y^2 - 3)[(y^2)^2 + (y^2)(3) + 3^2]$$
$$= 2(y^2 - 3)(y^4 + 3y^2 + 9).$$

Factoring can be a useful tool for solving equations, as Examples 4 and 5 show.

Example 4

Solve $x = ax + b$ for x

Solution

$$x = ax + b$$
$$x - ax = b \quad \text{[getting terms involving } x \text{ on one side]}$$
$$x(1 - a) = b \quad \text{[factoring]}.$$

Assuming that $1 - a \neq 0$, we divide both sides by $1 - a$.

$$\frac{x(1-a)}{1-a} = \frac{b}{1-a}$$

$$x = \frac{b}{1-a}.$$

Example 5

Solve $(a + c)x + x^2 = (x + a)^2$ for x.

Solution

$$(a + c)x + x^2 = (x + a)^2$$

$$ax + cx + x^2 = x^2 + 2ax + a^2$$

$$ax + cx = 2ax + a^2$$

$$ax + cx - 2ax = a^2$$

$$cx - ax = a^2$$

$$x(c - a) = a^2 \qquad \text{[factoring]}.$$

Assuming that $c - a \neq 0$, we divide both sides by $c - a$:

$$x = \frac{a^2}{c-a}.$$

Problem Set 10.3

In Problems **1–10**, *factor completely by grouping.*

1. $x + xy + 4 + 4y$.

2. $2ax - ay + 6bx - 3by$.

3. $xy - 2x - 4y + 8$.

4. $x^3 + x^2 + x + 1$.

5. $2x^3 - 3x^2 - 8x + 12$.

6. $x^4 - x^3 + x^2 - 1$.

7. $ax^2 - ay^2 + bx^2 - by^2$.

8. $x^2 + x - y^2 - y$.

9. $x^2 + 4x + 4 - a^2$.

10. $a^2 - b^2 + 2bc - c^2$.

In Problems **11–18**, *use the formulas for the sum or difference of cubes to factor completely.*

11. $x^3 + 27$.

12. $x^3 - 1$.

13. $8y^3 - 27$.

14. $64y^3 + 8$.

15. $24x^3 + 3y^6$.

16. $a^2 - b^2 + a^3 - b^3$.

17. $(a + 1)^3 - (b + 1)^3$.

18. $(1 - y^2)^3 - (1 - x^2)^3$.

In Problems **19–28**, *solve each equation for the given letter.*

19. $2R_1i + 3R_2i - 4R_4i = -(E_1 - E_2);$ *i.*

20. $ax + cx = 2ax + a^2;$ *x.*

21. $ax + b = cx + d;$ *x.*

22. $2at - t = 4bt + 6;$ *t.*

23. $mgh + \frac{1}{2}mv^2 = c;$ *m.*

24. $Q = mc(t_2 - t_1) + mL;$ *m.*

25. $a(s - t) + b(t + s) = 7;$ *s.*

26. $a(x - y) + b(x + y) = cxy;$ *x.*

27. $Q = mct_2 - mct_1 + mL;$ *c.*

28. $mgh = \frac{1}{2}mv^2 + \frac{1}{2}I\omega^2;$ *m.*

29. In an analysis of a negative feedback amplifier, the following equation was obtained:

$$V_0 = \alpha(V_s - \beta V_0),$$

where V_0 and V_s are output and input voltages, respectively. Find the ratio of V_0 to V_s.

30. If a ball of mass m is tied to a string and whirled around in a vertical circle, at the ball's lowest point Newton's second law indicates that

$$T - mg = m\frac{v^2}{r},$$

where T is the tension in the string, g is the acceleration due to gravity, v is the speed of the ball, and r is the radius of the circle. Solve for m.

10.4 REDUCING FRACTIONS

The fraction $\frac{-2}{3}$ can be thought of as

$$+ \frac{-2}{+3}.$$

Thus we can associate three signs with the fraction:

The sign in front of the fraction $(+)$,

The sign of the numerator $(-)$, and

The sign of the denominator $(+)$.

Furthermore, from the rules of signed numbers given in Chapter 1, we have

$$\frac{-2}{3} = -\frac{2}{3}.$$

Notice that $-\frac{2}{3}$ can be obtained from $\frac{-2}{3}$ by changing *two* signs of $\frac{-2}{3}$, namely, the sign in front of the fraction and the sign of the numerator. We also have

$$\frac{2}{-3} = -\frac{2}{3}.$$

Again, by changing *two* signs of $\frac{2}{-3}$, namely, the sign in front of the fraction and the sign of the denominator, we can obtain $-\frac{2}{3}$.

In general, **by changing any two of the three signs of a fraction, we obtain a fraction equal** (equivalent) **to the original one**. Thus $-\frac{-6}{7}$ can be written in the following ways:

$$+\frac{+6}{+7}, \qquad -\frac{+6}{-7}, \qquad +\frac{-6}{-7}.$$

Example 1

a. $\dfrac{-x}{y} = -\dfrac{x}{y}$.

b. $\dfrac{-x}{-y} = \dfrac{x}{y}$.

c. $-\dfrac{x(-y)}{z} = -\dfrac{-(xy)}{z} = \dfrac{xy}{z}$.

d. $\dfrac{w(-x)}{(-y)(-z)} = \dfrac{-(wx)}{yz} = -\dfrac{wx}{yz}$.

e. $\dfrac{-a-b}{-c} = \dfrac{-(a+b)}{-c} = \dfrac{a+b}{c}$.

f. $\dfrac{y}{(-1)ab} = \dfrac{y}{-ab} = -\dfrac{y}{ab}$.

If the numerator and denominator of a fraction have a *common factor*, then we can *reduce* the fraction. This is the same as canceling common factors.

$$\boxed{\dfrac{ab}{ac} = \dfrac{b}{c}} \quad \text{or} \quad \boxed{\dfrac{\cancel{a}b}{\cancel{a}c} = \dfrac{b}{c}}.$$

A fraction is **reduced**, or is in **simplest form**, when its numerator and denominator have no common factors except 1.

> To reduce a fraction, first completely factor both the numerator and denominator. Then cancel all common factors.

For example,

$$\frac{3ab - 3b^2}{a^2 - b^2} = \frac{3b\cancel{(a-b)}^{1}}{(a+b)\cancel{(a-b)}_{1}} = \frac{3b}{a+b}.$$

Keep in mind that you can cancel only **factors** of both the entire numerator and entire denominator. *Common terms **cannot** be canceled.* Here are some examples of incorrect canceling. In each case the error can be seen by replacing x by 2.

$$\frac{\overset{1}{\cancel{4}} + 3x}{\underset{1}{\cancel{4}}} \neq 1 + 3x \qquad [4 \text{ is } \textbf{\textit{not}} \textit{ a factor of the entire numerator}],$$

$$\frac{\overset{1}{\cancel{x}}}{3 + 4\underset{1}{\cancel{x}}} \neq \frac{1}{3 + 4} \qquad [x \text{ is } \textbf{\textit{not}} \textit{ a factor of the entire denominator}],$$

$$\frac{\overset{1}{\cancel{x}} + 3}{\underset{1}{\cancel{x}} + 5} \neq \frac{1 + 3}{1 + 5} \qquad [x \text{ is } \textbf{\textit{not}} \textit{ a factor}].$$

Example 2

Reduce the following fractions.

a. $\dfrac{3x - 9}{x - 3} = \dfrac{3\cancel{(x - 3)}}{\cancel{x - 3}} = 3.$

b. $\dfrac{4x^2}{8x^2 - 4x^3} = \dfrac{\cancel{4x^2}}{\cancel{4x^2}(2 - x)} = \dfrac{1}{2 - x}.$

c. $\dfrac{2x + 6}{6x}$. Here 2 is a common factor of the numerator and denominator.

$$\frac{2x + 6}{6x} = \frac{\cancel{2}(x + 3)}{\underset{3}{\cancel{6}}x} = \frac{x + 3}{3x}.$$

Example 3

Reduce the following fractions.

a. $\dfrac{x^2 - 1}{x^2 + 2x + 1} = \dfrac{(x - 1)\cancel{(x + 1)}}{\underset{x+1}{\cancel{(x + 1)^2}}} = \dfrac{x - 1}{x + 1}.$

b. $\dfrac{2x^2 - 2x - 12}{4x^2 - 8x - 12} = \dfrac{\cancel{2}(x^2 - x - 6)}{\underset{2}{\cancel{4}}(x^2 - 2x - 3)} = \dfrac{(x + 2)\cancel{(x - 3)}}{2(x + 1)\cancel{(x - 3)}} = \dfrac{x + 2}{2(x + 1)}.$

c. $\dfrac{(y - 2)^3(y - 4)^2}{(y - 2)^2(y - 4)^4} = \dfrac{y - 2}{(y - 4)^2}.$

The next example makes use of the following fact.

$$a - b = -(b - a) = (-1)(b - a).$$

This fact may sometimes be used to reduce a fraction containing containing factors with terms that differ in sign.

Example 4

Reduce the following.

a. $\dfrac{3-x}{x-3} = \dfrac{-(x-3)}{x-3} = \dfrac{(-1)\cancel{(x-3)}}{\cancel{x-3}} = -1.$

Another way to reduce the fraction is by rewriting the denominator:

$$\frac{3-x}{x-3} = \frac{\cancel{3-x}}{(-1)\cancel{(3-x)}} = \frac{1}{-1} = -1.$$

b. $\dfrac{x^2+3x-4}{1-x^2} = \dfrac{(x-1)(x+4)}{(1-x)(1+x)} = \dfrac{\cancel{(x-1)}(x+4)}{(-1)\cancel{(x-1)}(1+x)} = -\dfrac{x+4}{1+x}.$

Problem Set 10.4

In Problems **1–6***, write each fraction so that minus signs do not appear in the numerator and in the denominator. For example, write* $\dfrac{ab}{-c}$ *as* $-\dfrac{ab}{c}$.

1. $\dfrac{-x}{-y}.$

2. $-\dfrac{-wx}{-z}.$

3. $-\dfrac{a}{b(-c)}.$

4. $-\dfrac{ab}{(-c)(-d)}.$

5. $\dfrac{x}{-y-z}.$

6. $\dfrac{-x(y+z)}{(-a)(b)(-c)}.$

In Problems **7–36***, reduce the fractions.*

7. $\dfrac{(x+2)(x-1)}{(x-1)(3x+5)}.$

8. $\dfrac{(x-3)(x+4)(x+6)}{3x^2(x-3)}.$

9. $\dfrac{2x+4}{6}.$

10. $\dfrac{10x-15}{20}.$

11. $\dfrac{3y+12}{y+4}.$

12. $\dfrac{x^2-y^2}{x+y}.$

13. $\dfrac{2x+2y}{6ax+6ay}.$

14. $\dfrac{3x^2-12}{3x-6}.$

15. $\dfrac{12x^2+6x}{3x^2-9x}.$

16. $\dfrac{x^4-x^2}{x^2+x}.$

17. $\dfrac{x^2-81}{x^2+9x}.$

18. $\dfrac{x^2y - 4xy}{x^3y^2 - 2x^2y^2}.$

19. $\dfrac{6m + 2m^2}{8m - 8m^2}.$

20. $\dfrac{y^4 + y^2}{y^6 - 4y^5}.$

21. $\dfrac{x + 1}{x^2 + 7x + 6}.$

22. $\dfrac{z^2 - 9}{z^2 - 6z + 9}.$

23. $\dfrac{2x^2 + 11x + 12}{2x^2 + 8x}.$

24. $\dfrac{3z^2 + z - 2}{6z^2 - z - 2}.$

25. $\dfrac{25 - t^2}{t^2 - 2t - 15}.$

26. $\dfrac{x^2 + 3x - 4}{2 - x - x^2}.$

27. $\dfrac{(x + 3)(x^2 - 144)}{(x^2 - 9)(x + 12)}.$

28. $\dfrac{6a^2b^2 + 6ab^3y + 6a^2b^2c^2}{3abc}.$

29. $\dfrac{x^2 + 5x + 6}{x^2 - 2x - 8}.$

30. $\dfrac{x^2 + x - 12}{x^2 - 6x + 9}.$

31. $\dfrac{c - d}{d - c}.$

32. $\dfrac{(x - 2)(4 - x)}{(x - 4)(2 - x)}.$

33. $\dfrac{(x + 5)(2 - x)(x + 7)(6 - x)}{(x - 6)(7 - x)(x - 2)(x + 5)}.$

34. $\dfrac{(x + a)(-x - b)(c - x)(x + d)}{(-x - d)(x - c)(x + b)(x - a)}.$

35. $\dfrac{x^2 + x - 12}{-x - 4}.$

36. $\dfrac{16x^4 - 4b^4}{4x^2 - 2b^2}.$

*Reducing the fractions in Problems **37–44** is somewhat more challenging than the preceding problems.*

37. $\dfrac{(x - 2y)^2}{(2y - x)^3}.$

38. $\dfrac{(y - x)^2y}{(x^2 - xy)(x + y)}.$

39. $\dfrac{a^4 - b^4}{a + b}.$

40. $\dfrac{3(x + y) - 6x - 6y}{-3(x - y)}.$

41. $\dfrac{(2x^2 - x - 3)(4x^2 - 1)}{(2x^2 + 3x + 1)(6x^2 - 13x + 6)}.$

42. $\dfrac{(x^2 - xy)(x^2 - y^2)(xy - 2y^2)}{(x - y)^2(x^2 - 3xy + 2y^2)(x^2 + xy)}.$

43. $\dfrac{9w^2 - 4x^2 + 4x - 1}{3w + 2x - 1}.$

44. $\dfrac{(x + y)(y - x)(x + 1)}{x^3 - xy^2 + x^2 - y^2}.$

45. For the arrangement of particles shown in Fig. 10.1, it is shown in mechanics that the

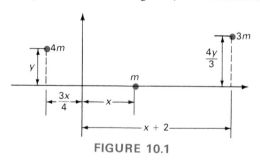

FIGURE 10.1

coordinates $\bar{x}$ and $\bar{y}$ of the center of mass of the particles are given by

$$\bar{x} = \frac{m(x) + 3m(x + 2) + 4m\left(-\dfrac{3x}{4}\right)}{m + 3m + 4m}$$

and

$$\bar{y} = \frac{m(0) + 3m\left(\dfrac{4y}{3}\right) + 4m(y)}{m + 3m + 4m}.$$

Simplify the expressions for $\bar{x}$ and $\bar{y}$.

10.5 MULTIPLICATION AND DIVISION OF FRACTIONS

To multiply fractions, we multiply their numerators, multiply their denominators, and simplify.

> **MULTIPLICATION OF FRACTIONS**
>
> $$\frac{a}{b} \cdot \frac{c}{d} = \frac{ac}{bd}.$$

Here a, b, c, and d can be replaced by other expressions, and the rule also applies to more than two fractions.

Example 1

Multiply the fractions.

a. $\dfrac{x}{x+2} \cdot \dfrac{x+3}{x-5} = \dfrac{x(x+3)}{(x+2)(x-5)}$. *We usually keep our answer in factored form.*

b. $\dfrac{3}{-x} \cdot \dfrac{x^2+4}{x} \cdot \dfrac{x+5}{x} = \dfrac{3(x^2+4)(x+5)}{-x^3} = -\dfrac{3(x^2+4)(x+5)}{x^3}.$

c. $z^3 \cdot \dfrac{z+2}{z+1} = \dfrac{z^3}{1} \cdot \dfrac{z+2}{z+1} = \dfrac{z^3(z+2)}{z+1}.$

Here we wrote z^3 as $\dfrac{z^3}{1}$ so that we could multiply fractions. More simply, we can directly

multiply the numerator of $\dfrac{z+2}{z+1}$ by z^3:

$$z^3 \cdot \frac{z+2}{z+1} = \frac{z^3(z+2)}{z+1}.$$

Before multiplying fractions, you should completely factor all numerators and denominators. This may simplify the final reducing process. As stated in Sec. 1.5, you can cancel common factors *before* multiplying fractions. Thus instead of writing

$$\frac{x}{x+1} \cdot \frac{x+1}{x-5} = \frac{x(x+1)}{(x+1)(x-5)} = \frac{x}{x-5},$$

you can write

$$\frac{x}{\cancel{x+1}} \cdot \frac{\cancel{x+1}}{x-5} = \frac{x}{x-5}.$$

Example 2

Multiply the fractions.

a. $\cancel{(x+2)} \cdot \dfrac{x-4}{\cancel{x+2}} = x - 4.$

b. $\dfrac{3x-3}{x} \cdot \dfrac{5}{x^2-1} = \dfrac{3(x-1)}{x} \cdot \dfrac{5}{(x-1)(x+1)}$ [factoring]

$$= \frac{3\cancel{(x-1)}}{x} \cdot \frac{5}{\cancel{(x-1)}(x+1)} = \frac{15}{x(x+1)}.$$

c. $\dfrac{x^3}{4x^2-9} \cdot \dfrac{2x+3}{x^2+x} = \dfrac{\overset{x^2}{\cancel{x^3}}}{(2x-3)\cancel{(2x+3)}} \cdot \dfrac{\cancel{2x+3}}{\cancel{x}(x+1)} = \dfrac{x^2}{(2x-3)(x+1)}.$

Example 3

Multiply the fractions.

a. $\dfrac{x^2+5x+4}{x^2+5x+6} \cdot \dfrac{x^2+3x+2}{x^2+2x-8} = \dfrac{(x+1)\cancel{(x+4)}}{(x+3)\cancel{(x+2)}} \cdot \dfrac{(x+1)\cancel{(x+2)}}{\cancel{(x+4)}(x-2)}$

$$= \frac{(x+1)^2}{(x+3)(x-2)}.$$

b. $\dfrac{6}{x^2-3x+2} \cdot \dfrac{2-x}{x+3} = \dfrac{6}{(x-1)\cancel{(x-2)}} \cdot \dfrac{(-1)\cancel{(x-2)}}{x+3}$

$$= \frac{-6}{(x-1)(x+3)} = -\frac{6}{(x-1)(x+3)}.$$

After factoring $x^2 - 3x + 2$, we wrote $2 - x$ as $(-1)(x-2)$ so that we could cancel.

c. $\dfrac{x^2+8x+16}{x^2+5x+6} \cdot \dfrac{x^2+3x+2}{x^2+2x-8} = \dfrac{(x+4)^2}{(x+3)(x+2)} \cdot \dfrac{(x+1)(x+2)}{(x+4)(x-2)}$

$$= \frac{\overset{x+4}{\cancel{(x+4)}^2}(x+1)\overset{1}{\cancel{(x+2)}}}{(x+3)\underset{1}{\cancel{(x+2)}}\underset{1}{\cancel{(x+4)}}(x-2)}$$

$$= \frac{(x+4)(x+1)}{(x+3)(x-2)}.$$

As stated in Chapter 1, to divide a/b by c/d we *multiply* a/b by the reciprocal of c/d, namely, d/c. This is commonly referred to as *inverting* the divisor and multiplying.

DIVISION OF FRACTIONS

$$\frac{\dfrac{a}{b}}{\dfrac{c}{d}} = \frac{a}{b} \cdot \frac{d}{c} = \frac{ad}{bc}.$$

Example 4

Perform the divisions and simplify.

a. $\dfrac{\dfrac{4y}{3z}}{\dfrac{z}{9}} = \dfrac{4y}{3z} \cdot \dfrac{9}{z} = \dfrac{4y}{\cancel{3}z} \cdot \dfrac{\overset{3}{\cancel{9}}}{z} = \dfrac{12y}{z^2}.$

b. $\dfrac{x}{x+2} \div \dfrac{x+3}{x-5} = \dfrac{x}{x+2} \cdot \dfrac{x-5}{x+3} = \dfrac{x(x-5)}{(x+2)(x+3)}.$

c. $\dfrac{\dfrac{4x}{x^2-1}}{\dfrac{2x^2+8x}{x-1}} = \dfrac{4x}{x^2-1} \cdot \dfrac{x-1}{2x^2+8x} = \dfrac{\overset{2}{\cancel{4x}}}{\cancel{(x-1)}(x+1)} \cdot \dfrac{\cancel{x-1}}{\cancel{2x}(x+4)}$

$$= \dfrac{2}{(x+1)(x+4)}.$$

Example 5

Perform the divisions and simplify.

a. $\dfrac{\dfrac{x}{x-3}}{2x} = \dfrac{\dfrac{x}{x-3}}{\dfrac{2x}{1}} = \dfrac{\cancel{x}}{x-3} \cdot \dfrac{1}{2\cancel{x}} = \dfrac{1}{2(x-3)}.$

b. $\dfrac{\dfrac{4x^2-4x+1}{x}}{2x-1} = (4x^2-4x+1) \cdot \dfrac{2x-1}{x}$

$$= (2x-1)^2 \cdot \dfrac{2x-1}{x} = \dfrac{(2x-1)^3}{x}.$$

Problem Set 10.5

*In Problems **1–52**, perform the indicated operations and simplify.*

1. $\dfrac{7}{y} \cdot \dfrac{1}{x}$.

2. $\dfrac{2}{x} \cdot \dfrac{y}{5}$.

3. $\dfrac{2x^2}{3} \cdot \dfrac{6}{x^5}$.

4. $\dfrac{1}{x} \cdot \dfrac{3xy^2z^3}{x^2y} \cdot \dfrac{x^2}{3yz^3}$.

5. $\dfrac{x^2y^2z}{abc} \cdot \dfrac{b^2c}{y^3z} \cdot \dfrac{a^2b}{x^3y}$.

6. $\dfrac{3x^2y}{a^4} \cdot \dfrac{2ab^2x}{y} \cdot \dfrac{a^2x^3}{6} \cdot \dfrac{a^2}{b^2x^6}$.

7. $\dfrac{y^2}{y-3} \cdot \dfrac{-1}{y+2}$.

8. $\dfrac{x-3}{x+4} \cdot \dfrac{x-5}{x-3}$.

9. $\dfrac{x-3}{x^2} \cdot \dfrac{x-3}{x^4}$.

10. $\dfrac{-x}{x+2} \cdot \dfrac{x+2}{x^2}$.

11. $\dfrac{2x-3}{x-2} \cdot \dfrac{2-x}{2x+3}$.

12. $\dfrac{x^2-3x+2}{x^2-7x+12} \cdot \dfrac{(x-3)(x+2)}{x^2+x-2}$.

13. $\dfrac{x^2-y^2}{x+y} \cdot \dfrac{x^2+2xy+y^2}{y-x}$.

14. $\dfrac{b(6a-6b)}{a^2+ab} \cdot \dfrac{a^3-ab^2}{2b^2}$.

15. $\dfrac{5x-10}{5(x+3)} \cdot \dfrac{x}{2x-4}$.

16. $\dfrac{4}{6x^2} \cdot \dfrac{3x+9}{x^2+3x}$.

17. $(x^2-9) \cdot \dfrac{x-3}{4x+12}$.

18. $\dfrac{5}{(x+1)^2} \cdot (x^2+2x+1)$.

19. $\dfrac{x^2-5x+6}{x-1} \cdot \dfrac{x-1}{x-3} \cdot \dfrac{x+3)}{(x-2)^2}$.

20. $\dfrac{(x-2)^2}{x+4} \cdot \dfrac{(x+4)^2}{x-2} \cdot \dfrac{(x-1)^2}{(x+4)^2}$.

21. $\dfrac{x^2+6x+8}{x^2-5x+6} \cdot \dfrac{x-2}{x^2+5x+4}$.

22. $\dfrac{2x^2+3x+1}{x^2-4x-5} \cdot \dfrac{x^2-25}{4x^2-1}$.

23. $\dfrac{8x^2+32}{x^2+2x} \cdot \dfrac{x^2+4x+4}{8x^2-32}$.

24. $\dfrac{x^3+8x^2+15x}{x^2+4x+3} \cdot \dfrac{x-1}{x^2+5x}$.

25. $\dfrac{x^2-8x+7}{(x+2)^2} \cdot \dfrac{x^2+x-2}{49-x^2}$.

26. $\dfrac{8-4x}{6x^3} \cdot \dfrac{4-x}{x^2-6x+8}$.

27. $\dfrac{2x^2y^3}{5w^2} \div \dfrac{2xy^2}{3wz^2}$.

28. $\dfrac{3xy}{2w^2z} \div \dfrac{4xyz}{5w}$.

29. $\dfrac{x+3}{x-4} \div \dfrac{x+3}{x+2}$.

30. $\dfrac{x+5}{x-3} \div \dfrac{x-5}{x-3}$.

31. $\dfrac{2x-2y}{3z} \div \dfrac{x-y}{6z^3}$.

32. $\dfrac{x^2-y^2}{xy} \div \dfrac{x+y}{xy}$.

33. $\dfrac{x-1}{x(x^2-y^2)} \div \dfrac{1-x}{x+y}$.

34. $\dfrac{3xz-15z}{2x^2y} \div \dfrac{5-x}{2(x+1)}$.

35. $6 \div \dfrac{x}{y}$.

36. $\dfrac{x}{6} \div y$.

37. $\dfrac{2}{9x} \div x.$

38. $\dfrac{x^2}{6} \div \dfrac{x}{3}.$

39. $\dfrac{\dfrac{4x^3}{9x}}{\dfrac{x}{18}}.$

40. $\dfrac{\dfrac{2m}{n^3}}{\dfrac{4m}{n^2}}.$

41. $\dfrac{\dfrac{c+d}{c}}{\dfrac{c-d}{2c}}.$

42. $\dfrac{\dfrac{4x}{3}}{\dfrac{2x}{}}.$

43. $\dfrac{-9x^3}{\dfrac{x}{3}}.$

44. $\dfrac{\dfrac{x-5}{x^2-7x+10}}{x-2}.$

45. $\dfrac{\dfrac{x^2+6x+9}{x}}{x+3}.$

46. $\dfrac{\dfrac{2x-4}{-6x}}{\dfrac{x-2}{3x^2}}.$

47. $\dfrac{\dfrac{x^2-4}{x^2+2x-3}}{\dfrac{x^2-x-6}{x^2-9}}.$

48. $\dfrac{\dfrac{x^2+7x+10}{x^2-2x-8}}{\dfrac{x^2+6x+5}{x^2-3x-4}}.$

49. $\dfrac{\dfrac{2x^2+5x-3}{4x^2-1}}{\dfrac{x^2+4x+3}{6x^2+x-1}}.$

50. $\dfrac{\dfrac{(x+2)^2}{3x-2}}{\dfrac{9x+18}{4-9x^2}}.$

51. $\dfrac{\dfrac{3}{5}(x^2+4x+4)}{\dfrac{9}{8}(x^2-4)}.$

52. $\dfrac{\dfrac{8}{3}(x^2+x-20)}{\dfrac{3x+15}{4}}.$

53. For a stretched steel wire, stress $= F/A$ and strain $= e/L$, where F is the tension in the wire, A is its cross-sectional area, e is its elongation, and L is its original length. Young's modulus for the material is stress/strain. Find an expression for this modulus and simplify.

54. Dimensional analysis of a fluid flow equation leads to the expression

$$L^3 \left[L \left(\dfrac{M}{L^3} \right) \left(\dfrac{L}{T^2} \right) \right] \left(\dfrac{T}{L^2} \right).$$

Simplify the expression.

55. The power P dissipated in a resistance R is given $P = V^2/R$, where V is the voltage across the resistor terminals. What happens to the power if the voltage is halved?

56. Under certain conditions, the pressure p of a vacuum system is given by the formula

$$p = \dfrac{ah^2}{V_0}.$$

Find p if $h = \tfrac{2}{3}d$.

57. In determining the moment of inertia of a flywheel, the expression for the ratio of natural frequencies of vibration of the flywheel with and without added loads is

$$\frac{\dfrac{JG}{IL}}{\dfrac{JG}{(I + 2mr^2)L}}.$$

Simplify the expression.

58. When a certain yo-yo is allowed to fall, the tension in the supporting string is given by the expression

$$\frac{\dfrac{1}{2}\left(\dfrac{W}{g}\right)\left(\dfrac{1}{6}\right)^2 (6a)}{\left(\dfrac{1}{6}\right)},$$

where W is weight and g and a are accelerations. Simplify this expression.

59. The power P dissipated in a resistor is given by $P = i^2R$, where i is current and R is resistance. In terms of time t, the current and resistance are given by

$$i = \frac{2t + 1}{3t^2} \quad \text{and} \quad R = \frac{3t}{2t + 1}.$$

Express P exclusively in terms of t and simplify the result.

10.6 ADDITION AND SUBTRACTION OF FRACTIONS

To add or subtract fractions with a *common denominator*, we use the next rules.

$$\frac{a}{c} + \frac{b}{c} = \frac{a + b}{c}.$$

$$\frac{a}{c} - \frac{b}{c} = \frac{a - b}{c}.$$

That is, the sum (or difference) is a fraction whose denominator is the common denominator and whose numerator is the sum (or difference) of the numerators of the fractions.

There are similar rules for handling sums and differences involving any number of fractions having a common denominator. For example,

$$\frac{a}{d} - \frac{b}{d} + \frac{c}{d} = \frac{a - b + c}{d}.$$

Example 1

Perform the indicated operations and simplify.

a. $\dfrac{x}{y} + \dfrac{3x^2}{y} = \dfrac{x + 3x^2}{y}$.

b. $\dfrac{3}{x-4} - \dfrac{2+3x}{x-4} = \dfrac{3-(2+3x)}{x-4}$ [note use of parentheses]

$$= \dfrac{3-2-3x}{x-4} = \dfrac{1-3x}{x-4}.$$

c. $\dfrac{x^2-5}{x-2} + \dfrac{2x-3}{x-2} = \dfrac{(x^2-5)+(2x-3)}{x-2}$

$$= \dfrac{x^2+2x-8}{x-2} \qquad \text{[simplifying numerator]}$$

$$= \dfrac{(x-2)(x+4)}{x-2} \qquad \text{[factoring and canceling]}$$

$$= x+4.$$

We emphasize that *only fractions with the **same** denominator can be directly combined under addition or subtraction.* In general,

$$\frac{a}{b} + \frac{a}{c} \neq \frac{a}{b+c} \quad \text{and} \quad \frac{a}{b} + \frac{c}{d} \neq \frac{a+c}{b+d}.$$

To add or subtract fractions with different denominators, we first use the fundamental principle of fractions (see Sec. 1.5) to rewrite the fractions as equivalent fractions that *do* have the same denominator. Then we add or subtract as we did before. For example, we shall find the sum

$$\frac{2}{x} + \frac{3}{x+4}.$$

Because the denominators are x and $x+4$, we shall rewrite each fraction as an equivalent fraction with denominator $x(x+4)$. The first fraction becomes

$$\frac{2(x+4)}{x(x+4)} \qquad \text{[multiplying numerator and denominator by } x+4],$$

and the second fraction becomes

$$\frac{3x}{(x+4)x} \qquad \text{[multiplying numerator and denominator by } x].$$

These fractions have a common denominator, so we can combine them.

$$\frac{2}{x} + \frac{3}{x+4} = \frac{2(x+4)}{x(x+4)} + \frac{3x}{(x+4)x}$$

$$= \frac{2(x+4) + 3x}{x(x+4)}$$

$$= \frac{2x + 8 + 3x}{x(x+4)}$$

$$= \frac{5x + 8}{x(x+4)}.$$

We could have rewritten the original fractions as equivalent fractions with other possible common denominators. But we chose to rewrite them as fractions with the denominator $x(x+4)$, which is called the **least common denominator** (L.C.D.) of the fractions $\frac{2}{x}$ and $\frac{3}{x+4}$.* It can be shown that if any other common denominator were chosen, it must have the L.C.D. as a factor. In this sense the L.C.D. is the *least* such common denominator. For this reason our work may be simplified by using the L.C.D. In general, we have the following rule for finding the L.C.D.

FINDING THE L.C.D.

To find the L.C.D. of two or more fractions, first factor the denominator of each fraction. Then, from the different factors that occur in the denominators of the fractions, form a product in which each factor is raised to the greatest power to which that factor occurs in any denominator. That product is the L.C.D.

Example 2
_____

Find the L.C.D. of the fractions

$$\frac{2x}{(x+1)(x-2)}, \qquad \frac{x-3}{x^2(x+1)}, \qquad \text{and} \qquad \frac{x^2+7}{x(x-2)^2}.$$

Solution   There are three different factors in the denominators:

$$x + 1, \qquad x - 2, \quad \text{and} \quad x.$$

The factor $x + 1$ occurs at most one time in any denominator. The factor $x - 2$ occurs at most two times (in the third fraction). The factor x occurs at most two times (in the second

---

* Section 0.4 discussed finding the L.C.D. of numerical fractions. A review of that section may be helpful.

fraction). The L.C.D. is the product of these factors, each raised to the highest power to which it occurs in any denominator. Thus the L.C.D. is

$$(x + 1)(x - 2)^2 x^2.$$

Example 3

Find the L.C.D. of the fractions

$$\frac{2}{x^2}, \quad \frac{x}{x + 1}, \quad \frac{3 + x}{x^3(x - 1)}, \quad \text{and} \quad \frac{5}{\underbrace{x^2 + 2x + 1}_{(x+1)^2}}.$$

Solution Note that we wrote the denominator $x^2 + 2x + 1$ in factored form: $(x + 1)^2$. In the denominators, the factor x occurs at most three times, the factor $x + 1$ occurs at most twice, and the factor $x - 1$ occurs at most once. Thus the L.C.D. is

$$x^3(x + 1)^2(x - 1).$$

To add or subtract fractions with different denominators, use the following rule.

ADDING OR SUBTRACTING FRACTIONS

For each fraction, multiply both its numerator and denominator by a quantity that makes its denominator equal to the L.C.D. of the fractions. Then combine and, if possible, simplify.

Example 4

Find $\dfrac{3}{x + 2} + \dfrac{x - 1}{x - 6}$.

Solution The L.C.D. is $(x + 2)(x - 6)$. To get the denominator of the first fraction equal to the L.C.D., we multiply the numerator and denominator by $x - 6$. In the second fraction, we multiply numerator and denominator by $x + 2$.

$$\frac{3}{x + 2} + \frac{x - 1}{x - 6}$$

$$= \frac{3(x - 6)}{(x + 2)(x - 6)} + \frac{(x - 1)(x + 2)}{(x - 6)(x + 2)} \qquad \text{[putting the L.C.D. in each fraction]}$$

$$= \frac{3(x - 6) + (x - 1)(x + 2)}{(x + 2)(x - 6)} \qquad \text{[combining fractions with common denominators]}$$

$$= \frac{3x - 18 + x^2 + x - 2}{(x + 2)(x - 6)}$$

$$= \frac{x^2 + 4x - 20}{(x + 2)(x - 6)}.$$

Example 5

Find $\dfrac{2}{x} - \dfrac{3}{xy} + \dfrac{4}{xz^2}$.

Solution The L.C.D. is xyz^2.

$$\dfrac{2}{x} - \dfrac{3}{xy} + \dfrac{4}{xz^2} = \dfrac{2(yz^2)}{xyz^2} - \dfrac{3(z^2)}{xyz^2} + \dfrac{4(y)}{xz^2y}$$

$$= \dfrac{2yz^2 - 3z^2 + 4y}{xyz^2}.$$

Example 6

$$(x^{-1} - y^{-1})^2 = \left(\dfrac{1}{x} - \dfrac{1}{y}\right)^2 = \left(\dfrac{y}{xy} - \dfrac{x}{yx}\right)^2 = \left(\dfrac{y - x}{xy}\right)^2$$

$$= \dfrac{(y - x)^2}{(xy)^2} = \dfrac{y^2 - 2xy + x^2}{x^2y^2}.$$

Example 7

Find $3x - 4 + \dfrac{2}{x - 1}$.

Solution Because $3x - 4 = \dfrac{3x - 4}{1}$, the L.C.D. of $3x - 4$ and $\dfrac{2}{x - 1}$ is $x - 1$.

$$3x - 4 + \dfrac{2}{x - 1} = \dfrac{(3x - 4)(x - 1)}{x - 1} + \dfrac{2}{x - 1}$$

$$= \dfrac{(3x - 4)(x - 1) + 2}{x - 1}$$

$$= \dfrac{3x^2 - 7x + 4 + 2}{x - 1}$$

$$= \dfrac{3x^2 - 7x + 6}{x - 1}.$$

Example 8

Find $\dfrac{6x - 17}{x^2 - 5x + 6} - \dfrac{1}{x - 3} + 3$.

Solution The first denominator factors into $(x - 3)(x - 2)$. Thus the denominators of the three terms are $(x - 3)(x - 2)$, $x - 3$, and 1. The L.C.D. is $(x - 3)(x - 2)$.

$$\underbrace{\frac{6x - 17}{x^2 - 5x + 6}}_{(x-3)(x-2)} - \frac{1}{x - 3} + 3 = \frac{6x - 17}{(x - 3)(x - 2)} - \frac{x - 2}{(x - 3)(x - 2)} + \frac{3(x - 3)(x - 2)}{(x - 3)(x - 2)}$$

$$= \frac{6x - 17 - (x - 2) + 3\overbrace{(x - 3)(x - 2)}^{x^2-5x+6}}{(x - 3)(x - 2)}$$

$$= \frac{6x - 17 - x + 2 + 3x^2 - 15x + 18}{(x - 3)(x - 2)}$$

$$= \frac{3x^2 - 10x + 3}{(x - 3)(x - 2)} = \frac{(3x - 1)\cancel{(x-3)}}{\cancel{(x-3)}(x - 2)}$$

$$= \frac{3x - 1}{x - 2}.$$

Example 9

Find $\dfrac{6}{x - 2} + \dfrac{7}{2 - x}$.

Solution You might be tempted to say that the L.C.D. is $(x - 2)(2 - x)$. However, we can rewrite the second fraction so that its denominator is $x - 2$:

$$\frac{6}{x - 2} + \frac{7}{2 - x} = \frac{6}{x - 2} + \frac{7}{-(x - 2)}$$

$$= \frac{6}{x - 2} - \frac{7}{x - 2}$$

$$= \frac{6 - 7}{x - 2} = \frac{-1}{x - 2} = -\frac{1}{x - 2}.$$

Example 10

Find $\dfrac{x - 2}{x^2 + 6x + 9} - \dfrac{x + 2}{2(x^2 - 9)}$.

Solution Because $x^2 + 6x + 9 = (x + 3)^2$ and $2(x^2 - 9) = 2(x + 3)(x - 3)$, the L.C.D. is $2(x + 3)^2(x - 3)$.

$$\frac{x - 2}{(x + 3)^2} - \frac{x + 2}{2(x + 3)(x - 3)} = \frac{(x - 2)(2)(x - 3)}{(x + 3)^2(2)(x - 3)} - \frac{(x + 2)(x + 3)}{2(x + 3)(x - 3)(x + 3)}$$

$$= \frac{(x - 2)(2)(x - 3) - (x + 2)(x + 3)}{2(x + 3)^2(x - 3)}$$

$$= \frac{2(x^2 - 5x + 6) - [x^2 + 5x + 6]}{2(x + 3)^2(x - 3)}$$

$$= \frac{2x^2 - 10x + 12 - x^2 - 5x - 6}{2(x + 3)^2(x - 3)}$$

$$= \frac{x^2 - 15x + 6}{2(x + 3)^2(x - 3)}.$$

Example 11

When two springs are connected in series as shown in Fig. 10.2, the reciprocal of the effective spring constant k is given by

$$\frac{1}{k} = \frac{1}{k_1} + \frac{1}{k_2},$$

where k_1 and k_2 are the spring constants for the two springs. Express $1/k$ as a single fraction.

FIGURE 10.2

Solution

$$\frac{1}{k} = \frac{1}{k_1} + \frac{1}{k_2} = \frac{k_2}{k_1 k_2} + \frac{k_1}{k_1 k_2} = \frac{k_2 + k_1}{k_1 k_2}.$$

Problem Set 10.6

In Problems **1-8**, *find the L.C.D.*

1. $\dfrac{6}{(x-4)^2}, \dfrac{7}{(x-4)^5}.$

2. $\dfrac{x}{x+1}, \dfrac{2}{x-3}.$

3. $\dfrac{4}{x^2y}, \dfrac{5}{xy^3}.$

4. $\dfrac{x+1}{x^2y^3z}, \dfrac{y-1}{xyz^4}.$

5. $\dfrac{3x}{x^2+6x+9}, \dfrac{x^2}{x^2-9}.$

6. $\dfrac{2}{x^2+3x-4}, \dfrac{1}{x-1}.$

7. $\dfrac{1}{2x+2}, \dfrac{x}{x^2+x}, \dfrac{2}{x+1}.$

8. $\dfrac{x}{4x+2}, \dfrac{4}{4x^2-1}, \dfrac{x}{3}.$

In Problems **9-48**, *perform the indicated operations and simplify.*

9. $\dfrac{x+1}{x-3} + \dfrac{4}{x-3}.$

10. $\dfrac{x^2}{x-2} + \dfrac{x-6}{x-2}.$

11. $\dfrac{3x}{x+1} + \dfrac{4}{x+1} - \dfrac{x+2}{x+1}.$

12. $\dfrac{2x}{x^2-1} - \dfrac{2}{x^2-1}.$

13. $\dfrac{3x^2+6x}{x^2+x-2} - \dfrac{x^2+2x+1}{x^2-1}.$

14. $\dfrac{3x+4}{x+2} + \dfrac{x^2-9}{x^2+5x+6}.$

15. $\dfrac{2}{x} + \dfrac{3}{y}.$

16. $3 + \dfrac{x}{y}.$

17. $\dfrac{x-4}{6} - \dfrac{x-2}{9}.$

18. $\dfrac{x-2}{3} + 1.$

19. $\dfrac{3}{2x} - \dfrac{2}{xy}.$

20. $\dfrac{4}{x} + \dfrac{2}{y} - \dfrac{x+1}{xy}.$

21. $\dfrac{5}{x-2} + \dfrac{3}{x-3}.$

22. $\dfrac{y}{y-2} - \dfrac{3}{y}.$

23. $\dfrac{5y}{x^2} - \dfrac{2}{xy} + \dfrac{3}{y}.$

24. $\dfrac{x}{a^2} + \dfrac{y}{ab}.$

25. $\dfrac{x+3}{x-1} + 4.$

26. $\dfrac{a}{b} + \dfrac{c}{d}.$

27. $\dfrac{x}{x-y} + \dfrac{y}{x+y}.$

28. $\dfrac{2}{x+1} - \dfrac{3}{x-1}.$

29. $\dfrac{x+3}{x-3} - \dfrac{x-3}{2(x+3)}.$

30. $\dfrac{4}{2x-1} + \dfrac{x}{x+2}.$

31. $\dfrac{6x+12}{x^2+5x+4} + \dfrac{x}{x+4}.$

32. $\dfrac{5}{x^2+3x-4} + \dfrac{1}{x+4}.$

33. $\dfrac{1}{x^2-1} - \dfrac{1}{x-1} + \dfrac{1}{x+1}.$

34. $\dfrac{x}{x+1} - \dfrac{2x}{x^2+3x+2}.$

35. $\dfrac{x-1}{x^2+6x+9} + \dfrac{2}{x^2-9}.$

36. $x^2+2-\dfrac{x^4}{x^2-2}.$

37. $\dfrac{x+1}{x^2+7x+10} - \dfrac{2x}{x^2+6x+5}.$

38. $\dfrac{y}{3y^2-5y-2} - \dfrac{2}{3y^2-7y+2}.$

39. $\dfrac{2x-6}{x^2-5x+6} + \dfrac{x^2+8x+16}{x^2+6x+8}.$

40. $\dfrac{2}{x^3(x-3)} + \dfrac{3}{x(x-3)^2}.$

41. $2x+3+\dfrac{2}{x-1}.$

42. $\dfrac{3}{x-1} - \dfrac{4}{1-x}.$

43. $\dfrac{x-2}{x^2+x} + \dfrac{3}{x^3+2x^2} - \dfrac{2x-3}{x^2+3x+2}.$

44. $\dfrac{2}{x^2-5x+6} - \dfrac{1}{x^2-3x+2} + \dfrac{4}{x^2-4x+3}.$

45. $\dfrac{y}{2y^2+7y+3} - \dfrac{2}{4y^2+4y+1}.$

46. $\dfrac{3}{x+3} + \dfrac{1}{x-3} - \dfrac{4}{x+2}.$

47. $\dfrac{y}{x^2+2xy+y^2} + \dfrac{3x}{x^2-y^2} - \dfrac{2}{x+y}.$

48. $1 - \dfrac{2y^2}{x^2-y^2} + \dfrac{2xy}{x^2+y^2}.$

Problems 49–52 are more challenging than the previous ones. Again, perform the indicated operations and simplify.

49. $\left(2x - \dfrac{x+3}{2} - 2a\right) + \dfrac{a-4}{3} - a.$

50. $\dfrac{ay}{a-y} + \dfrac{a^2}{y-a} + a.$

51. $\dfrac{2y+3}{2-y} - \dfrac{2-3y}{y+2} + \dfrac{16y-y^2}{(y+2)(y-2)}.$

52. $\dfrac{1}{8-6x+x^2} - \dfrac{x+2}{(x-3)(4-x)} + \dfrac{x+2}{x^2-5x+6}.$

In Problems 53–56, perform the operations and simplify. Give all answers as single fractions with positive exponents only.

53. $(x+y^{-1})^2.$

54. $x^{-2} - y^{-2}.$

55. $[x^{-1}(x+y^{-1})]^{-1}.$

56. $(x^{-1}+y^{-1})^{-1}.$

In Problems 57 and 58, solve each equation for the given letter.

57. $U = \left(\dfrac{f-f_0}{f_0}\right)\dfrac{\epsilon}{v_0};\ f.$

58. $\sigma = \dfrac{n_0-n_e}{\lambda} L;\ n_e.$

59. For the arrangement in Fig. 10.3, it is shown in physics that the potential at point O due to the point charges q_1, q_2, and q_3 is given by

$$\dfrac{kq_1}{x} + \dfrac{kq_2}{x+2} + \dfrac{kq_3}{x+3}.$$

Express the potential as a single fraction. Do not simplify your answer.

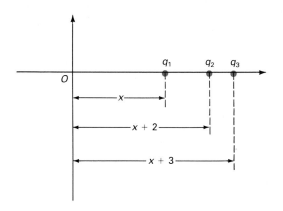

FIGURE 10.3

60. For the arrangement in Fig. 10.3, if the distance of charge q_2 from point O were doubled, theory shows that the potential at point O would be

$$\frac{kq_1}{x} + \frac{kq_2}{2(x+2)} + \frac{kq_3}{x+3}.$$

Express the potential as a single fraction. Do not simplify your answer.

61. In studies of electric potential, the expression

$$\frac{1}{4\pi\epsilon_0} \left(\frac{q_1}{r_1} - \frac{q_2}{r_2} \right)$$

was encountered. Write this expression as a single fraction.

62. When three springs with spring constants of k_1 k_2, k_3 are connected in series (see Example 11), the reciprocal of the effective spring constant k is given by

$$\frac{1}{k} = \frac{1}{k_1} + \frac{1}{k_2} + \frac{1}{k_3}.$$

Express $1/k$ as a single fraction.

63. The expression for the viscosity of a liquid has the form

$$\frac{a}{kR_1^2} - \frac{a}{kR_2^2}.$$

Combine these fractions and give your answer in factored form.

64. The formula for the stress factor f of a particular coil spring is

$$f = \frac{4k-1}{4k-4} + \frac{0.5}{k}.$$

Express f as a single fraction and simplify.

65. When two batteries with emf and internal resistance $\mathcal{E}_1$, r_1 and $\mathcal{E}_2$, r_2, respectively, are connected in parallel, the effective emf $\mathcal{E}$ of the combination is given by

$$\mathcal{E} = \left(\frac{1}{r_1} + \frac{1}{r_2} \right)\left(\frac{\mathcal{E}_1}{r_1} + \frac{\mathcal{E}_2}{r_2} \right).$$

Express $\mathcal{E}$ as a single fraction. Do not simplify your answer.

66. The reciprocal of the square of the impedance of an ac circuit containing a resistor, capacitor, and inductor connected in parallel is given by

$$\frac{1}{R^2} + \left(\frac{1}{X_L^2} - \frac{1}{X_C^2} \right)^2.$$

Express this quantity as a single fraction.

10.7 COMPLEX FRACTIONS

A fraction whose numerator or denominator (or both) contains a fraction is called a **complex fraction**. To simplify a complex fraction, there are two methods that are commonly used.

Method 1. Perform any indicated operations in the numerator and in the denominator. Then divide the numerator by the denominator.

Method 2. Multiply the numerator and denominator of the given fraction by the L.C.D. of all the fractions that appear. This gives an equivalent fraction (by the fundamental principle of fractions) in which the numerator and denominator do not contain fractions.

Both methods are shown in Example 1.

Example 1

Simplify $\dfrac{1 + \dfrac{1}{x}}{2 - \dfrac{1}{y}}$.

Solution **Method 1.** We separately combine the terms in the numerator and denominator and then divide.

$$\frac{1 + \dfrac{1}{x}}{2 - \dfrac{1}{y}} = \frac{\dfrac{x}{x} + \dfrac{1}{x}}{\dfrac{2y}{y} - \dfrac{1}{y}} = \frac{\dfrac{x+1}{x}}{\dfrac{2y-1}{y}}$$

$$= \frac{x+1}{x} \cdot \frac{y}{2y-1}$$

$$= \frac{y(x+1)}{x(2y-1)}.$$

Method 2. We multiply the numerator and denominator by the L.C.D. of all the fractions that appear. The fractions are $1/x$ and $1/y$, so the L.C.D. is xy.

$$\frac{1 + \dfrac{1}{x}}{2 - \dfrac{1}{y}} = \frac{xy\left(1 + \dfrac{1}{x}\right)}{xy\left(2 - \dfrac{1}{y}\right)}$$

$$= \frac{xy(1) + xy\left(\dfrac{1}{x}\right)}{xy(2) - xy\left(\dfrac{1}{y}\right)} \qquad \text{[distributive law]}$$

$$= \frac{xy + y}{2xy - x} = \frac{y(x + 1)}{x(2y - 1)}.$$

Example 2

Simplify $\dfrac{3 - \dfrac{1}{2x}}{6x + \dfrac{11x}{x - 2}}$.

Solution Using Method 1 we have

$$\frac{3 - \dfrac{1}{2x}}{6x + \dfrac{11x}{x - 2}} = \frac{\dfrac{3(2x)}{2x} - \dfrac{1}{2x}}{\dfrac{6x(x - 2)}{x - 2} + \dfrac{11x}{x - 2}}$$

$$= \frac{\dfrac{6x - 1}{2x}}{\dfrac{6x^2 - 12x + 11x}{x - 2}}$$

$$= \frac{6x - 1}{2x} \cdot \frac{x - 2}{6x^2 - x}$$

$$= \frac{(6x - 1)(x - 2)}{2x(x)(6x - 1)}$$

$$= \frac{x - 2}{2x^2}.$$

Example 3

Simplify $\dfrac{\dfrac{x}{x - 1}}{\dfrac{1}{x - 1} + \dfrac{1}{x + 1}}$.

Solution   Using Method 2, we multiply the numerator and denominator by the L.C.D. of the fractions that appear. The L.C.D. is $(x - 1)(x + 1)$.

$$\frac{\dfrac{x}{x-1}}{\dfrac{1}{x-1} + \dfrac{1}{x+1}} = \frac{(x-1)(x+1)\left[\dfrac{x}{x-1}\right]}{(x-1)(x+1)\left[\dfrac{1}{x-1} + \dfrac{1}{x+1}\right]}$$

$$= \frac{(x+1)(x)}{(x-1)(x+1)\left(\dfrac{1}{x-1}\right) + (x-1)(x+1)\left(\dfrac{1}{x+1}\right)}$$

$$= \frac{(x+1)(x)}{(x+1) + (x-1)} = \frac{(x+1)(x)}{2x} = \frac{x+1}{2}.$$

Problem Set 10.7

In Problems **1–26**, *simplify.*

1. $\dfrac{\frac{1}{2} + \frac{1}{3}}{\frac{1}{2}}.$

2. $\dfrac{2^{-2} + 3^{-2}}{13}.$

3. $\dfrac{\dfrac{1}{x} + \dfrac{3}{x}}{4}.$

4. $\dfrac{\dfrac{x}{x-1} - \dfrac{1}{x-1}}{x-1}.$

5. $\dfrac{4 - \dfrac{6}{x}}{2}.$

6. $\dfrac{x-1}{1 - \dfrac{1}{x}}.$

7. $\dfrac{7}{3x - \frac{1}{2}}.$

8. $\dfrac{x - \dfrac{1}{x}}{x+1}.$

9. $\dfrac{3 - \dfrac{1}{y}}{2 + \dfrac{1}{x}}.$

10. $\dfrac{\dfrac{a}{b} + 2}{\dfrac{b}{a} - 2}.$

11. $\dfrac{\dfrac{1}{x} + \dfrac{x}{2x-3}}{\dfrac{x-1}{x}}.$

12. $\dfrac{a - \dfrac{b^2}{a}}{b - \dfrac{a^2}{b}}.$

13. $\dfrac{\dfrac{x}{y} - \dfrac{y}{x}}{\dfrac{x+y}{xy}}.$

14. $\dfrac{\dfrac{x^2 - 9}{4}}{\dfrac{1}{x} - \dfrac{1}{3}}.$

15. $\dfrac{\dfrac{x}{y} + \dfrac{y}{x}}{\dfrac{x}{y} - \dfrac{y}{x}}.$

16. $\dfrac{\dfrac{x}{x^2 + 4x + 3}}{\dfrac{1}{x^2 - 1} + 1}.$

17. $\dfrac{x + 1 + \dfrac{1}{x+3}}{\dfrac{x+2}{3}}.$

18. $\dfrac{3x + \dfrac{x}{x-3}}{3x - \dfrac{x}{x-3}}.$

19. $\dfrac{\dfrac{1}{x^4} + \dfrac{1}{x^2} + 1}{\dfrac{1}{x^3} + \dfrac{1}{x} + x}.$

20. $\dfrac{\dfrac{x}{y} + x}{y - \dfrac{1}{y}} - 1.$

21. $\dfrac{2 + x^{-1}}{x}.$

22. $\dfrac{x + x^{-1}}{x^{-1} - x}.$

23. $\left(\dfrac{x^{-1} + y^{-1}}{x^{-2} - y^{-2}}\right)^{-2}.$

24. $\left(\dfrac{1 + y^{-2}}{y + y^{-1}}\right)^{-2}.$

25. $\dfrac{a}{1 + \dfrac{c}{d + \frac{1}{2}}}.$

26. $\dfrac{\dfrac{a + b}{a - b} - \dfrac{a - b}{a + b}}{1 - \dfrac{a^2 + b^2}{(a + b)^2}}.$

27. The theory of relativity shows that the mass m of an object varies with its speed v according to the equation

$$m^2 = \frac{m_0^2}{1 - \left(\dfrac{v}{c}\right)^2},$$

where m_0 is the rest mass of the object and c is the speed of light. Simplify the expression for m^2.

28. When the system of masses m_1 and m_2 shown in Fig. 10.4 is released from rest, the acceleration a of the masses is given by

$$a = \frac{g}{1 + \dfrac{m_1}{m_2}},$$

where g is a constant. Simplify the right side of the equation.

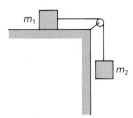

FIGURE 10.4

29. In a model of traffic flow on a lane of a freeway,* the number N of cars the lane can carry per unit time is given by

$$N = \frac{-2a}{-2at_r + v - \dfrac{2al}{v}},$$

where a is acceleration of a car when stopping, t_r is reaction time to begin braking, v is average speed of the cars, and l is length of a car. Simplify the right side of the equation.

* J. I. Shonle, *Environmental Applications of General Physics* (Reading, Mass.: Addison-Wesley, 1975.)

30. The effective spring constant k for a certain spring system is given by

$$k = \frac{1}{4\left(\dfrac{1}{k_1} + \dfrac{1}{k_2}\right)}.$$

Simplify the expression for k.

31. The effective spring constant k of three springs connected in series is given by

$$k = \frac{1}{\dfrac{1}{k_1} + \dfrac{1}{k_2} + \dfrac{1}{k_3}}.$$

Simplify the expression for k.

32. The lens-maker's equation gives the focal length f of a lens with spherical surfaces of radii R_1 and R_2. The equation is

$$f = \frac{1}{\left(\dfrac{n_1}{n_2} - 1\right)\left(\dfrac{1}{R_1} + \dfrac{1}{R_2}\right)},$$

where n_1 and n_2 are the indices of refraction of the lens material and surrounding medium, respectively. Simplify the expression for f.

33. If capacitors C_1 and C_2 are connected in series, the total capacitance C_T is given by

$$\frac{1}{C_T} = \frac{1}{C_1} + \frac{1}{C_2}.$$

Taking reciprocals of both sides yields the equation

$$C_T = \frac{1}{\dfrac{1}{C_1} + \dfrac{1}{C_2}}.$$

Show that this equation can be written as

$$C_T = \frac{C_1 C_2}{C_1 + C_2}.$$

10.8 FRACTIONAL EQUATIONS

An equation containing a fraction in which the unknown occurs in the denominator is called a **fractional equation**. An example is

$$\frac{6}{x - 3} = \frac{5}{x - 4}.$$

One way to solve a fractional equation is to first clear fractions by multiplying each side by the L.C.D. of all the fractions involved. Then we solve the resulting equation and check that all its solutions satisfy the *original* equation.

To illustrate, we shall solve

$$\frac{6}{x - 3} = \frac{5}{x - 4}.$$ (1)

To clear fractions we multiply both sides by the L.C.D., which is $(x - 3)(x - 4)$.

$$(x - 3)(x - 4)\left(\frac{6}{x - 3}\right) = (x - 3)(x - 4)\left(\frac{5}{x - 4}\right)$$

$$6(x - 4) = 5(x - 3)$$ (2)

$$6x - 24 = 5x - 15$$

$$x = 9.$$

In the first step we multiplied each side by an expression involving the variable x. As mentioned in Sec. 4.1, this means that we must check whether or not 9 satisfies the *original* equation. If 9 is substituted for x in Eq. 1, the left side is

$$\frac{6}{9 - 3} = \frac{6}{6} = 1$$

and the right side is

$$\frac{5}{9 - 4} = \frac{5}{5} = 1.$$

Because both sides are equal, 9 is a solution. It can be shown that when both sides of an equation (with variable x) are multiplied by the same polynomial (in x), then the resulting equation has *all* the solutions of the original and perhaps *more*. Thus any other solution of the original equation has to be a solution of Eq. 2, which has 9 as its only solution. Hence the only solution of the original equation is 9.

Example 1

Solve $\dfrac{4}{x - 1} = \dfrac{7}{x - 1} - 2.$

Solution   To clear fractions, we multiply both sides by the L.C.D., which is $x - 1$.

$$(x - 1)\left(\frac{4}{x - 1}\right) = (x - 1)\left(\frac{7}{x - 1} - 2\right)$$

$$(x - 1)\left(\frac{4}{x - 1}\right) = (x - 1)\left(\frac{7}{x - 1}\right) - (x - 1)(2) \qquad \begin{array}{l}\text{[distributive law} \\ \text{and cancellation]}\end{array}$$

$$4 = 7 - 2x + 2$$

$$2x = 5$$

$$x = \frac{5}{2}.$$

You may verify by substitution that $\frac{5}{2}$ is indeed a solution.

Example 2

Solve $\dfrac{9}{x - 3} = 0$.

Solution   Multiplying both sides by the L.C.D., which is $x - 3$, gives

$$9 = 0.$$

Because it is never true that $9 = 0$, we conclude that **there are no solutions**. The solution set has no elements in it. Recall that we denote this set by $\varnothing$, the empty set. There is another way of handling the given equation that is useful. The *only* way a fraction can equal zero is if the numerator is zero and the denominator is different from zero. Because the numerator, 9, is never zero, the solution set is $\varnothing$.

Example 3

Solve

$$\frac{1}{x + 2} + \frac{1}{x - 2} = \frac{4}{x^2 - 4}. \tag{3}$$

Solution   We clear fractions by multiplying both sides by the L.C.D. of all fractions involved. Since $x^2 - 4 = (x + 2)(x - 2)$, clearly the L.C.D. is $(x + 2)(x - 2)$.

$$\frac{1}{x + 2} + \frac{1}{x - 2} = \frac{4}{(x + 2)(x - 2)}$$

$$(x + 2)(x - 2)\left[\frac{1}{x + 2} + \frac{1}{x - 2}\right] = (x + 2)(x - 2)\left[\frac{4}{(x + 2)(x - 2)}\right]$$

$$\frac{(x + 2)(x - 2)}{x + 2} + \frac{(x + 2)(x - 2)}{x - 2} = \frac{(x + 2)(x - 2) \cdot 4}{(x + 2)(x - 2)} \qquad \begin{array}{l}\text{[distributive law}\\ \text{and cancellation]}\end{array}$$

$$(x - 2) + (x + 2) = 4 \tag{4}$$

$$2x = 4$$

$$x = 2.$$

Because we multiplied both sides by the polynomial $(x + 2)(x - 2)$, we must see if 2 is a solution of the original equation. When $x = 2$ in that equation, there is a denominator of 0. But we cannot divide by 0. Thus we conclude that **there is no solution**. Although 2 is a solution of Eq. 4, it is not a solution of the original equation and is sometimes called an **extraneous solution** of Eq. 3. In general, *multiplying both sides of an equation by a polynomial may lead to extraneous solutions.*

Example 4

Solve $\dfrac{1}{a} + \dfrac{1}{x} = \dfrac{1}{b}$ for x.

Solution Multiplying both sides by the the L.C.D., which is abx, we have

$$abx\left(\frac{1}{a} + \frac{1}{x}\right) = abx\left(\frac{1}{b}\right)$$

$$abx\left(\frac{1}{a}\right) + abx\left(\frac{1}{x}\right) = abx\left(\frac{1}{b}\right) \quad \text{[distributive law]}$$

$$bx + ab = ax$$

$$bx - ax = -ab$$

$$x(b - a) = -ab \quad \text{[factoring]}$$

$$x = \frac{-ab}{b - a} = \frac{ab}{a - b},$$

provided $b - a \neq 0$.

Example 5

If n cells, each having an internal resistance r and electromotive force $\mathcal{E}$, are connected in series to a load resistance R, the current i in the circuit is given by

$$i = \frac{n\mathcal{E}}{R + nr}.$$

Solve for r.

Solution

$$i = \frac{n\mathcal{E}}{R + nr}$$

$$i(R + nr) = n\mathcal{E} \quad \text{[multiplying both sides by } R + nr\text{]}$$

$$iR + nir = n\mathcal{E}$$

$$nir = n\mathcal{E} - iR$$

$$r = \frac{n\mathcal{E} - iR}{ni}.$$

Example 6

An important equation for lenses is

$$\frac{1}{f} = \frac{1}{p} + \frac{1}{q},$$

where f is focal length, p is object distance, and q is image distance. Suppose that for a converging lens the focal length is 12 cm and the object distance is 24 cm. Find the image distance.

Solution   Substituting 12 for f and 24 for p gives

$$\frac{1}{12} = \frac{1}{24} + \frac{1}{q}$$

$$24q\left(\frac{1}{12}\right) = 24q\left(\frac{1}{24} + \frac{1}{q}\right) \qquad \text{[multiplying both sides by } 24q\text{]}$$

$$2q = q + 24$$

$$q = 24 \text{ cm.}$$

Problem Set 10.8

In Problems 1–28, solve the fractional equations.

1. $\dfrac{3}{x} = 12.$

2. $\dfrac{1}{x} + \dfrac{1}{5} = \dfrac{4}{5}.$

3. $\dfrac{x}{3x - 4} = 3.$

4. $\dfrac{4}{x - 1} = 2.$

5. $\dfrac{10}{3r} - \dfrac{9r + 2}{6r} = 3.$

6. $\dfrac{4x}{7 - x} = 1.$

7. $\dfrac{3}{7 - x} = 0.$

8. $\dfrac{2x - 3}{4x - 5} = 6.$

9. $1 + \dfrac{2}{x} = \dfrac{2(x + 1)}{x}.$

10. $\dfrac{5x - 2}{x + 1} = 0.$

11. $\dfrac{3}{4y} - \dfrac{5}{6y} = \dfrac{1}{6}.$

12. $\dfrac{x + 3}{x} = \dfrac{2}{5}.$

13. $\dfrac{1}{p - 1} = \dfrac{2}{p - 2}.$

14. $\dfrac{2x - 3}{4x - 5} = 6.$

15. $\dfrac{1}{10} + \dfrac{1}{x} = \dfrac{4}{5}.$

16. $\dfrac{1}{2} - \dfrac{3}{x} = \dfrac{2}{3}.$

17. $\dfrac{3x - 2}{2x + 3} = \dfrac{3x - 1}{2x + 1}.$

18. $\dfrac{x + 2}{x - 1} + \dfrac{x + 1}{2 - x} = 0.$

19. $\dfrac{y - 6}{y} - \dfrac{6}{y} = \dfrac{y + 6}{y - 6}.$

20. $\dfrac{y - 1}{y + 3} = 4 + \dfrac{2}{y + 3}.$

21. $\dfrac{-4}{x - 1} = \dfrac{7}{2 - x} + \dfrac{3}{x + 1}.$

22. $\dfrac{1}{x - 3} - \dfrac{3}{x - 2} = \dfrac{4}{1 - 2x}.$

23. $\dfrac{9}{x - 3} = \dfrac{3x}{x - 3}.$

24. $\dfrac{x}{x + 3} - \dfrac{x}{x - 3} = \dfrac{3x - 4}{x^2 - 9}.$

25. $\dfrac{2x}{x - 1} - \dfrac{3}{x + 2} = \dfrac{4x}{(x + 2)(x - 1)} + 2.$

26. $\dfrac{2}{x} + \dfrac{3}{x + 1} = \dfrac{x}{x + 1} - \dfrac{x + 1}{x}.$

27. $\dfrac{3x+4}{x+2} - \dfrac{3x-5}{x-4} = \dfrac{12}{x^2-2x-8}$.

28. $\dfrac{4}{x-4} - \dfrac{3}{x-3} = \dfrac{1}{x-5}$.

*In Problems **29-42**, express the indicated letter(s) in terms of the remaining letters.*

29. $2 + \dfrac{b}{abx} = \dfrac{4}{x} - \dfrac{a}{b}$; x.

30. $\dfrac{x-a}{b-x} = \dfrac{x-b}{a-x}$; x.

31. $\dfrac{P_1 V_1}{T_1} = \dfrac{P_2 V_2}{T_2}$; T_1.

32. $h = kat\left(\dfrac{T}{L}\right)$; L.

33. $V = V_0\left(\dfrac{P_1}{P_2}\right)\left(\dfrac{T_2}{T_1}\right)$; P_2.

34. $F = \dfrac{1}{2\pi}\left(\dfrac{e}{m}\right)b$; m.

35. $\dfrac{1}{p} + \dfrac{1}{q} = \dfrac{1}{f}$; q, f.

36. $\dfrac{1}{C_t} = \dfrac{1}{C_1} + \dfrac{1}{C_2} + \dfrac{1}{C_3}$; C_t.

37. $R_t = \dfrac{R_1 R_2}{R_1 + R_2}$; R_1.

38. $\dfrac{x}{a} + \dfrac{y}{b} = 1$; a.

39. $\dfrac{1}{2}mv^2 - \dfrac{p^2}{2m} = 0$; m^2.

40. $S = \dfrac{\dfrac{W_s}{V}}{\dfrac{W_w}{V}}$; W_w, W_s.

41. $\dfrac{1}{f} = (n-1)\left(\dfrac{1}{R_1} + \dfrac{1}{R_2}\right)$; R_1.

42. $V = \dfrac{1}{4\pi\epsilon_0}\left(\dfrac{q_1}{r_1} + \dfrac{q_2}{r_2}\right)$; r_1.

43. When three cells, each having an internal resistance of $0.2\,\Omega$ and emf $\mathcal{E} = 5\,\text{V}$, are connected in series with a load resistance R, the current in the circuit is $0.1\,\text{A}$. Find the load resistance R, in ohms. (See Example 5.)

44. When an object is placed in front of a converging lens with a focal length of $20\,\text{cm}$, the image distance is $40\,\text{cm}$. Find the object distance. (See Example 6.)

45. The combined resistance R of two resistors R_1 and R_2 connected in parallel is given by

$$R = \dfrac{1}{\dfrac{1}{R_1} + \dfrac{1}{R_2}}.$$

If one resistor has a resistance of $60\,\Omega$ and the combined resistance is $10\,\Omega$, find the resistance of the other resistor.

46. The total capacitance C_T of a circuit network containing two capacitors C_1 and C_2 in series is given by

$$\dfrac{1}{C_T} = \dfrac{1}{C_1} + \dfrac{1}{C_2}.$$

If $C_1 = 2\,\mu\text{F}$ and $C_T = \tfrac{2}{3}\,\mu\text{F}$, find C_2.

47. If a sound source moves towards you, the pitch of the sound seems to change. This is known as the Doppler effect. For example, if you were standing near a railroad track, you would hear an increase of pitch in the train's whistle as the train approaches. If the actual frequency of a sound source is f, and the sound source has speed v, and V is the velocity of sound, then the apparent frequency heard, f' (read f prime), is given by

$$f' = \left(\frac{V}{V - v}\right)f.$$

Suppose a train whistle has a frequency of 480 vibrations per second and approaches you at 44 ft/s. Also, suppose sound travels at 1100 ft/s.
(a) Find the apparent frequency you would hear.
(b) At what speed would the train have to approach you so that you would hear an apparent frequency of 507 vibrations per second?

48. If a space capsule of mass m were launched from the surface of the earth with speed v, it would rise, neglecting the effects of air resistance, to a height h such that

$$\frac{1}{2}mv^2 = \frac{mghR}{R + h},$$

where R is the radius of the earth and g is the acceleration due to gravity at the earth's surface. Solve for h.

49. The reciprocal of the focal length f of a thin lens with spherical surfaces is given by

$$\frac{1}{f} = (n - 1)\left(\frac{1}{R_1} + \frac{1}{R_2}\right)$$

(called the lens-maker's equation), where n is the index of refraction of the lens material and R_1 and R_2 are radii of curvature of the surfaces. For a certain double-convex lens, $n = 1.5$, $R_1 = 10$ cm, and $R_2 = 20$ cm. Find f to one decimal place.

50. The combined resistance R of two resistors R_1 and R_2 connected in parallel is given by

$$\frac{1}{R} = \frac{1}{R_1} + \frac{1}{R_2}.$$

Show that

$$R = \frac{R_1 R_2}{R_1 + R_2}.$$

10.9   REVIEW

Important Terms

Section 10.1	Square of a binomial, cube of a binomial, product of the sum and difference, FOIL method.
Section 10.2	Factoring, common factor, completely factored, prime polynomial, difference of two squares.
Section 10.3	Factoring by grouping, sum or difference of two cubes.
Section 10.4	Reduced fraction, simplest form of a fraction.
Section 10.6	Least common denominator (L.C.D.).
Section 10.7	Complex fraction.
Section 10.8	Fractional equation, extraneous solution.

Summary of Rules and Formulas

SPECIAL PRODUCTS

$(x + a)(x + b) = x^2 + (a + b)x + ab.$

$(a + b)^2 = a^2 + 2ab + b^2$ square of a binomial.

$(a - b)^2 = a^2 - 2ab + b^2$ square of a binomial.

$(a + b)(a - b) = a^2 - b^2$ product of sum and difference.

$(a + b)^3 = a^3 + 3a^2b + 3ab^2 + b^3$ cube of a binomial.

$(a - b)^3 = a^3 - 3a^2b + 3ab^2 - b^3$ cube of a binomial.

FACTORING RULES

$ab + ac = a(b + c)$ common factor.

$x^2 + 2ax + a^2 = (x + a)^2$ perfect-square trinomial.

$x^2 - 2ax + a^2 = (x - a)^2$ perfect-square trinomial.

$a^2 - b^2 = (a + b)(a - b)$ difference of two squares.

$a^3 + b^3 = (a + b)(a^2 - ab + b^2)$ sum of two cubes.

$a^3 - b^3 = (a - b)(a^2 + ab + b^2)$ difference of two cubes.

RULES FOR FRACTIONS

$\dfrac{ab}{ac} = \dfrac{b}{c}$ cancellation property.

$\dfrac{a}{b} \cdot \dfrac{c}{d} = \dfrac{ac}{bd}$ multiplication of fractions.

$\dfrac{\frac{a}{b}}{\frac{c}{d}} = \dfrac{a}{b} \cdot \dfrac{d}{c} = \dfrac{ad}{bc}$ division of fractions.

$\dfrac{a}{c} + \dfrac{b}{c} = \dfrac{a + b}{c}$ addition of fractions.

$\dfrac{a}{c} - \dfrac{b}{c} = \dfrac{a - b}{c}$ subtraction of fractions.

To find the L.C.D. of two or more fractions, first factor the denominator of each fraction. Then, from the different factors that occur in the denominators of the fractions, form a product in which each factor is raised to the greatest power to which that factor occurs in any denominator. That product is the L.C.D.

Review Questions

1. $x^2 + 6x + 9$ __(is) (is not)__ the square of a binomial.
2. To factor $2x^2 + 10x^2 + 12x$, you should first remove the common factor _____.
3. If $x^4 - y^4$ is written as $(x^2 + y^2)(x^2 - y^2)$, is it completely factored? _____.
4. $(53)(47) = (50 + 3)(50 - 3) = 2500 -$ _____.
5. $(22)^2 = (20 + 2)^2 = 400 +$ _____ $+ 4$.
6. Because $x^2 + 36$ cannot be factored, it is said to be _____.
7. Insert $+$ or $-$ in the parentheses to make the equalities true in the following.

$$\frac{a - b}{c - d} = (\;)\frac{-(a - b)}{c - d} = (\;)\frac{b - a}{d - c} = (\;)\frac{b - a}{-(c - d)}.$$

Write true *if the following are, in general, true and write* false *if otherwise.*

$$\frac{a + bx}{a + cy} = \frac{bx}{cy}. \qquad \text{(8)} \underline{\hspace{2cm}}$$

$$\frac{1}{x} + \frac{1}{y} = \frac{xy}{x + y}. \qquad \text{(9)} \underline{\hspace{2cm}}$$

$$\frac{abx}{acy} = \frac{bx}{cy}. \qquad \text{(10)} \underline{\hspace{2cm}}$$

$$\frac{1}{x} - \frac{x - y}{x^2} = \frac{x - x - y}{x^2}. \qquad \text{(11)} \underline{\hspace{2cm}}$$

$$\frac{1}{a} + \frac{1}{b} = \frac{2}{ab}. \qquad \text{(12)} \underline{\hspace{2cm}}$$

$$\frac{a^2 - b^2}{a - b} = a - b \qquad \text{(13)} \underline{\hspace{2cm}}$$

$$\frac{\frac{1}{x}}{y} = \frac{1}{xy}. \qquad \text{(14)} \underline{\hspace{2cm}}$$

$$\frac{1}{x} - \frac{1}{y} = \frac{x - y}{xy}. \qquad \text{(15)} \underline{\hspace{2cm}}$$

$$\frac{x - y}{y + x} + \frac{y - x}{x + y} = 0. \qquad \text{(16)} \underline{\hspace{2cm}}$$

$$\frac{1}{x} + \frac{1}{y} = \frac{1}{x + y}. \qquad \text{(17)} \underline{\hspace{2cm}}$$

$$\frac{1}{x} \cdot \frac{1}{y} = \frac{1}{xy}. \qquad \text{(18)} \underline{\hspace{2cm}}$$

19. The L.C.D. of the fractions

$$\frac{6}{x^2 - 1}, \quad \frac{8}{x + 1}, \quad \text{and} \quad \frac{7}{2(x + 1)}$$

is _____

20. If $p_1/p_2 = t_1/t_2$, then $t_2 = $ _____.

Answers to Review Questions

1. Is.   2. $2x$.   3. No.   4. 9.   5. 80.   6. Prime   7. $-, +, +$.   8. False.
9. False.   10. True.   11. False.   12. False.   13. False.   14. True.

15. False.   16. True.   17. False.   18. True.   19. $2(x + 1)(x - 1)$.   20. $\dfrac{t_1 p_2}{p_1}$.

Review Problems

In Problems 1–20, find the special products.

1. $(x + 6)^2$.
2. $(x - 7)^2$.
3. $(x - 5)^2$.
4. $(x + 3y)^2$.
5. $(2x + 4y)^2$.
6. $(3x - 6)^2$.
7. $(x - 8)(x + 8)$.
8. $(9 - 2x)(9 + 2x)$.
9. $(3x + 2)(3x - 2)$.
10. $(1 - 3x)(3x + 1)$.
11. $(2x - 4y)(2x + 4y)$.
12. $(a - 2b)(a + 2b)$.
13. $(x - 6)(x + 4)$.
14. $(x + 3)(x - 2)$.
15. $(x - 6)(x - 7)$.
16. $(x + 5)(x + 8)$.
17. $(2x - 3)(2x - 4)$.
18. $(3x - 2)(2x + 3)$.
19. $(y^2 + 4)(y^2 - 4)$.
20. $(y^3 + 1)(y^3 - 1)$.

In Problems 21–26, perform the indicated operations.

21. $2x^2(x - 3)(x + 4)$.
22. $-3x(x + 3)^2$.
23. $(y - \sqrt{2})(y + \sqrt{2}) - (y - 4)^2$.
24. $5y(1 + 2y)^2$.
25. $(4x + 3)(4x - 3)(x + 2)$.
26. $3(3x - 1)^2 + 5x(x + 2)^2$.

In Problems 27–50, factor completely.

27. $6x^3y^4 + 4xy^6$.
28. $10abc^8 - 15ab^2c$.
29. $x^2 - 11x + 30$.
30. $x^2 + 6x + 8$.
31. $16 - y^2$.
32. $2x^3 - 5x^2 + 6x - 15$.
33. $x^3 - x^2 - 56x$.
34. $2x^2 + 4x - 96$.
35. $3x^2 + 10x - 8$.
36. $2x^2 + 9x - 35$.
37. $8x^2 - 50$.
38. $15y^2 + 2y - 8$.
39. $2z^3 - 128$.
40. $8y^2 + 6y + 1$.
41. $x^4 - 2x^2 - 8$.
42. $x^4 + 10x^2 + 25$.
43. $x^3(x - 6)^2 + x^4(x - 6)$.
44. $(x + 4)^3(x + 6)^4 + (x + 4)^4(x + 6)^3$.
45. $2x^3 + 8x^2 - 3x - 12$.
46. $8x^6 + 27$.
47. $x^2 - y^2 + x + y$.
48. $x^8 - y^8$.
49. $9z^2 - 4x^2 + 4x - 1$.
50. $x^3 - x^2 + x - 1$.

*In Problems **51–80**, perform the indicated operations and simplify your answers.*

51. $\dfrac{2}{x} + \dfrac{4}{x-6}$.

52. $\dfrac{x-7}{x+4} - \dfrac{6}{2x+8}$.

53. $\dfrac{x^2-64}{x^3} \cdot \dfrac{x^2}{2x+16}$.

54. $\dfrac{x-2}{4} \cdot \dfrac{8x+4}{x^2+2x-8}$.

55. $\dfrac{x+2}{\dfrac{2x+4}{3}}$.

56. $\dfrac{\dfrac{x^3}{x-1}}{x^5}$.

57. $\dfrac{3}{x-2} - \dfrac{x+2}{x-3}$.

58. $\dfrac{2}{3x} + \dfrac{3}{x} - \dfrac{4}{5x}$.

59. $\dfrac{9x}{x^2+2x+1} \cdot \dfrac{(x+1)^3}{-3}$.

60. $\dfrac{-(x+3)}{x^2+x} \cdot \dfrac{x}{-(x^2-9)}$.

61. $2 + \dfrac{x}{x-1} - \dfrac{x-1}{x^2-1}$.

62. $\dfrac{2}{x-y} + \dfrac{2}{y-x}$.

63. $\dfrac{x^2+5x+6}{x^2-2x-8} \cdot \dfrac{x^2-16}{x^2+7x+12}$.

64. $\dfrac{4x^2+4x+1}{x^2-2x-3} \cdot \dfrac{x^2+2x+1}{2x^2+3x+1}$.

65. $\dfrac{\dfrac{8-4x}{2x}}{\dfrac{x^2-4x+4}{x^2-2x}}$.

66. $\dfrac{\dfrac{4x^2-9}{(x+1)^3}}{\dfrac{4x+6}{(x+1)^2}}$.

67. $\dfrac{x+2}{x^2+4x+4} + \dfrac{x-3}{x+2}$.

68. $\dfrac{6}{y+3} - \dfrac{2y}{y-3} + \dfrac{3}{y^2-9}$.

69. $\dfrac{2x+2}{x^3-x} \div \dfrac{x-1}{x^2}$.

70. $\dfrac{3x^2-12x-15}{x^2+5x+4} \div \dfrac{30-6x}{x+4}$.

71. $\dfrac{x+1}{x^2+x-12} \cdot \dfrac{9-x^2}{x^2+3x+2}$.

72. $\dfrac{4-x^2}{25-x^2} \cdot \dfrac{x-5}{x-2}$.

73. $\dfrac{x+2}{\dfrac{x}{x+1} + \dfrac{4}{x}}$.

74. $\dfrac{\dfrac{1}{x+2} - \dfrac{1}{x-2}}{\dfrac{2}{x-2}}$.

75. $\dfrac{1 - \dfrac{7}{x^2-9}}{\dfrac{x-4}{3-x}}$.

76. $\dfrac{\dfrac{2}{x+3} + 1}{1 - \dfrac{2}{x+4}}$.

77. $(x^{-1} + x)^{-1}$.

78. $\dfrac{x}{y^{-1}} + \left(\dfrac{y}{x^{-1}}\right)^{-2}$.

79. $\dfrac{x^{-2}-4}{x^{-1}+2}$.

80. $\left(\dfrac{x+y^{-1}}{x^{-1}+y}\right)^{-3}$.

In Problems 81–84, solve the equations.

81. $\dfrac{2}{x+5} = \dfrac{4}{x-5}$.

82. $\dfrac{2x}{x-3} - \dfrac{x+1}{x+2} = 1$.

83. $\dfrac{x+2}{x-5} - \dfrac{7}{x-5} = 0$.

84. $\dfrac{5}{p} - \dfrac{2}{3p} = 6$.

In Problems 85–88, use the given formulas to express the given symbols in terms of the remaining symbols.

85. $n - 1 = C + \dfrac{C'}{\lambda^2};\quad C, C'$.

86. $\sigma = \dfrac{n_0 - n_e}{\lambda} L;\quad n_0, n_e$.

87. $y = \dfrac{(B+D)\lambda}{2B(N-1)\alpha};\quad B$.

88. $P = \dfrac{E^2}{R+r} - \dfrac{E^2 r}{(R+r)^2};\quad E^2$.

In Problems 89 and 90, solve for x.

89. $a(x+2) = b(x+3)$.

90. $ax + b = c - \dfrac{bx}{2}$.

91. The image distance for a simple lens is given by the expression

$$\frac{fp}{p-f},$$

where p is the object distance and f is the focal length.
(a) Simplify this expression, if possible.
(b) Simplify the reciprocal of this expression, if possible.

92. When a flexible chain l units long and of linear density w is released with a length c overhanging a smooth table, energy considerations result in the following expression for the square of the velocity of the chain as it leaves the table:

$$\frac{2g}{wl}\left[wc(l-c) + \frac{1}{2} w(l-c)^2 \right].$$

Show that this can be expressed in completely factored form as

$$\frac{g}{l}(l+c)(l-c).$$

93. According to the Bohr theory of the hydrogen atom, the frequency of the energy radiated when an electron goes from orbit n_1 to orbit n_2 is given by the expression

$$\frac{2\pi^2 m e^4}{h^3 n_2^2} - \frac{2\pi^2 m e^4}{h^3 n_1^2},$$

where m is the mass of the electron, e is the charge of the electron, and h is Planck's constant. Express this frequency in factored form. (Do not combine any fractions.)

94. Under certain conditions, the force that the rear wheels of an automobile of weight W must exert on the ground to cause an acceleration a is given by

$$\frac{\mu W d}{c+d} + \frac{\mu W h a}{gc + gd},$$

where μ (mu) is the coefficient of friction, g is the acceleration due to gravity, and c, d, and h are constants that specify the location of the center of gravity of the automobile. Express this force in factored form. (Do not combine any fractions.)

95. An expression for the energy stored in the magnetic field of an inductor is

$$\left(\frac{it}{2}\right)\left(\frac{Li}{t}\right).$$

Simplify this expression.

96. The static displacement of an object connected to a certain complex system of springs is given by

$$\frac{W}{k} + \frac{1}{2}\left(\frac{W}{2k} + \frac{W}{3k}\right).$$

Write this expression as a single fraction.

97. The equivalent resistance R of three resistances R_1, R_2, and R_3 connected in parallel is given by

$$R = \frac{1}{\dfrac{1}{R_1} + \dfrac{1}{R_2} + \dfrac{1}{R_3}}.$$

Simplify this complex fraction.

98. The Doppler effect for light can be used to determine the speeds of heavenly bodies that are moving toward or receding from the earth. The frequency f' received from such a body is given by

$$f' = f - \frac{fv}{c} + \frac{f}{2}\left(\frac{v}{c}\right)^2,$$

where f is the emitted frequency, v is the speed of the body, and c is the speed of light. Write the expression for f' as a single fraction.

99. In studies of photosynthesis, the formula

$$P = \frac{bE}{1 + aE}$$

occurs. Solve for a.

100. The total capacitance C_T of two capacitors C_1 and C_2 connected in series is given by

$$\frac{1}{C_T} = \frac{1}{C_1} + \frac{1}{C_2}.$$

Show that $C_T = \dfrac{C_1 C_2}{C_1 + C_2}$.

Operations with Radicals

11.1 CHANGING THE FORM OF A RADICAL

In Chapter 2 we discussed the following basic rules for radicals (where a, b, $\sqrt[n]{a}$, and $\sqrt[n]{b}$ are real numbers):

1. $(\sqrt[n]{a})^n = \sqrt[n]{a^n} = a$.

2. $\sqrt[n]{ab} = \sqrt[n]{a}\ \sqrt[n]{b}$.

3. $\sqrt[n]{\dfrac{a}{b}} = \dfrac{\sqrt[n]{a}}{\sqrt[n]{b}}$.

4. $a^{1/n} = \sqrt[n]{a}$.

5. $a^{m/n} = (\sqrt[n]{a})^m = \sqrt[n]{a^m}$.

We can take a radical with index n and rewrite it as an expression that does not have a radical, provided the radicand is a perfect nth power of an expression. For example, just as the rule $\sqrt[n]{a^n} = a$ allows us to write $\sqrt[3]{x^3} = x$, we have

$$\sqrt[3]{x^6 y^9} = \sqrt[3]{(x^2 y^3)^3} = x^2 y^3.$$

407

We were able to do this because the exponents 6 and 9 in $x^6 y^9$ are multiples of the index 3. This guarantees that $x^6 y^9$ is a perfect cube. Other examples of applying Rule 1 are given in Example 1.

Example 1

a. $\sqrt{x^6 y^8} = \sqrt{(x^3 y^4)^2} = x^3 y^4.$

b. $\sqrt[4]{\dfrac{x^{16}}{y^8}} = \sqrt[4]{\left(\dfrac{x^4}{y^2}\right)^4} = \dfrac{x^4}{y^2}.$

c. $\sqrt[5]{32 x^5 y^{15}} = \sqrt[5]{(2xy^3)^5} = 2xy^3.$

One way to change the form of a radical with index n is to remove from the radicand all factors whose nth roots can easily be found. The rule $\sqrt[n]{ab} = \sqrt[n]{a}\,\sqrt[n]{b}$ is used. For example,

$$\sqrt{50} = \sqrt{25 \cdot 2} = \sqrt{5^2}\,\sqrt{2} = 5\sqrt{2}.$$

In general, radicals should be expressed in a form such that the exponent of any factor in the radicand is less than the index of the radical. For example, in $\sqrt[3]{x^7 y^3}$ the exponents 7 and 3 are not less than the index 3. Therefore, we can write the radicand as a product of two factors such that one of the factors is a perfect cube.

$$\sqrt[3]{x^7 y^3} = \sqrt[3]{(x^6 y^3)x} = \sqrt[3]{(x^2 y)^3}\,\sqrt[3]{x} = x^2 y \sqrt[3]{x}.$$

Example 2

Remove as many factors as possible from the radicand.

a. $\sqrt{8} = \sqrt{4 \cdot 2} = \sqrt{2^2}\,\sqrt{2} = 2\sqrt{2}.$

b. $\sqrt[4]{48} = \sqrt[4]{16 \cdot 3} = \sqrt[4]{2^4}\,\sqrt[4]{3} = 2\sqrt[4]{3}.$

c. $\sqrt[3]{x^3 y} = \sqrt[3]{x^3}\,\sqrt[3]{y} = x\sqrt[3]{y}.$

d. $\sqrt{25 x^7} = \sqrt{(25 x^6)(x)} = \sqrt{25 x^6}\,\sqrt{x} = 5x^3\sqrt{x}.$

e. $\sqrt[5]{64 x^6 y^{14} z^2} = \sqrt[5]{(32 x^5 y^{10})(2 x y^4 z^2)} = 2xy^2\sqrt[5]{2 x y^4 z^2}$, because $\sqrt[5]{32 x^5 y^{10}}$ is $2xy^2$.

f. $\sqrt{x^2 + 2x + 1} = \sqrt{(x+1)^2} = x + 1$, provided that $x + 1 \geq 0$. Otherwise we must write $\sqrt{(x+1)^2} = |x+1|$. (Recall from Sec. 2.5 that $\sqrt{x^2} = |x|$.)

Example 3

The frequency of vibration of a string of length L, fixed at both ends and vibrating in its fundamental mode, is given by

$$f = \frac{1}{2L}\sqrt{\frac{T}{\mu}},$$

where f is frequency, μ (mu) is mass per unit length, and T is the tension in the string. If the tension in the string is quadrupled, what happens to the frequency?

Solution   Let f_0 denote the frequency when the tension is quadrupled, that is, when it is $4T$. Then

$$f_0 = \frac{1}{2L}\sqrt{\frac{4T}{\mu}} = \frac{1}{2L}(2)\sqrt{\frac{T}{\mu}} = 2\left(\frac{1}{2L}\sqrt{\frac{T}{\mu}}\right) = 2f.$$

Thus the frequency is doubled.

When a radicand is a fraction, we may use the previous rule $\sqrt[n]{\dfrac{a}{b}} = \dfrac{\sqrt[n]{a}}{\sqrt[n]{b}}$ to get an equivalent expression in which the radicand is not a fraction. For example,

$$\sqrt{\frac{7}{16}} = \frac{\sqrt{7}}{\sqrt{16}} = \frac{\sqrt{7}}{4}.$$

In the answer the radicand does not contain a fraction because 16 is a perfect square $(16 = 4^2)$ and $\sqrt{4^2} = 4$. When an expression involves radicals, we use the phrase **rationalizing the denominator** to refer to the process of rewriting the expression so that *no fraction appears in a radicand* and *no radical appears in the denominator of a fraction.** Thus we have just rationalized the denominator of $\sqrt{\frac{7}{16}}$.

Sometimes the denominator of a radical of index 2 is not a perfect square. To rationalize the denominator, we multiply the numerator and denominator of the radicand by a number that makes the denominator a perfect square. For example, to rationalize

$$\sqrt{\frac{3}{7}}$$

we change the denominator to 7^2 by multiplying the numerator and denominator of the fraction $\frac{3}{7}$ by 7 (using the fundamental principle of fractions).

$$\sqrt{\frac{3}{7}} = \sqrt{\frac{3\cdot 7}{7\cdot 7}} = \sqrt{\frac{21}{7^2}} = \frac{\sqrt{21}}{\sqrt{7^2}} = \frac{\sqrt{21}}{7}.$$

* This process was discussed in Sec. 2.4.

The denominator is now rationalized. Actually, multiplying the numerator and denominator by 7 is equivalent to multiplying $\frac{3}{7}$ by $\frac{7}{7}$. That is,

$$\sqrt{\frac{3}{7}} = \sqrt{\frac{3}{7}\cdot\frac{7}{7}} = \sqrt{\frac{21}{7^2}} = \frac{\sqrt{21}}{\sqrt{7^2}} = \frac{\sqrt{21}}{7}.$$

Looking at this problem yet another way, we can multiply $\sqrt{\frac{3}{7}}$ by $\sqrt{7}/\sqrt{7}$. Here are the steps.

$$\sqrt{\frac{3}{7}} = \frac{\sqrt{3}}{\sqrt{7}} \qquad \text{[Rule 3]}$$

$$= \frac{\sqrt{3}}{\sqrt{7}}\cdot\frac{\sqrt{7}}{\sqrt{7}} \qquad \text{[multiplying by 1]}$$

$$= \frac{\sqrt{21}}{\sqrt{7^2}} \qquad \text{[Rule 2]}$$

$$= \frac{\sqrt{21}}{7} \qquad \text{[Rule 1]}.$$

To rationalize the denominator of $\dfrac{2}{\sqrt[3]{5}}$, we want the radicand to be a perfect cube. This will be the case if we multiply the numerator and denominator by $\sqrt[3]{5^2}$, because $\sqrt[3]{5}\,\sqrt[3]{5^2} = \sqrt[3]{5^3}$. Thus

$$\frac{2}{\sqrt[3]{5}} = \frac{2}{\sqrt[3]{5}}\cdot\frac{\sqrt[3]{5^2}}{\sqrt[3]{5^2}} = \frac{2\sqrt[3]{5^2}}{\sqrt[3]{5^3}} = \frac{2\sqrt[3]{25}}{5}.$$

Example 4

Rationalize the denominator.

a. $\sqrt[4]{\dfrac{y}{x^8}} = \dfrac{\sqrt[4]{y}}{\sqrt[4]{(x^2)^4}} = \dfrac{\sqrt[4]{y}}{x^2}.$

b. $\sqrt{\dfrac{21}{x}} = \dfrac{\sqrt{21}}{\sqrt{x}} = \dfrac{\sqrt{21}}{\sqrt{x}}\cdot\dfrac{\sqrt{x}}{\sqrt{x}} = \dfrac{\sqrt{21x}}{(\sqrt{x})^2} = \dfrac{\sqrt{21x}}{x}.$

c. $\sqrt[5]{\dfrac{x}{y^2}} = \dfrac{\sqrt[5]{x}}{\sqrt[5]{y^2}}.$

Because the index is 5, we shall change the denominator to $\sqrt[5]{y^5}$ by multiplying the numerator and denominator by $\sqrt[5]{y^3}$

$$\sqrt[5]{\frac{x}{y^2}} = \frac{\sqrt[5]{x}}{\sqrt[5]{y^2}} = \frac{\sqrt[5]{x}}{\sqrt[5]{y^2}}\cdot\frac{\sqrt[5]{y^3}}{\sqrt[5]{y^3}} = \frac{\sqrt[5]{xy^3}}{\sqrt[5]{y^5}} = \frac{\sqrt[5]{xy^3}}{y}.$$

d. $\sqrt[3]{\dfrac{2}{3x^4 y^2}} = \dfrac{\sqrt[3]{2}}{\sqrt[3]{3x^4 y^2}}.$

Because the index is 3, we want each factor in the denominator to be a perfect cube of a number. That is, the exponent of each such factor must be a 3 or a multiple of 3. This will be the case if we multiply the 3 by 3^2, the x^4 by x^2, and the y^2 by y. Thus we multiply both numerator and denominator by $\sqrt[3]{3^2 x^2 y}$.

$$\sqrt[3]{\dfrac{2}{3x^4 y^2}} = \dfrac{\sqrt[3]{2}}{\sqrt[3]{3x^4 y^2}} \cdot \dfrac{\sqrt[3]{3^2 x^2 y}}{\sqrt[3]{3^2 x^2 y}} = \dfrac{\sqrt[3]{18x^2 y}}{\sqrt[3]{3^3 x^6 y^3}} = \dfrac{\sqrt[3]{18x^2 y}}{3x^2 y}.$$

Example 5

Rationalize the denominator.

a. $\dfrac{3}{\sqrt{2x}} = \dfrac{3}{\sqrt{2x}} \cdot \dfrac{\sqrt{2x}}{\sqrt{2x}} = \dfrac{3\sqrt{2x}}{\sqrt{2^2 x^2}} = \dfrac{3\sqrt{2x}}{2x}.$

b. $\dfrac{6}{\sqrt[4]{2}} = \dfrac{6}{\sqrt[4]{2}} \cdot \dfrac{\sqrt[4]{2^3}}{\sqrt[4]{2^3}} = \dfrac{6\sqrt[4]{8}}{\sqrt[4]{2^4}} = \dfrac{6\sqrt[4]{8}}{2} = 3\sqrt[4]{8}.$

c. $\dfrac{7xy^2}{2\sqrt[5]{3x^3}} = \dfrac{7xy^2}{2\sqrt[5]{3x^3}} \cdot \dfrac{\sqrt[5]{3^4 x^2}}{\sqrt[5]{3^4 x^2}} = \dfrac{7xy^2 \sqrt[5]{81x^2}}{2\sqrt[5]{3^5 x^5}} = \dfrac{7xy^2 \sqrt[5]{81x^2}}{2(3x)} = \dfrac{7y^2 \sqrt[5]{81x^2}}{6}.$

We can obtain another rule for radicals that involves a root of a root of a number. Because $\sqrt[m]{\sqrt[n]{a}} = (a^{1/n})^{1/m} = a^{1/(mn)} = \sqrt[mn]{a}$, we have the following rule:

$$\boxed{6. \quad \sqrt[m]{\sqrt[n]{a}} = \sqrt[mn]{a}.}$$

Observe that the final index is the product of the original indices. For example,

$$\sqrt[3]{\sqrt{64}} = \sqrt[6]{64} = 2.$$

Alternatively, without using Rule 6, we have

$$\sqrt[3]{\sqrt{64}} = \sqrt[3]{8} = 2.$$

Example 6

a. $\sqrt[3]{\sqrt[4]{2}} = \sqrt[12]{2}.$

b. $\sqrt{\sqrt[4]{x}} = \sqrt[8]{x}.$

c. $\sqrt[4]{81} = \sqrt{\sqrt{81}} = \sqrt{9} = 3.$

Sometimes it is possible to **reduce the index** of a radical, as the following shows.

$$\sqrt[6]{x^3} = x^{3/6} = x^{1/2} = \sqrt{x}.$$

Here the index 6 was reduced to 2. The rule $\sqrt[mn]{a} = \sqrt[m]{\sqrt[n]{a}}$ can also be used for this type of problem. Since $6 = 2 \cdot 3$ we have

$$\sqrt[6]{x^3} = \sqrt{\sqrt[3]{x^3}} = \sqrt{x}.$$

Example 7

Reduce the index.

a. $\sqrt[4]{25} = \sqrt[4]{5^2} = 5^{2/4} = 5^{1/2} = \sqrt{5}$ or, alternatively,

$$\sqrt[4]{25} = \sqrt{\sqrt{25}} = \sqrt{5}.$$

b. $\sqrt[6]{16x^2}$.

Solution   We cannot conveniently take the sixth root of the radicand, but we can take the square root. Because $6 = 3 \cdot 2$, we have

$$\sqrt[6]{16x^2} = \sqrt[3]{\sqrt{16x^2}} = \sqrt[3]{4x}.$$

Another approach is

$$\sqrt[6]{16x^2} = \sqrt[6]{(4x)^2} = (4x)^{2/6} = (4x)^{1/3} = \sqrt[3]{4x}.$$

c. $\sqrt[12]{8x^6y^9}$

Solution   Here the radicand is the cube $(2x^2y^3)^3$ and 3 is a factor of the index 12. Thus we have

$$\sqrt[12]{8x^6y^9} = \sqrt[4]{\sqrt[3]{8x^6y^9}} = \sqrt[4]{2x^2y^3}.$$

An expression that contains radicals is said to be in **simplest radical form** when all the following conditions are met:

1. *As many factors as possible are removed from all radicands (Example 2).*
2. *All denominators are rationalized (Examples 4 and 5).*
3. *The index of each radical cannot be reduced (Example 7).*

Example 8

Express in simplest radical form.

a. $\sqrt{\dfrac{x^5}{z}} = \dfrac{\sqrt{x^5}}{\sqrt{z}} = \dfrac{\sqrt{x^5}}{\sqrt{z}} \cdot \dfrac{\sqrt{z}}{\sqrt{z}} = \dfrac{\sqrt{x^5 z}}{\sqrt{z^2}} = \dfrac{\sqrt{x^5 z}}{z} = \dfrac{\sqrt{x^4(xz)}}{z} = \dfrac{x^2\sqrt{xz}}{z}.$

b. $\sqrt[4]{x^6 y^{10}} = \sqrt[4]{x^4 y^8 (x^2 y^2)} = xy^2 \sqrt[4]{x^2 y^2} = xy^2 \sqrt[4]{(xy)^2}$

$= xy^2 (xy)^{2/4} = xy^2 (xy)^{1/2} = xy^2 \sqrt{xy}.$

c. $\sqrt[6]{\dfrac{x^3}{y^9}} = \dfrac{\sqrt[6]{x^3}}{\sqrt[6]{y^9}} = \dfrac{\sqrt[6]{x^3}}{\sqrt[6]{y^9}} \cdot \dfrac{\sqrt[6]{y^3}}{\sqrt[6]{y^3}} = \dfrac{\sqrt[6]{x^3 y^3}}{\sqrt[6]{y^{12}}} = \dfrac{\sqrt[6]{(xy)^3}}{y^2} = \dfrac{\sqrt{xy}}{y^2}.$

d. $\sqrt[3]{x^{-6} y^6} = \sqrt[3]{\dfrac{y^6}{x^6}} = \sqrt[3]{\left(\dfrac{y^2}{x^2}\right)^3} = \dfrac{y^2}{x^2}.$

Problem Set 11.1

In Problems **1–14**, *find the root.*

1. $\sqrt[4]{x^8}.$

2. $\sqrt[3]{x^9}.$

3. $\sqrt[3]{8x^{12}}.$

4. $\sqrt[6]{9^{12}}.$

5. $\sqrt{9x^{16}y^{18}}.$

6. $\sqrt[3]{x^3 y^3 z^6}.$

7. $\sqrt[3]{x^3 y^6 z^9}.$

8. $\sqrt{(x^3 y^7)(x^7 y^{13})}.$

9. $\sqrt[5]{\dfrac{x^{15}}{y^{20}}}.$

10. $\sqrt[4]{\dfrac{16x^8}{y^{16}}}.$

11. $\sqrt{\sqrt{x^8}}.$

12. $\sqrt[3]{\sqrt{x^{18}}}.$

13. $\sqrt[4]{\sqrt[3]{x^{12}}}.$

14. $\sqrt{\sqrt[6]{x^{24}}}.$

In Problems **15–72**, *express in simplest radical form.*

15. $\sqrt{12}.$

16. $\sqrt{18}.$

17. $\sqrt{32}.$

18. $\sqrt{20}.$

19. $\sqrt[3]{16}.$

20. $\sqrt[3]{54}.$

21. $\sqrt{x^7}.$

22. $\sqrt[3]{x^7}.$

23. $\sqrt[3]{24x^6}.$

24. $\sqrt{8x^3}.$

25. $\sqrt[4]{x^9 y^2}.$

26. $\sqrt[5]{x^{17} y^{20}}.$

27. $\sqrt[3]{x^6 yz^4}.$

28. $\sqrt{x^5 y^4 z}.$

29. $\sqrt[3]{8a^3 y^5}.$

30. $\sqrt[3]{24(a+b)^7}.$

31. $\sqrt[5]{x^{23} y^{10} z^6}.$

32. $\sqrt[4]{x^3 y^3 z^7}.$

33. $\sqrt{81xy^2 z^3 w^4}.$

34. $\sqrt[4]{32x^{17}}.$

35. $\sqrt{\dfrac{1}{2}}.$

36. $\sqrt{\dfrac{1}{5}}.$

37. $\sqrt[3]{\dfrac{2}{5}}.$

38. $\sqrt[3]{\dfrac{1}{3}}.$

39. $\sqrt[3]{\dfrac{x^2}{y^3}}.$

40. $\sqrt[4]{\dfrac{3}{x^4}}.$

41. $\sqrt{\dfrac{x}{y^4}}.$

42. $\sqrt[3]{\dfrac{y^2}{x^{12}}}.$

43. $\sqrt{\dfrac{2x}{y}}.$

44. $\sqrt{\dfrac{y}{x^9}}.$

45. $\sqrt[3]{\dfrac{2}{xy^2}}.$

46. $\sqrt[3]{\dfrac{2y}{xz}}.$

47. $\sqrt[4]{\dfrac{3}{2x^7yz^2}}.$

48. $\sqrt[4]{\dfrac{1}{x^3yz^5}}.$

49. $\sqrt[6]{x^2}.$

50. $\sqrt[8]{(xy)^4}.$

51. $\sqrt[4]{9}.$

52. $\sqrt[6]{8}.$

53. $\sqrt[4]{16x^4y^2}.$

54. $\sqrt[6]{27x^3y^3z^3}.$

55. $\sqrt[8]{\dfrac{x^4}{y^4}}.$

56. $\sqrt[6]{\dfrac{16x^2}{y^2}}.$

57. $\sqrt[12]{x^2y^2z^{10}}.$

58. $\sqrt[15]{x^3y^{15}z^6}.$

59. $\sqrt[4]{x^{10}y^2}.$

60. $\sqrt[12]{x^{15}y^{27}}.$

61. $\sqrt[6]{x^{20}x^{26}}.$

62. $\sqrt[8]{x^2z^{10}}.$

63. $\sqrt[3]{\sqrt{x^{12}y^5w^{25}}}.$

64. $\sqrt{\sqrt[3]{64x^{12}y^{11}w^7}}.$

65. $\sqrt[4]{\dfrac{16x^5}{y^8}}.$

66. $\sqrt[3]{\dfrac{x^6}{y^{-3}}}.$

67. $\sqrt[4]{\dfrac{1}{4}}.$

68. $\sqrt[6]{\dfrac{x^9}{y^{15}}}.$

69. $\sqrt[8]{\dfrac{x^4}{y^{12}}}.$

70. $\sqrt{x^4(2+x)}.$

71. $\sqrt[3]{\sqrt{x^3}}.$

72. $\sqrt[4]{\sqrt[3]{\sqrt{x^{48}}}}.$

In Problems **73–84**, *rationalize the denominators and simplify.*

73. $\dfrac{3}{\sqrt{7}}.$

74. $\dfrac{5}{\sqrt{11}}.$

75. $\dfrac{4}{\sqrt{2x}}.$

76. $\dfrac{y}{\sqrt{2y}}.$

77. $\dfrac{1}{\sqrt[3]{2}}.$

78. $\dfrac{3}{\sqrt[4]{2}}.$

79. $\dfrac{1}{\sqrt[3]{3x}}.$

80. $\dfrac{4}{3\sqrt[3]{x^2}}.$

81. $\dfrac{2xy}{3\sqrt[5]{xy^3z^6}}.$

82. $\dfrac{-3y}{x\sqrt{x^3y}}.$

83. $\dfrac{4ab^2}{\sqrt[4]{2ab^3}}.$

84. $\dfrac{3a^2b^2c}{\sqrt[3]{9abc^2}}.$

85. When a solid sphere of radius r and mass m is released from the top of an inclined plane of vertical height h, its speed v at the bottom, if the sphere rolls without slipping, can be shown to be given by

$$v = \sqrt{\dfrac{10gh}{7}}.$$

Express v in rationalized form.

86. When the mass m shown in Fig. 11.1 is released from rest and falls through a distance h, the wheel can be shown to have an angular speed ω, given by

$$\omega = \sqrt{\frac{2mgh}{mr^2 + I}},$$

where I is the moment of inertia of the wheel and g is the acceleration due to gravity. Express ω with a rationalized denominator.

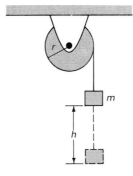

FIGURE 11.1

87. In the formula of Example 3, if the mass per unit length is quadrupled, what happens to the frequency?

88. The grating spacing, the distance between ions, of a sodium chloride crystal can be shown to be given by the expression

$$\sqrt[3]{\frac{M}{2\rho N_0}}.$$

Express this distance in rationalized form.

89. If a circular ring of radius r has a charge per unit length of λ, it can be shown that the electric field intensity E at a point a distance y from the center of the ring on an axis perpendicular to the plane of the ring (Fig. 11.2), is given by

$$E = \frac{\lambda r y}{2\epsilon_0 \sqrt{(r^2 + y^2)^3}},$$

where ϵ_0 is a constant. Express E in rationalized form.

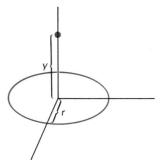

FIGURE 11.2

11.2 ADDITION AND SUBTRACTION OF RADICALS

We can add or subtract radicals that have the *same index* and the *same radicand*. Such radicals are called **like radicals**. The distributive law is used just as it was in combining like terms in Chapter 3. For example,

$$8\sqrt{xy} - 4\sqrt{xy} + 7\sqrt{xy} = (8 - 4 + 7)\sqrt{xy} = 11\sqrt{xy}.$$

Sometimes, radicals that are unlike can be combined by simplifying them first. For example, the radicands in $\sqrt[3]{81} - \sqrt[3]{24}$ are different. But we can write

$$\sqrt[3]{81} - \sqrt[3]{24} = \sqrt[3]{27 \cdot 3} - \sqrt[3]{8 \cdot 3}$$
$$= 3\sqrt[3]{3} - 2\sqrt[3]{3} = \sqrt[3]{3}.$$

We caution you: *Do not hastily add unlike radicals.* Students frequently make this error. For example,

$$\sqrt{16} + \sqrt{9} \neq \sqrt{16 + 9} = \sqrt{25} = 5,$$

but

$$\sqrt{16} + \sqrt{9} = 4 + 3 = 7.$$

Example 1

Perform the indicated operations.

a. $(16\sqrt[3]{x} + 15) - (12 - 2\sqrt[3]{x}) = 16\sqrt[3]{x} + 15 - 12 + 2\sqrt[3]{x}$
$$= 18\sqrt[3]{x} + 3.$$

b. $3\sqrt{75} - 2\sqrt{12} + \sqrt{7} = 3\sqrt{25 \cdot 3} - 2\sqrt{4 \cdot 3} + \sqrt{7}$
$$= 3(5\sqrt{3}) - 2(2\sqrt{3}) + \sqrt{7}$$
$$= 15\sqrt{3} - 4\sqrt{3} + \sqrt{7} = 11\sqrt{3} + \sqrt{7}.$$

c. $\sqrt[3]{\dfrac{3}{4}} + \sqrt[3]{\dfrac{2}{9}} - \sqrt[3]{\dfrac{1}{36}} = \sqrt[3]{\dfrac{3}{4} \cdot \dfrac{2}{2}} + \sqrt[3]{\dfrac{2}{9} \cdot \dfrac{3}{3}} - \sqrt[3]{\dfrac{1}{6^2} \cdot \dfrac{6}{6}}$

$$= \dfrac{\sqrt[3]{6}}{2} + \dfrac{\sqrt[3]{6}}{3} - \dfrac{\sqrt[3]{6}}{6}$$

$$= \left(\dfrac{1}{2} + \dfrac{1}{3} - \dfrac{1}{6}\right)\sqrt[3]{6} = \dfrac{2}{3}\sqrt[3]{6}.$$

Problem Set 11.2

*In Problems **1–20**, perform the indicated operations and simplify.*

1. $4\sqrt[3]{3} - 2\sqrt[3]{3} + \sqrt[3]{3}.$

2. $6\sqrt{7} - (2\sqrt{7} + 3\sqrt{7}).$

3. $x^2\sqrt{2x} - 3x^2\sqrt{2x} + 4x^2\sqrt{2x}.$

4. $xy\sqrt{x} + 4xy\sqrt{x} - 2xy\sqrt{x}.$

5. $3\sqrt{75} - 2\sqrt{12}.$

6. $5\sqrt{18} - (\sqrt{2} + 1).$

7. $5\sqrt{8} - (2\sqrt{18} - 4\sqrt{32}).$

8. $\sqrt{75} - (\sqrt{27} - 2\sqrt{3}).$

9. $2y\sqrt{16x} - 3y\sqrt{9x}.$

10. $3\sqrt[3]{16} - \sqrt[3]{54}.$

11. $\sqrt[3]{128} - 6\sqrt[3]{16}.$

12. $2(\sqrt[3]{54} - 2\sqrt[3]{128}) - 2\sqrt[3]{16}.$

13. $30\sqrt{\dfrac{1}{15}} - 72\sqrt{\dfrac{5}{12}} + 50\sqrt{\dfrac{3}{5}}.$

14. $20\sqrt{\dfrac{2}{5}} - 3\sqrt{40} - 4\sqrt{\dfrac{5}{2}}.$

15. $4\sqrt[3]{\dfrac{1}{2}} + \sqrt[3]{32} - 3\sqrt[3]{4}.$

16. $\sqrt{\dfrac{1}{6}} + \sqrt{\dfrac{2}{3}} + \sqrt{\dfrac{3}{2}}.$

17. $\sqrt[4]{4x^4} - 3\sqrt[6]{8x^6}.$

18. $\sqrt[4]{4x^2} - 3\sqrt[6]{8x^3}.$

19. $5x\sqrt[3]{3000y} - 2x\sqrt[3]{81y} - \sqrt[3]{24x^3y}.$

20. $4x^2\sqrt{27y} - (\sqrt{75x^4y} - 2x^2\sqrt{3y}) + 3x^2\sqrt{12}.$

11.3 MULTIPLICATION OF RADICALS

Radicals having the *same index* can be multiplied by means of the previous rule, $\sqrt[n]{a} \cdot \sqrt[n]{b} = \sqrt[n]{ab}$. For example,

$$\sqrt{2x^3} \cdot \sqrt{8x} = \sqrt{(2x^3)(8x)} = \sqrt{16x^4} = 4x^2.$$

Example 1

Determine the given products and simplify.

a. $\sqrt{yz} \cdot \sqrt{3z} = \sqrt{3yz^2} = z\sqrt{3y}.$

b. $\sqrt[4]{\dfrac{3}{2}} \cdot \sqrt[4]{6} = \sqrt[4]{\dfrac{3}{2} \cdot 6} = \sqrt[4]{9} = \sqrt[4]{3^2} = 3^{2/4} = 3^{1/2} = \sqrt{3}.$

c. $\sqrt{3}(\sqrt{3} - \sqrt{6}) = \sqrt{3}(\sqrt{3}) - \sqrt{3}(\sqrt{6})$ [distributive law]

$$= 3 - \sqrt{18} = 3 - \sqrt{9 \cdot 2}$$

$$= 3 - 3\sqrt{2}.$$

d. $(x\sqrt[3]{x^2y^3})^4 = x^4(\sqrt[3]{x^2y^3})^4$ [using $(ab)^n = a^nb^n$]

$\qquad\qquad = x^4\sqrt[3]{(x^2y^3)^4}$ [using $(\sqrt[n]{a})^m = \sqrt[n]{a^m}$]

$\qquad\qquad = x^4\sqrt[3]{x^8y^{12}}$ [using $(ab)^n = a^nb^n$]

$\qquad\qquad = x^4(x^2y^4\sqrt[3]{x^2})$ [simplifying]

$\qquad\qquad = x^6y^4\sqrt[3]{x^2}.$

Example 2

Determine the given products and simplify.

a. $(\sqrt{8} - \sqrt{3})(\sqrt{18} - \sqrt{48})$.

Solution   Simplifying the radicals and multiplying, we have

$(2\sqrt{2} - \sqrt{3})(3\sqrt{2} - 4\sqrt{3})$

$\qquad = 2\sqrt{2}\cdot3\sqrt{2} - 2\sqrt{2}\cdot4\sqrt{3} - \sqrt{3}\cdot3\sqrt{2} + \sqrt{3}\cdot4\sqrt{3}$ [multiplying binomials]

$\qquad = 6(\sqrt{2})^2 - 8\sqrt{6} - 3\sqrt{6} + 4(\sqrt{3})^2$

$\qquad = 12 - 11\sqrt{6} + 12$

$\qquad = 24 - 11\sqrt{6}.$

b. $(\sqrt{5} - 2\sqrt{7})(\sqrt{5} + 2\sqrt{7})$.

Solution   This is a special product of the form $(a - b)(a + b)$. Thus the product is a difference of squares.

$$(\sqrt{5} - 2\sqrt{7})(\sqrt{5} + 2\sqrt{7}) = (\sqrt{5})^2 - (2\sqrt{7})^2$$
$$= 5 - 4(7) = 5 - 28 = -23.$$

c. $(\sqrt{x} - \sqrt{y})^2$.

Solution   This is the square of a binomial.

$$(\sqrt{x} - \sqrt{y})^2 = (\sqrt{x})^2 - 2\sqrt{x}\sqrt{y} + (\sqrt{y})^2$$
$$= x - 2\sqrt{xy} + y.$$

To find the product of two radicals with different indices, we first rewrite them so that they do have the same index. We make use of fractional exponents, as Example 3 shows.

Example 3

Determine the given products and simplify.

a. $\sqrt{2} \cdot \sqrt[3]{3}$.

Solution We first express each radical in terms of fractional exponents.

$$\sqrt{2} \cdot \sqrt[3]{3} = 2^{1/2} \cdot 3^{1/3}.$$

Next we rewrite the exponents so that their denominators are the same. Because $\frac{1}{2} = \frac{3}{6}$ and $\frac{1}{3} = \frac{2}{6}$, we have

$$2^{1/2} \cdot 3^{1/3} = 2^{3/6} \cdot 3^{2/6}.$$

Going back to radical form and multiplying gives

$$2^{3/6} \cdot 3^{2/6} = \sqrt[6]{2^3} \, \sqrt[6]{3^2} = \sqrt[6]{2^3 \cdot 3^2} = \sqrt[6]{8 \cdot 9} = \sqrt[6]{72}.$$

Summarizing, we can write

$$\sqrt{2} \cdot \sqrt[3]{3} = 2^{1/2} \cdot 3^{1/3} = 2^{3/6} \cdot 3^{2/6} = \sqrt[6]{2^3 \cdot 3^2} = \sqrt[6]{72}.$$

b. $\sqrt[4]{3x^2} \, \sqrt[3]{2x^2} = (3x^2)^{1/4}(2x^2)^{1/3}$

$$= (3x^2)^{3/12}(2x^2)^{4/12} = \sqrt[12]{(3x^2)^3} \cdot \sqrt[12]{(2x^2)^4}$$

$$= \sqrt[12]{(3x^2)^3(2x^2)^4} = \sqrt[12]{3^3 x^6 \cdot 2^4 x^8}$$

$$= \sqrt[12]{432x^{14}} = x \sqrt[12]{432x^2}.$$

Problem Set 11.3

In Problems 1–34, perform the operations and simplify.

1. $\sqrt{3}\,\sqrt{4}$.

2. $\sqrt{9}\,\sqrt{2}$.

3. $(2\sqrt{6})(3\sqrt{3})$.

4. $(5\sqrt{27})(2\sqrt{3})$.

5. $\sqrt[3]{3}\,\sqrt[3]{9}\,\sqrt[3]{12}$.

6. $\sqrt[3]{4}\,\sqrt[3]{16}\,\sqrt[3]{-1}$.

7. $\sqrt{2x}\,\sqrt{x}\,\sqrt{3x}$.

8. $\sqrt{5y}\,\sqrt{2y}\,\sqrt{y}$.

9. $(-\sqrt{3})^2$.

10. $(-\sqrt{5})^3$.

11. $(2\sqrt[3]{x})^4$.

12. $(\frac{1}{3}\sqrt{x})^3$.

13. $\sqrt{30}\,\sqrt{\dfrac{2}{3}}$.

14. $\sqrt{\dfrac{5}{8}}\,\sqrt{\dfrac{16}{3}}$.

15. $\sqrt{3}(2\sqrt{6} - 4\sqrt{3})$.

16. $\sqrt{2}(\sqrt{2} + 2\sqrt{18})$.

17. $(\sqrt{6} + \sqrt{2})(\sqrt{2} - 2\sqrt{6})$.

18. $(\sqrt{3} - 1)(\sqrt{3} + 2)$.

19. $(2 + \sqrt{7})(2 - \sqrt{7})$.

20. $(\sqrt{7} - \sqrt{2})(\sqrt{7} + \sqrt{2})$.

21. $(\sqrt{5} + 2)^2.$ **22.** $(1 - \sqrt{5})^2.$

23. $(\sqrt{x} - 1)(2\sqrt{x} + 5).$ **24.** $\sqrt{ab}\,\sqrt{a^2bc^2}\,\sqrt{abc^2}.$

25. $\sqrt{3xy^2}\,\sqrt{2xy}\,\sqrt{3xy^3}.$ **26.** $5\sqrt[4]{ab}(1 - 2\sqrt[4]{ab}).$

27. $(\sqrt{6} - 5)^2.$ **28.** $(\sqrt{3} + 4)^2.$

29. $(3\sqrt{8} + \sqrt{3})(8\sqrt{3} - \sqrt{8}).$ **30.** $(5\sqrt{2} + \sqrt{5})(2\sqrt{5} - \sqrt{2}).$

31. $(\sqrt{2y} + 3)^2.$ **32.** $(\sqrt{xy} - \sqrt{z})(2\sqrt{x} + \sqrt{y}).$

33. $(\sqrt{2x^2y})^5.$ **34.** $(\sqrt[4]{2x^3y^2z})^5.$

In Problems **35–46,** *perform the operations and simplify.*

35. $\sqrt{5} \cdot \sqrt[4]{5}.$ **36.** $\sqrt[3]{2} \cdot \sqrt[6]{3}.$ **37.** $\sqrt{3x} \cdot \sqrt[3]{x^2}.$

38. $\sqrt[4]{x} \cdot \sqrt[3]{x^2}.$ **39.** $\sqrt[3]{9x} \cdot \sqrt{3x}.$ **40.** $\sqrt{8x^2} \cdot \sqrt[3]{4x^3}.$

41. $(3\sqrt[3]{x^2y})(2\sqrt{2x}).$ **42.** $5\sqrt{ab^2} \cdot \sqrt[3]{ab}.$ **43.** $\sqrt[5]{x^2y^3} \cdot \sqrt[4]{xy^2}.$

44. $\sqrt[6]{xy^5} \cdot \sqrt[3]{x^2y}.$ **45.** $\sqrt{x} \cdot \sqrt[3]{x^2y} \cdot \sqrt[4]{x^4y^2}.$ **46.** $\sqrt{x} \cdot \sqrt[4]{y} \cdot \sqrt[8]{xy}.$

11.4   DIVISION OF RADICALS

Division of radicals with the *same index* can be handled by using the rule $\dfrac{\sqrt[n]{a}}{\sqrt[n]{b}} = \sqrt[n]{\dfrac{a}{b}}$ and simplifying the result. For example,

$$\frac{\sqrt{7}}{\sqrt{3}} = \sqrt{\frac{7}{3}} = \sqrt{\frac{7 \cdot 3}{3 \cdot 3}} = \frac{\sqrt{21}}{\sqrt{3^2}} = \frac{\sqrt{21}}{3}.$$

Alternatively, instead of using the preceding rule, we can rationalize the denominator immediately:

$$\frac{\sqrt{7}}{\sqrt{3}} = \frac{\sqrt{7}}{\sqrt{3}} \cdot \frac{\sqrt{3}}{\sqrt{3}} = \frac{\sqrt{21}}{3}.$$

Example 1

Perform the indicated operations and simplify.

a. $\dfrac{\sqrt[3]{3x}}{\sqrt[3]{2x}} = \sqrt[3]{\dfrac{3x}{2x}} = \sqrt[3]{\dfrac{3}{2}} = \sqrt[3]{\dfrac{3 \cdot 2^2}{2 \cdot 2^2}} = \dfrac{\sqrt[3]{12}}{2}.$

b. $\dfrac{\sqrt[4]{x^2y}}{\sqrt[4]{z}} = \sqrt[4]{\dfrac{x^2y}{z}} = \sqrt[4]{\dfrac{x^2y}{z} \cdot \dfrac{z^3}{z^3}} = \dfrac{\sqrt[4]{x^2yz^3}}{z}.$

c. $\dfrac{\sqrt{5xy}}{\sqrt{18xy^3}} = \dfrac{\sqrt{5xy}}{3y\sqrt{2xy}} = \dfrac{1}{3y}\sqrt{\dfrac{5xy}{2xy}} = \dfrac{1}{3y}\sqrt{\dfrac{5}{2}} = \dfrac{1}{3y}\sqrt{\dfrac{5}{2}\cdot\dfrac{2}{2}} = \dfrac{\sqrt{10}}{6y}.$

Sometimes the denominator of a fraction is the sum or difference of two terms, where one or both of the terms involve square roots, such as $\sqrt{5} + \sqrt{2}$. In such cases the denominator can be rationalized by multiplying the numerator and denominator of the fraction by an expression that makes the denominator a difference of two squares. For example, if the denominator has the form $\sqrt{a} + \sqrt{b}$, the expression we use is $\sqrt{a} - \sqrt{b}$ (and vice versa) because

$$(\sqrt{a} + \sqrt{b})(\sqrt{a} - \sqrt{b}) = (\sqrt{a})^2 - (\sqrt{b})^2 = a - b.$$

To illustrate, we have

$$\dfrac{4}{\sqrt{5} + \sqrt{2}} = \dfrac{4}{\sqrt{5} + \sqrt{2}} \cdot \dfrac{\sqrt{5} - \sqrt{2}}{\sqrt{5} - \sqrt{2}}$$

$$= \dfrac{4(\sqrt{5} - \sqrt{2})}{(\sqrt{5})^2 - (\sqrt{2})^2} = \dfrac{4(\sqrt{5} - \sqrt{2})}{5 - 2}$$

$$= \dfrac{4(\sqrt{5} - \sqrt{2})}{3}.$$

Example 2

Rationalize the denominators.

a. $\dfrac{2}{x - \sqrt{3}} = \dfrac{2}{x - \sqrt{3}} \cdot \dfrac{x + \sqrt{3}}{x + \sqrt{3}} = \dfrac{2(x + \sqrt{3})}{x^2 - (\sqrt{3})^2} = \dfrac{2(x + \sqrt{3})}{x^2 - 3}.$

b. $\dfrac{\sqrt{2}}{\sqrt{2} - \sqrt{3}} = \dfrac{\sqrt{2}}{\sqrt{2} - \sqrt{3}} \cdot \dfrac{\sqrt{2} + \sqrt{3}}{\sqrt{2} + \sqrt{3}}$

$$= \dfrac{\sqrt{2}(\sqrt{2} + \sqrt{3})}{2 - 3}$$

$$= \dfrac{2 + \sqrt{6}}{-1} = -2 - \sqrt{6}.$$

c. $\dfrac{\sqrt{5} - \sqrt{2}}{\sqrt{5} + \sqrt{2}} = \dfrac{\sqrt{5} - \sqrt{2}}{\sqrt{5} + \sqrt{2}} \cdot \dfrac{\sqrt{5} - \sqrt{2}}{\sqrt{5} - \sqrt{2}}$

$$= \dfrac{(\sqrt{5} - \sqrt{2})^2}{5 - 2} = \dfrac{5 - 2\sqrt{5}\sqrt{2} + 2}{3}$$

$$= \dfrac{7 - 2\sqrt{10}}{3}.$$

To find the quotient of two radicals with different indices, we first use fractional exponents to rewrite each radical so that the indices are the same.

Example 3

Perform the indicated operations and simplify.

a. $\dfrac{\sqrt[3]{10}}{\sqrt{10}} = \dfrac{10^{1/3}}{10^{1/2}} = \dfrac{10^{2/6}}{10^{3/6}} = \dfrac{\sqrt[6]{10^2}}{\sqrt[6]{10^3}} = \sqrt[6]{\dfrac{10^2}{10^3}} = \sqrt[6]{\dfrac{10^2}{10^3} \cdot \dfrac{10^3}{10^3}} = \dfrac{\sqrt[6]{10^5}}{10}.$

b. $\dfrac{\sqrt{2}}{\sqrt[3]{3}} = \dfrac{2^{1/2}}{3^{1/3}} = \dfrac{2^{3/6}}{3^{2/6}} = \dfrac{\sqrt[6]{2^3}}{\sqrt[6]{3^2}} = \sqrt[6]{\dfrac{2^3}{3^2} \cdot \dfrac{3^4}{3^4}} = \dfrac{\sqrt[6]{8 \cdot 81}}{3} = \dfrac{\sqrt[6]{648}}{3}.$

c. $\dfrac{\sqrt[4]{2}}{\sqrt[3]{xy^2}} = \dfrac{2^{1/4}}{x^{1/3}y^{2/3}} = \dfrac{2^{3/12}}{x^{4/12}y^{8/12}} = \sqrt[12]{\dfrac{2^3}{x^4y^8} \cdot \dfrac{x^8y^4}{x^8y^4}} = \dfrac{\sqrt[12]{8x^8y^4}}{xy}.$

Problem Set 11.4

In Problems 1–28, perform the indicated operations and simplify.

1. $\dfrac{\sqrt{32}}{\sqrt{2}}.$

2. $\dfrac{\sqrt{18}}{\sqrt{2}}.$

3. $\dfrac{\sqrt{3}}{\sqrt{7}}.$

4. $\dfrac{\sqrt{8}}{\sqrt{3}}.$

5. $\dfrac{\sqrt{18}}{\sqrt{3}}.$

6. $\dfrac{\sqrt{15}}{\sqrt{5}}.$

7. $\dfrac{\sqrt{2a^3}}{\sqrt{a}}.$

8. $\dfrac{\sqrt{3x^5}}{\sqrt{x}}.$

9. $\dfrac{2x\sqrt[4]{x^7}}{\sqrt[4]{x^{12}}}.$

10. $\dfrac{3a^2b\sqrt[5]{a^3}}{\sqrt[5]{a^5}}.$

11. $\dfrac{\sqrt[3]{6}}{\sqrt[3]{4x}}.$

12. $\dfrac{\sqrt[3]{3y}}{\sqrt[3]{x}}.$

13. $\dfrac{\sqrt[3]{2x}}{\sqrt[3]{5xy^2}}.$

14. $\dfrac{\sqrt{7xy}}{\sqrt{14xy^3}}.$

15. $\dfrac{1}{2 + \sqrt{3}}.$

16. $\dfrac{1}{1 - \sqrt{2}}.$

17. $\dfrac{\sqrt{2}}{\sqrt{3} - \sqrt{6}}.$

18. $\dfrac{5}{\sqrt{6} + \sqrt{7}}.$

19. $\dfrac{2\sqrt{2}}{\sqrt{2} - \sqrt{3}}.$

20. $\dfrac{2\sqrt{3}}{\sqrt{5} - \sqrt{2}}.$

21. $\dfrac{1}{x + \sqrt{5}}.$

22. $\dfrac{3 - \sqrt{5}}{\sqrt{2} + \sqrt{4}}.$

23. $\dfrac{1 + \sqrt{2}}{\sqrt{3} + \sqrt{6}}.$

24. $\dfrac{\sqrt{3} - 5\sqrt{2}}{\sqrt{3} + \sqrt{2}}.$

25. $\dfrac{\sqrt{6}-\sqrt{3}}{\sqrt{6}+\sqrt{3}}.$

26. $\dfrac{x-3}{\sqrt{x-1}}+\dfrac{4}{\sqrt{x-1}}.$

27. $\dfrac{5}{1+\sqrt{3}}-\dfrac{4}{2-\sqrt{2}}.$

28. $\dfrac{4}{\sqrt{x+2}}\cdot\dfrac{x^2}{3}.$

In Problems **29–38**, *perform the indicated operations and simplify.*

29. $\dfrac{\sqrt{6}}{\sqrt[4]{6}}.$

30. $\dfrac{\sqrt[3]{5}}{\sqrt[6]{5}}.$

31. $\dfrac{\sqrt[3]{3}}{\sqrt{2}}.$

32. $\dfrac{\sqrt[8]{3}}{\sqrt[4]{2}}.$

33. $\dfrac{\sqrt{2x}}{\sqrt[3]{x}}.$

34. $\dfrac{\sqrt{2x}}{\sqrt[3]{y}}.$

35. $\dfrac{\sqrt[9]{2y}}{\sqrt[3]{y}}.$

36. $\dfrac{\sqrt[3]{3y}}{\sqrt[9]{x}}.$

37. $\dfrac{\sqrt{xy}}{\sqrt[4]{xy}}.$

38. $\dfrac{2x\sqrt{x}}{\sqrt[3]{2xy}}.$

39. If a block of mass m were released at the top of a frictionless inclined plane with the same vertical height h as in Problem 85 of Sec. 11.1, its speed v at the bottom of the incline is $\sqrt{2gh}$. Find the ratio of the speed of the block to the speed of the sphere in Problem 85 and simplify.

40. One result of Einstein's theory of relativity is that the mass of an object increases as the speed of the object increases. If the mass at speed v is m_v and the rest mass of the object is m_0, then the formula for m_v is

$$m_v = \dfrac{m_0}{\sqrt{1-\dfrac{v^2}{c^2}}},$$

where c is the speed of light. If the object's speed is half the speed of light, find its mass m_v in terms of its rest mass.

41. For a piano tuned to the equally tempered scale, the frequency of a halftone above any note can be obtained by multiplying the frequency of the note by $\sqrt[12]{2}$. The standard concert frequency for A above middle C is 440 vibrations per second.
    (a) What is the frequency for the note B that is two halftones above standard A? Give your answer in simplest radical form.
    (b) What is the frequency for the note D-sharp that is six halftones below standard A? Give your answer in simplified radical form and then approximate your answer to the nearest full vibration per second.

11.5 RADICAL EQUATIONS

An equation in which the unknown occurs under a radical sign or has a fractional exponent is called a **radical equation**. One way to solve a radical equation is to raise both sides to the same appropriate power to remove the radical. This operation, you may recall from Chapter 4, does not guarantee that the resulting equation is equivalent to the original one. The resulting equation may have more roots. For that reason it is essential that all "solutions" be checked so that any extraneous roots that may have been introduced can be rejected.

To illustrate, we shall solve

$$\sqrt{x - 7} = 4.$$

Squaring both sides gives

$$(\sqrt{x - 7})^2 = 4^2$$

$$x - 7 = 16$$

$$x = 23.$$

Although 23 appears to be the solution, we must check it. Substituting 23 for x in $\sqrt{x - 7} = 4$ gives $\sqrt{23 - 7} = 4$, or $\sqrt{16} = 4$, which is true. Thus 23 is the solution.

Example 1

Solve $\sqrt[3]{x - 4} - 3 = 0$.

Solution   When a radical equation has only one radical term, it is best to rewrite the equation so that the radical is isolated on one side. Doing this, we have

$$\sqrt[3]{x - 4} = 3$$

$$x - 4 = 27 \qquad [cubing \text{ both sides}]$$

$$x = 31.$$

You may check that 31 is indeed the solution.

Example 2

Solve $\sqrt{(x - 3)^3} = 64$.

Solution   This equation can be rewritten as

$$(x - 3)^{3/2} = 64.$$

Raising both sides to the two-thirds power gives

$$[(x - 3)^{3/2}]^{2/3} = (64)^{2/3}$$

$$x - 3 = (\sqrt[3]{64})^2 = 4^2 = 16$$

$$x = 19.$$

You should verify that 19 is the solution.

In some cases we have to raise both sides of an equation to the same power more than once. Example 3 shows this.

Example 3

Solve $\sqrt{y - 3} - \sqrt{y} = -3$.

Solution   When an equation has two terms involving radicals, we first rewrite the equation so that one radical is on each side. This often simplifies the work.

$$\sqrt{y - 3} = \sqrt{y} - 3$$

$$(\sqrt{y - 3})^2 = (\sqrt{y} - 3)^2 \qquad \text{[squaring both sides]}$$

$$y - 3 = y - 6\sqrt{y} + 9$$

$$6\sqrt{y} = 12$$

$$\sqrt{y} = 2$$

$$y = 4 \qquad \text{[squaring both sides]}.$$

Now we check our result. Replacing y by 4 in the left side of the original equation gives $\sqrt{1} - \sqrt{4}$, which is -1. Since this does not equal the right side, -3, there is **no solution**.

Problem Set 11.5

In Problems **1–18**, *solve the radical equations.*

1. $\sqrt{x - 2} = 5$.

2. $\sqrt{x + 7} = 9$.

3. $\sqrt{2y - 5} - 6 = 0$.

4. $\sqrt{2x - 6} - 16 = 0$.

5. $(x^2 + 33)^{1/2} = x + 3$.

6. $(y + 6)^{1/2} = 7$.

7. $\left(\dfrac{x}{2} + 1\right)^{1/4} = \dfrac{1}{2}$.

8. $\sqrt[4]{2x + 1} = 3$.

9. $(z - 3)^{3/2} = 8$.

10. $\sqrt{x - 3} + 4 = 1$.

11. $2\sqrt{2x + 1} = 3\sqrt{3x - 8}$.

12. $\sqrt{4x - 6} - \sqrt{x} = 0$.

13. $\sqrt{x} - \sqrt{x + 1} = 1$.

14. $4\sqrt{8y + 1} = 5\sqrt{5y + 1}$.

15. $\sqrt{x + 1} = \sqrt{x} + 1$.

16. $\sqrt{7 - 2x} - \sqrt{x - 1} = 0$.

17. $\sqrt{y} + \sqrt{y + 2} = 3$.

18. $\sqrt{\dfrac{1}{x}} - \sqrt{\dfrac{2}{5x - 2}} = 0$.

19. The time T, in seconds, for one complete oscillation of a simple pendulum is given by

$$T = 2\pi \sqrt{\dfrac{L}{g}},$$

where L is the length of the pendulum and g is the acceleration due to gravity. Solve the equation for g.

20. The formula $f_r = \dfrac{1}{2\pi\sqrt{LC}}$ occurs in the study of alternating current. Here f_r is resonant frequency, L is inductance, and C is capacitance. Solve for C.

21. Two sources, A and B, are 8 m apart as shown in Fig. 11.3. The sources emit waves of wavelength $\lambda = 5$ m. *Constructive* interference of the waves from A and B will occur at point C, located a distance x to the right of A, when

$$\overline{BC} - \overline{AC} = \lambda.$$

How far is point C from A when constructive interference of the waves from A and B occurs at C?

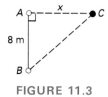

FIGURE 11.3

22. In Problem 21, *destructive* interference of the waves from A and B occurs at C when

$$\overline{BC} - \overline{AC} = \frac{\lambda}{2}.$$

How far is point C from source A when such destructive interference occurs?

23. In the study of the flow of water through a sloped rectangular channel, the equation $AR^{2/3} = 0.892$ occurs. For a certain channel, $A = b^2/2$ and $R = b/4$, where b is the width of the channel in meters. Find b.

24. An analysis of the structure of the hydrogen atom involves expressions for the angular momentum L and the total energy E. These can be written

$$L = \frac{nh}{2\pi} = \sqrt{\frac{me^2 r}{4\pi\epsilon_0}}$$

and

$$E = -\frac{e^2}{8\pi\epsilon_0 r}.$$

Solve the first equation for r and, by substituting into the second equation, show that

$$E = -\frac{me^4}{8\epsilon_0^2 h^2 n^2}.$$

25. Police have used the formula $s = \sqrt{30fd}$ to estimate the speed s (in miles per hour) of a car if it skidded d feet when stopping. The literal number f is the coefficient of friction determined by the kind of road (such as concrete, asphalt, gravel, or tar) and whether the

road is wet or dry. Some values of f are given in Table 11.1. At 40 mi/h, about how many feet will a car skid on a dry concrete road? Give your answer to the nearest foot.

TABLE 11.1

	Concrete	Tar
Wet	0.4	0.5
Dry	0.8	1.0

11.6 REVIEW

Important Terms

Section 11.1   Removing factors from the radicand, rationalizing the denominator, reducing the index, simplest radical form.

Section 11.2   Like radicals.

Section 11.5   Radical equation.

Summary of Rules

LAWS OF RADICALS

$$(\sqrt[n]{a})^n = \sqrt[n]{a^n} = a. \qquad \sqrt[n]{ab} = \sqrt[n]{a}\,\sqrt[n]{b}.$$

$$\sqrt[n]{\frac{a}{b}} = \frac{\sqrt[n]{a}}{\sqrt[n]{b}}. \qquad a^{1/n} = \sqrt[n]{a}.$$

$$a^{m/n} = (\sqrt[n]{a})^m = \sqrt[n]{a^m}. \qquad \sqrt[m]{\sqrt[n]{a}} = \sqrt[mn]{a}.$$

SIMPLEST RADICAL FORM

1.   As many factors as possible are removed from all radicands.

2.   All denominators are rationalized.

3.   The index of each radical cannot be reduced.

Review Questions

1. $\sqrt[5]{x^{10}y^{15}} = \sqrt[5]{(\underline{\qquad\qquad})^5} = \underline{\qquad\qquad}.$

2. $\sqrt[3]{x^5y^4} = \sqrt[3]{(\underline{\qquad\qquad})x^2y} = \sqrt[3]{\underline{\qquad\qquad}} \cdot \sqrt[3]{x^2y} = \underline{\qquad\qquad}\sqrt[3]{x^2y}.$

3. $\sqrt[5]{\dfrac{x^3}{y^2}} = \sqrt[5]{\dfrac{x^3(\underline{\hspace{2cm}})}{y^5}} = \dfrac{\sqrt[5]{x^3(\underline{\hspace{2cm}})}}{\sqrt[5]{y^5}} = \dfrac{\sqrt[5]{x^3(\underline{\hspace{2cm}})}}{y}.$

4. $6\sqrt[4]{3x} + 2\sqrt[4]{3x} - 3\sqrt[4]{3x} = \underline{\hspace{2cm}}\sqrt[4]{3x}.$

5. $\sqrt{x^3} + 2x\sqrt{x} = \sqrt{x^2 \cdot x} + 2x\sqrt{x} = \underline{\hspace{2cm}}\sqrt{x} + 2x\sqrt{x} = \underline{\hspace{2cm}}\sqrt{x}.$

6. $\sqrt[3]{y^2} \cdot \sqrt[3]{y} = \sqrt[3]{(\underline{\hspace{2cm}})} = \underline{\hspace{2cm}}.$

7. $\dfrac{1}{\sqrt[3]{x}} = \dfrac{1}{\sqrt[3]{x}} \cdot \dfrac{(\underline{\hspace{2cm}})}{\sqrt[3]{x^2}} = \dfrac{\sqrt[3]{(\underline{\hspace{2cm}})}}{x}.$

8. $\dfrac{\sqrt{10xy}}{\sqrt{5x}} = \sqrt{\dfrac{10xy}{5x}} = \sqrt{(\underline{\hspace{2cm}})}.$

9. $\dfrac{1}{3+\sqrt{2}} = \dfrac{1}{3+\sqrt{2}} \cdot \dfrac{3-\sqrt{2}}{3-\sqrt{2}} = \dfrac{3-\sqrt{2}}{(\underline{\hspace{2cm}})}.$

Answers to Review Questions

1. x^2y^3, x^2y^3.   2. x^3y^3, x^3y^3, xy.   3. y^3, y^3, y^3.   4. 5.   5. x, $3x$.   6. y^3, y.   7. $\sqrt[3]{x^2}$, x^2.
8. $2y$.   9. 7.

Review Problems

In Problems 1–36, perform the operations and simplify.

1. $\sqrt{32}.$

2. $\sqrt[3]{24}.$

3. $\sqrt[3]{2x^3}.$

4. $\sqrt{4x}.$

5. $\sqrt{16x^4}.$

6. $\sqrt[4]{\dfrac{x}{16}}.$

7. $\sqrt{7}\,\sqrt{4}\,\sqrt{14}.$

8. $\dfrac{2}{\sqrt{x^3}}.$

9. $\sqrt{\sqrt[3]{t^4}}.$

10. $\dfrac{\sqrt{3}\,\sqrt{6}}{\sqrt{2}}.$

11. $\sqrt[4]{\dfrac{3}{2x^2y^3}}.$

12. $\dfrac{\sqrt[3]{t^5}}{\sqrt[3]{t^2}}.$

13. $2\sqrt{8} - (5\sqrt{2} - \sqrt{18}).$

14. $(\sqrt{3} - \sqrt{2})(\sqrt{3} + 2\sqrt{2}).$

15. $\sqrt{2}(1 - \sqrt{6}).$

16. $\sqrt{75k^4}.$

17. $(\sqrt[5]{2})^{10}.$

18. $\sqrt{x}\,\sqrt{x^2y^3}\,\sqrt{xy^2}.$

19. $\dfrac{2}{\sqrt{7}}.$

20. $\dfrac{8}{\sqrt[3]{4}}.$

21. $\dfrac{3}{\sqrt[4]{x}}.$

22. $\sqrt[5]{\sqrt[3]{x^{10}}}.$

23. $\sqrt[4]{81x^6}.$

24. $(\sqrt[5]{x^2y})^{10}.$

25. $\sqrt[3]{\sqrt{\sqrt[3]{x^{36}}}}.$

26. $\sqrt[4]{2}\,\sqrt[4]{24}.$

27. $\sqrt{x}\,\sqrt{3x}.$

28. $\sqrt[6]{x^7y^{13}z^{12}}.$

29. $\sqrt[6]{\dfrac{x^6}{y^9}}.$

30. $\sqrt[9]{x^3y^6}.$

31. $\dfrac{3}{\sqrt[3]{xy^2}}.$

32. $\dfrac{1}{\sqrt{x}\sqrt{y}}.$

33. $\dfrac{\sqrt[3]{3x^2}}{\sqrt[3]{2x}}.$

34. $\sqrt[5]{\dfrac{xy^2}{x^2y}}.$

35. $\dfrac{1}{\sqrt{6}-2}.$

36. $\dfrac{4\sqrt{6}}{\sqrt{6}+\sqrt{4}}.$

*In Problems **37–44**, solve the equations.*

37. $\sqrt{2x+5}=5.$

38. $\sqrt{3x-4}=\sqrt{2x+5}.$

39. $\sqrt[3]{11x+9}=4.$

40. $\sqrt{x^2+5x+25}=x+4.$

41. $\sqrt{y}+6=5.$

42. $2\sqrt{x-2}=\sqrt{3}(\sqrt{x}-2).$

43. $\sqrt{x-1}+\sqrt{x+6}=7.$

44. $\sqrt{z^2+2z}=3+z.$

45. Figure 11.4 shows an arrangement of charges q_1 and q_2. Theory shows that the magnitudes of the electric field vectors $\mathbf{E}_1$ and $\mathbf{E}_2$ due to charges q_1 and q_2, respectively, are given by

$$E_1 = \frac{kq_1}{r_1^2} \quad \text{and} \quad E_2 = \frac{kq_2}{r_2^2}.$$

Find expressions for the magnitudes of the *vertical components* of $\mathbf{E}_1$ and $\mathbf{E}_2$ and simplify each. *Hint*: For $\mathbf{E}_1$ it is given by $E_1 \cos \theta$.

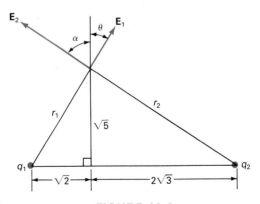

FIGURE 11.4

46. For the arrangement in Problem 45, find an expression for the magnitude of the *horizontal component* of $\mathbf{E}_1$ and simplify.

47. In a lab experiment dealing with the oscillation of a spring, a student encountered an equation of the form

$$T = 2\pi \sqrt{\frac{M_1 + M_2}{k}}$$

and had to show that the graph of T^2 (output) plotted against M_1 (input) is a straight line with slope $4\pi^2/k$. Verify that this is indeed the case.

12

Quadratic Equations and Functions

12.1 SOLUTION BY FACTORING AND SQUARE ROOT

If a ball is thrown vertically upward from ground level at a speed of 19.6 m/s, then (neglecting air resistance) its height s (in meters) after t seconds have elapsed is given by the equation of motion

$$s = -4.9t^2 + 19.6t.$$

For example, after 1 s its height is $-4.9(1)^2 + 19.6(1) = 14.7$ m. When does the ball hit the ground?

The ball hits the ground when s is 0. Thus we want to solve

$$0 = -4.9t^2 + 19.6t. \tag{1}$$

In Eq. 1, note that the highest power of the variable t that occurs is the second power. For this reason, Eq. 1 is called a **second-degree equation** or an **equation of degree 2**. It is most commonly referred to as a *quadratic equation.*

A **quadratic equation** in the variable x is an equation that can be written in the form

$$ax^2 + bx + c = 0$$

where a, b, and c are constants and $a \neq 0$.

Example 1

The following are quadratic equations of the form $ax^2 + bx + c = 0$.

a. $3x^2 - 5x + 2 = 0$ $a = 3, b = -5, c = 2$.   *Note that $b \neq 5$.*

b. $y^2 - 4 = 0$ The variable is y, and $a = 1, b = 0, c = -4$.

c. $x^2 - x = 0$ $a = 1, b = -1, c = 0$.

d. $2x^2 = x - 4$ By rewriting this as $2x^2 - x + 4 = 0$, we have $a = 2, b = -1, c = 4$.

A simple way of solving quadratic equations involves factoring $ax^2 + bx + c$. For example, let us return to the ball problem. We shall rewrite Eq. 1 as

$$4.9t^2 - 19.6t = 0, \tag{2}$$

because it is convenient that the coefficient of t^2 be positive. Factoring the left side,

$$t(4.9t - 19.6) = 0.$$

Think of this as two numbers, t and $4.9t - 19.6$, whose product is zero. Now we use an important fact:

> If the product of two or more numbers is *zero*, then at least one of the numbers must be zero.

Applying this to the two factors of our equation, we must have either

$$t = 0 \quad \text{or} \quad 4.9t - 19.6 = 0,$$

from which

$$t = 0 \quad \text{or} \quad t = \frac{19.6}{4.9} = 4.$$

You can easily check that the values $t = 0$ and $t = 4$ both satisfy the given equation. Thus *a quadratic equation can have two different solutions.* However, due to the physical situation involved in the problem, we must reject $t = 0$, because that is the time the ball *left* the ground. Thus the ball *strikes* the ground 4 s later. In summary, 0 and 4 are solutions of Eq. 2, but $t = 4$ s is the solution to the problem posed.

Solving a quadratic equation by factoring is practical only when the factoring is easy to peform. When factoring is difficult, other techniques can be used. These will be discussed later.

Example 2

Solve $x^2 - 3x + 2 = 0$ by factoring.

Solution   Factoring gives

$$x^2 - 3x + 2 = 0$$

$$(x - 1)(x - 2) = 0.$$

Setting each factor equal to zero and solving for x,

$$x - 1 = 0 \quad \bigg| \quad x - 2 = 0$$

$$x = 1. \quad \bigg| \quad x = 2.$$

The solutions are 1 and 2. The equation $x^2 - 3x + 2 = 0$ can be solved graphically by sketching the graph of $y = x^2 - 3x + 2$. The real solutions correspond to the x-intercepts, that is, those values of x for which $y = 0$. A sketch is shown in Fig. 12.1.

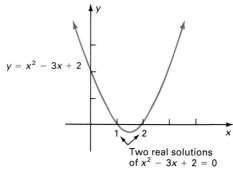

FIGURE 12.1

Example 3

Solve $6w^2 = 4w$ by factoring.

Solution   First we rewrite the equation so that one side is 0.

$$6w^2 - 4w = 0.$$

Factoring gives

$$2w(3w - 2) = 0.$$

Setting each factor equal to zero and solving for w,

$$2 = 0. \quad \bigg| \quad w = 0. \quad \bigg| \quad 3w - 2 = 0$$

$$\text{Impossible} \qquad\qquad\qquad 3w = 2$$

$$w = \frac{2}{3}.$$

Thus the solutions are 0 and $\frac{2}{3}$.

In Example 3, initially we could have divided both sides of the given equation ($6w^2 = 4w$) by 2 to make the coefficients easier to work with. This gives the equivalent equation $3w^2 = 2w$, which is then solved in the same manner as in Example 3 (but the factor 2 would not appear). *This type of simplifying operation is a great aid in solving equations and should not be overlooked.* Moreover, you may have noticed that w is also a factor of both sides. However, we must avoid dividing both sides by w (a variable) because equivalence is not guaranteed and we may lose a solution (see the discussion of Operation 4 in Sec. 4.1). Let's see the trouble we can get into by dividing by w.

$$\frac{6w^2}{w} = \frac{4w}{w}$$

$$6w = 4$$

$$w = \frac{2}{3}. \tag{3}$$

Here the only solution is $\frac{2}{3}$. Thus Eq. 3 is *not* equivalent to the original equation, which has solutions 0 and $\frac{2}{3}$. We lost a solution! This occurred because when we divided both sides by w, we had to assume that w was not zero (we cannot divide by 0). Thus zero had no chance of being a solution to the equations that followed. *Moral: Avoid dividing both sides of an equation by an expression involving the variable.*

Example 4

Solve $x - 3 = \dfrac{x^2}{3} + 3x$ by factoring.

Solution   First, to clear fractions we multiply both sides by 3.

$$3(x - 3) = 3\left(\frac{x^2}{3} + 3x\right)$$

$$3x - 9 = x^2 + 9x.$$

Next, we rewrite the equation so that one side is zero. Then we factor

$$0 = x^2 + 6x + 9$$

$$0 = (x + 3)(x + 3).$$

$$
\begin{array}{c|c}
x + 3 = 0 & x + 3 = 0 \\
x = -3. & x = -3.
\end{array}
$$

The solution is -3. Because two factors gave rise to the same root, -3, we say that there are two *equal roots* of -3 or that -3 is a **double root** (or **repeated root**) of the equation. The graph of $y = x^2 + 6x + 9$ meets, but does not cross, the x-axis when $x = -3$ (see Fig. 12.2).

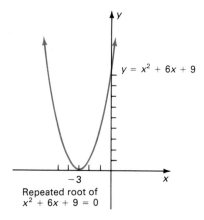

$y = x^2 + 6x + 9$

-3

Repeated root of
$x^2 + 6x + 9 = 0$

FIGURE 12.2

Example 5

Solve $(3x - 4)(x + 1) = -2$.

Solution   You should approach a problem like this with caution. If the product of two numbers is -2, this does *not* mean that one of the numbers must be -2. You should *not* set each factor equal to -2 because the equations $3x - 4 = -2$ and $x + 1 = -2$ do not provide solutions to the given equation. To solve the equation, we first multiply the factors in the left side. Then we combine terms so that one side is zero.

$$(3x - 4)(x + 1) = -2$$
$$3x^2 - x - 4 = -2$$
$$3x^2 - x - 2 = 0$$
$$(3x + 2)(x - 1) = 0.$$

$3x + 2 = 0$	$x - 1 = 0$
$3x = -2$	$x = 1.$
$x = -\frac{2}{3}.$	

The solutions are $-\frac{2}{3}$ and 1.

Example 6

Solve $x^2 = 3$ by factoring.

434   CHAP. 12   Quadratic Equations and Functions

Solution

$$x^2 = 3$$

$$x^2 - 3 = 0. \tag{4}$$

Because $3 = (\sqrt{3})^2$, the left side of Eq. 4 is a difference of squares, so we can factor it.

$$(x + \sqrt{3})(x - \sqrt{3}) = 0.$$

(In this chapter we allow factors involving coefficients that are not integers.) Thus

$$x + \sqrt{3} = 0 \qquad \Big| \qquad x - \sqrt{3} = 0$$

$$x = -\sqrt{3}. \qquad \Big| \qquad x = \sqrt{3}.$$

The pair of numbers $\sqrt{3}$ and $-\sqrt{3}$ may be written as $\pm\sqrt{3}$, which is read *plus or minus* $\sqrt{3}$. Thus the solutions are $\pm\sqrt{3}$.

A more general form of the equation $x^2 = 3$ in Example 6 is $u^2 = k$. In the same manner as in Example 6, we can show the following:

$$\boxed{\text{If } u^2 = k, \quad \text{then} \quad u = \pm\sqrt{k}.}$$

The formula $u = \pm\sqrt{k}$ is called the **square-root method** of solving $u^2 = k$. It easily allows us to solve quadratic equations having no first-degree term. We simply solve for x^2 and apply the rule. For example, to solve $5x^2 - 11 = 0$ we have

$$5x^2 = 11$$

$$x^2 = \frac{11}{5}$$

$$x = \pm\sqrt{\frac{11}{5}} = \pm\sqrt{\frac{11}{5} \cdot \frac{5}{5}} = \pm\frac{\sqrt{55}}{5}.$$

Do not forget the $\pm$ sign.

If k is negative, the equation $u^2 = k$ has no real solutions. For example, consider the equation $x^2 + 2 = 0$, or

$$x^2 = -2.$$

If x is a solution, its square must be -2. Because the square of any real number must be nonnegative, there are no real solutions.* Geometrically, this means that the graph of $y = x^2 + 2$ does not meet the x-axis (see Fig. 12.3).

* In Chapter 13, when complex numbers are considered, we shall write the solutions as $\pm\sqrt{2}j$, where $j = \sqrt{-1}$.

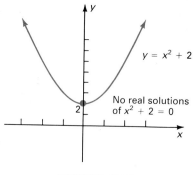

FIGURE 12.3

Example 7

Solve $(x + 2)^2 = 7$ by the square-root method.

Solution This equation has the special form $u^2 = k$, where u is $x + 2$ and k is 7. Thus

$$(x + 2)^2 = 7$$
$$x + 2 = \pm\sqrt{7}$$
$$x = -2 \pm \sqrt{7}.$$

Here we have two solutions: one when we use the $+$ sign, $-2 + \sqrt{7}$, and another when we use the $-$ sign, $-2 - \sqrt{7}$.

From the examples presented thus far, it is reasonable to infer that a quadratic equation may have two real solutions, only one, or none.

Sometimes factoring can be used to solve equations that are not quadratic, as Example 8 shows.

Example 8

Solve $4x^4 - 16x^2 = 0$.

Solution This is called a fourth-degree equation. (Why?) Dividing both sides by 4, we have

$$x^4 - 4x^2 = 0$$
$$x^2(x^2 - 4) = 0$$
$$x^2(x + 2)(x - 2) = 0$$
$$x \cdot x(x + 2)(x - 2) = 0.$$

Setting each of the *four* factors equal to 0 and solving the resulting equations give

$$x = 0. \quad | \quad x = 0. \quad | \quad x + 2 = 0 \quad | \quad x - 2 = 0$$
$$x = -2. \qquad x = 2.$$

Thus the roots are 0 and ± 2. Here 0 is a double root.

Example 9

A rectangular observation deck overlooking a scenic valley is to be built. It is to have dimensions 6 m by 12 m. A rectangular shelter of area 40 m² is to be centered over the deck. The uncovered part of the deck is to serve as a walkway of uniform width. How wide should this walkway be?

Solution   A diagram of the deck is shown in Fig. 12.4. Let w be the width (in meters) of the

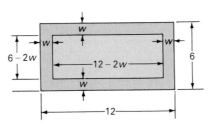

FIGURE 12.4

walkway. Then the part of the deck for the shelter has dimensions $12 - 2w$ by $6 - 2w$. Since this area must be 40 m², where area = (length)(width), we have

$$(12 - 2w)(6 - 2w) = 40$$
$$72 - 36w + 4w^2 = 40 \qquad \text{[multiplying]}$$
$$4w^2 - 36w + 32 = 0$$
$$w^2 - 9w + 8 = 0 \qquad \text{[dividing both sides by 4]}$$
$$(w - 8)(w - 1) = 0$$
$$w = 8, 1.$$

Although 8 is a solution to the equation, it is *not* a solution to our problem, because one of the dimensions of the deck itself is only 6 m. Thus the only possible solution is 1 m.

Problem Set 12.1

In Problems 1–50, solve each equation by using the methods of this section.

1. $x^2 + 3x + 2 = 0$.
2. $(x + 2)(x - 5) = 0$.
3. $(x - 3)(2x + 1) = 0$.
4. $x^2 + 8x + 15 = 0$.
5. $t^2 - 7t + 12 = 0$.
6. $t^2 - 4t + 4 = 0$.

7. $z^2 + 2z - 3 = 0$.

8. $z^2 + z - 12 = 0$.

9. $x^2 - 12x + 36 = 0$.

10. $x^2 - 1 = 0$.

11. $x^2 - 8x = 0$.

12. $t^2 = 16$.

13. $2x^2 + 10x = 0$.

14. $0 = 3x - 4x^2$.

15. $0 = 3t^2 - 7t$.

16. $x^2 = 32$.

17. $4 - x^2 = 0$.

18. $10x^2 - x - 3 = 0$.

19. $x^2 = 25$.

20. $x^2 = 8$.

21. $x^2 = 6$.

22. $9x^2 = 36$.

23. $\dfrac{x^2}{3} = 4$.

24. $\dfrac{x^2}{7} = 1$.

25. $9z^2 = 81$.

26. $3x^2 - 12x + 12 = 0$.

27. $6x^2 + 7x - 3 = 0$.

28. $3 = z^2$.

29. $\dfrac{4}{3}t^2 = 5$.

30. $7t^2 = 7$.

31. $2x^2 - 14 = 0$.

32. $x^2 + 2 = 18$.

33. $2x^2 + 7x = 4$.

34. $x^2 = 2x + 3$.

35. $-x^2 + 3x + 10 = 0$.

36. $2x^2 + 3x - 2 = 0$.

37. $4x^2 + 4x = -1$.

38. $9x^2 - 1 = 0$.

39. $6(x^2 + 2x) + 6 = 0$.

40. $4x^2 - 12x + 9 = 0$.

41. $t(t + 4) = 5$.

42. $2y^2 = 4y$.

43. $3x^2 + 5(2 - x) = 2x^2 + 4$.

44. $2(x^2 - 5) - 3(x^2 - 7) = 3$.

45. $4 - x^2 = (x + 1)^2 + 3$.

46. $(2x - 5)(x + 5) = -22$.

47. $\dfrac{x^2}{2} - x - 4 = 0$.

48. $x^2 + \dfrac{7}{2}x - 2 = 0$.

49. $6y^2 + \dfrac{5}{2}y + \dfrac{1}{4} = 0$.

50. $\dfrac{x^2}{2} + \dfrac{10}{3}x + 2 = 0$.

In Problems 51–56, solve each equation by using the square-root method.

51. $(x - 3)^2 = 16$.

52. $(x + 5)^2 = 6$.

53. $(x + 4)^2 = 8$.

54. $(w - 6)^2 = 1$.

55. $(y + \frac{1}{2})^2 = 1$.

56. $(x - 4)^2 = \frac{9}{4}$.

In Problems 57–68, solve each equation.

57. $x(x - 1)(x + 2) = 0$.

58. $x^2(x - 4) = 0$.

59. $(x - 2)^2(x + 1)^2 = 0$.

60. $x(x - 1)(x + 1) = 0$.

61. $7x^2(x - 2)^2(x + 3)(x - 4) = 0$.

62. $x(x^2 - 1)(x^2 - 4) = 0$.

63. $x(x^2 - 1)(x^2 - 1) = 0$.

64. $x^3 - x = 0$.

65. $x^3 - 64x = 0$.

66. $x^3 - 4x^2 - 5x = 0$.

67. $3y^3 + 18y^2 + 24y = 0$.

68. $3x^4 + 11x^3 - 4x^2 = 0$.

In Problems 69–74, solve each equation by factoring.

69. $x^4 - 10x^2 + 9 = 0$.

70. $x^4 - 29x^2 + 100 = 0$.

71. $x^2(x^4 - 2x^2 + 1) = 0$.

72. $x^5 - 6x^3 + 8x = 0$.

73. $x^4 - 13x^2 + 36 = 0$.

74. $x^5 - 4x^3 + 4x = 0$.

In Problems 75–78, solve the given equation for the indicated letter.

75. $T^2 = 4\pi^2\left(\dfrac{L}{g}\right); \quad T$.

76. $s = \dfrac{1}{2}at^2; \quad t$.

77. $mgh = \dfrac{1}{2}mv^2 + \dfrac{1}{2}I\omega^2; \quad \omega$.

78. $P = \dfrac{E^2}{R + r} - \dfrac{E^2 r}{(R + r)^2}; \quad E$.